INFECTIOUS DISEASE CONTROL IN ANIMAL SHELTERS: STRATEGIES AND PROTOCOLS

INFECTIOUS DISEASE CONTROL IN ANIMAL SHELTERS: STRATEGIES AND PROTOCOLS

Edited by

Dr. Balaji Marotrao Kondre

Delve Publishing

www.delvepublishing.com

Infectious Disease Control in Animal Shelters: Strategies and Protocols

Dr. Balaji Marotrao Kondre

Delve Publishing
4164 Lakeshore Road
Burlington, ON L7L 1A4
Canada
www.delvepublishing.com
Email: orders@arclereducation.com

ISBN: 978-1-77956-404-7 (Hardcover)

ABOUT THE EDITOR

Dr. Balaji Marotrao Kondre is presently working as Assistant Professor, Department of Veterinary Pathology, College of Veterinary and Animal Sciences, Udgir Dist. Latur, Maharashtra. He did his B. V. Sc & A. H. in 2003 from College of Veterinary and Animal Sciences, Udgir Dist. Latur and M. V. Sc. (Veterinary Pathology) in 2006 from College of Veterinary and Animal Sciences, Parbhani, Ph. D. (Veterinary Pathology) in 2021 from College of Veterinary Science & Animal Husbandry, Anand, (Gujarat). He started his career as a Livestock Development Officer in 2007 and as Assistant Professor (Veterinary Pathology) in 2009. During his total 17 year's tenure of services, he has Handled 7 externally funded project/Private Agency Schemes as PI / COPI. He has guided 25+ M. V. Sc students as a member of advisory Committee. He has published 12 research papers in international and national reputed journals. He had published research abstracts more than 31 in International, National, State Seminar, workshops and symposiums. Besides these activities he also delivered one Radio talks, guest Lectures in farmer training. In addition to above he also wrote popular articles in local language.

TABLE OF CONTENTS

List of Figures .... *xi*
List of Tables .... *xiii*
Preface .... *xv*

Chapter 1 Introduction to Infectious Disease Control in Animal Shelters 1
Introduction 2
1.1. Fundamentals of Disease Control in Shelters 2
1.2. Shelter Missions and Goals 7
1.3. Wellness 20
1.4. The Problem Oriented Approach to Small Animal Medicine 25
1.5. Wellness: Physical Health 30
1.6. Wellness: Behavioral Health 39
1.7. Environmental Wellness 46
References 52

Chapter 2 Dermatological Disease in Animal Shelters 55
Introduction 56
2.1. Dermatophytosis 56
2.2. Etiology 57
2.3. Pathogenesis 59
2.4. Disease 59
2.5. Transmission 61
2.6. Clinical Presentation and Differential Diagnoses 62
2.7. Diagnosis 64
2.8. Initial Risk Evaluation and Response 82
2.9. Treatment 84
2.10. Outbreak Management and Response 95
References 97

Chapter 3 **Small Rodents: Common Diseases** **101**
Introduction 102
3.1. Overview of Rodents 103
3.2. Husbandry 106
3.3. The Diagnostic Challenge 108
3.4. Diseases of Small Rodents 109
3.5. Disorders and Diseases of Rats 111
3.6. Clinical Techniques 121
References 129

Chapter 4 **Canine and Feline Vaccinations and Immunology** **133**
Introduction 134
4.1. Vaccinations and Immunity 134
4.2. Vaccine Characteristics: Efficacy, Onset, and Duration of Immunity For Modified Live, Killed, and Other Vaccine Types 136
4.3. Vaccination Protocols for Shelters 138
4.4. Vaccine Handling, Administration, and Adverse Reactions 141
4.5. Core Vaccines for Shelter Dogs and Cats 143
4.6. Optional Vaccines for Cats or Dogs in a Shelter 153
4.7. Juvenile Animals 155
4.8. Use of Hyperimmune Serum for Disease Prevention 160
4.9. Risk Assessment and Evaluation Of Serologic Immunity for Canine Distemper, Canine Parvovirus, and Feline Panleukopenia 161
4.10. Vaccine Effects on Diagnostic Testing 165
References 169

Chapter 5 **Feline Upper Respiratory Disease** **175**
Introduction 176
5.1. Agent Characteristics and Epidemiology 176
5.2. Clinical Signs 181
5.3. Therapy 183
5.4. Modes of Transmission 185
5.5. Diagnosis 188
5.6. Disease Control and Prevention 188
5.7. Disease Surveillance 201
5.8. Staff and Volunteer Education and Motivation 202

5.9. Adopting a Systematic Approach to Management of URTD 203
References 207

Chapter 6 **Avian Infections Disease Control and Care in the Animal Shelter 217**
Introduction 218
6.1. Species 218
6.2. Husbandry 220
6.3. History and Physical Examination 224
6.4. Infectious Disease 227
6.5. Nutritional/Metabolic Disorders 247
6.6. Behavior 252
6.7. Control of Avian Infectious Diseases 260
References 266

Chapter 7 **Infectious Diseases of Domestic Rabbits 269**
Introduction 270
7.1. Overview of Rabbits 270
7.2. Diseases of Domestic Rabbits 285
7.3. Bacterial Diseases 306
7.4. Viral Diseases 317
References 337

Chapter 8 **Infectious Diseases of Ferrets 345**
Introduction 346
8.1. Overview of Ferrets 346
8.2. Infectious Diseases of Ferrets 358
8.3. Neoplasia of Ferrets 364
8.4. Hormonal Disorders of Ferrets 369
References 373

Chapter 9 **Infectious Diseases of Guinea Pigs 397**
Introduction 398
9.1. Overview of Guinea Pigs 398
9.2. Diseases in Guinea Pigs 407
References 427

Index 433

LIST OF FIGURES

Figure 2.1. Puppy with an area of hair loss caudal to the ear

Figure 2.2. Ear margin crusting due to dermatophytosis. This was caused by a *Trichophyton* spp.

Figure 2.3. Cat studies

Figure 2.4. A kitten with *M. canis* dermatophytosis

Figure 3.1. Rodent health

Figure 5.1. Typical clinical signs after FHV infection. (A) FHV-A1 isolate; (B) negative control group

Figure 5.2. Microscopic pathological changes of the lung after FHV infection. (A) Control; (B) mild interstitial pneumonia; (C) severe interstitial pneumonia. Scale bar = 50 μm

Figure 6.1. Grey African parrot

Figure 6.2. Indian ringneck parakeet (*Psittacula krameri*)

Figure 6.3. Cage setup with perching options and foraging toys

Figure 6.4. Bornaviruses specific to psittacine birds

Figure 6.5. These are the main organs that should be investigated using histopathology, immunohistochemistry, and RT-PCR for confirmation of PDD

Figure 6.6. Outdoor enclosure for parrots with lots of space, and foraging toys

Figure 7.1. Rabbit breeds

Figure 7.2. The rabbit ear mite, *Psoroptes cuniculi*, causes crusting and inflammation of the external ear canal, which often extends up the pinna. Lesions are sometimes found on other parts of the body such as the perineal skin folds. Mites can just be seen with the naked eye in exudate from the lesions. Large numbers of *P. cuniculi* are visible on microscopic examination of the exudate, which can be softened in liquid paraffin before placing on a glass slide

Figure 7.3. An example of a Cheyletiella species. *Cheyletiella parasitovorax* can be found in the fur of healthy rabbits. It is not always associated with skin lesions. In large numbers, *C. parasitovorax* mites cause pruritus and areas of white, flaky skin. Heavy infestation is usually linked to some underlying problem with grooming, such as dental disease, obesity or spinal disorders. Mites may be seen moving among skin flakes that are combed out and placed under a bright light. *Cheyletiella parasitovorax* can also be detected by combing out the flakes and applying acetate strips to the exposed underlying skin. The acetate strip is placed on a microscope slide and examined on low power. In heavy infestations, a variety of nymphal stages, eggs, and adult mites are seen

Figure 7.4. *L. gibbus* male, from cat (a) and from rabbit (b), showing finely striated cuticle (black arrows), sclerotised "dark hooded" gnathostoma (red arrow), two elongated adanal processes (blue arrows)

Figure 7.5. *Cysticercus pisiformis* (It is the larval stage of *Taenia pisiformis*, which is a tapeworm that affects dogs and foxes, with rabbits acting as the intermediate host)

Figure 7.6. *Encephalitozoon cuniculi*

Figure 7.7. Myxomatosis and rabbit hemorrhagic disease (RHD)

Figure. 8.1. A pet. This is the basic (wild type) color of the ferret, but many color variations exist

Figure 8.2. Ferrets can detect dimensions

Figure 8.3. The utility in virology

Figure 8.4. An age-related significantly greater cerebellar cortex volume in males than in female ferrets

Figure 8.5. Microscopic appearance of lung sample from ferret lung congestion and edema are present; hematoxylin–eosin stain, 100× magnification

Figure 8.6. Survival and petechial rash following EBOV mucosal challenge. (a) Kaplan–Meier survival curve following oronasal, oral, and ocular virus challenge

Figure 8.7. *M. xenopi* mycobacteriosis in a pet ferret

Figure 9.1. Lateral radiographs of the lumbosacral vertebral column demonstrating morphological variants of the sacral (S) bone-in guinea pigs: (a) Complete fusion of the vertebral bodies and fusion of spinous processes S1–S4 creates the median sacral crest; (b) Fusion of the vertebral bodies with vestigial intervertebral space (arrows) and fusion of spinous processes S1–S4 represents the most common variant; (c) Fusion of vertebral bodies and separation of all spinous process es (arrows); (d) Fusion of vertebral bodies and separation of first and fourth spinous processes

Figure 9.2. Morphology of guinea pig spermatozoa with Hemacolor® staining (Merck, Darmstadt, Germany). Type 1: intact acrosome (a), Type 2: expansion by acrosomal reaction (b), Type 3: the onset of decomposition (c), and Type 4: loss of acrosomal contents and matrix (d). Bar represents 10 μm

Figure 9.3. Atypical forms in guinea pig spermatozoa with Hemacolor® staining (Merck, Darmstadt, Germany). Normal (a), spermatozoa with macronucleus (b), micronucleus (c), pyriform (d), elongated (e), and lanceolate (f). Bar represents 10 μm

Figure 9.4. Dystocia

LIST OF TABLES

Table 2.1. Clinicopathological features of cutaneous and extracutaneous invasive dermatophyte lesions in cats

Table 6.1. Virus names, abbreviations, and genus/species designations are listed according to the taxonomy and nomenclature approved by the International Committee on Taxonomy of Viruses (ICTV)

Table 6.2. Key recommendations to strengthen surveillance of wild birds and collaboration

Table 6.3. A list of non-native bird species reported as lost (missing), found, or sighted by pet bird breeders, keepers, and traders

Table. 7.1. Summary of some production and health benefits of supplementing rabbit diets with nutraceutical products

Table 7.2. Answers provided to "Please choose how much you agree or disagree with the following statements: Rabbits can form strong bonds with other rabbits; Rabbits can form strong bonds with humans; Rabbits are happy to live alone" $n = 1516$

Table 7.3. Nutraceuticals' potential to alleviate heat stress impacts in rabbit production

Table 8.1. Infection with influenza virus types A and B show variable, strain-dependent clinical pictures in ferrets

Table 8.2. Predominant histopathological findings following EBOV mucosal challenge in ferrets

Table 9.1. Cardiorespiratory and metabolic measurements in guinea pigs

Table 9.2. Association between patient and diagnostic features with the guinea pigs' sex (female versus male) in 117 clinical cases diagnosed with lower urinary tract disease

Table 9.3. Association between diagnosis and therapeutic features with the guinea pigs' response to treatment in 114 clinical cases diagnosed with lower urinary tract disease

PREFACE

Veterinary medicine is a crucial pillar of public health, agriculture, and food security, playing an indispensable role in safeguarding both human and animal health. With the rise in pet ownership, millions of dogs and cats, including those in shelters and households, are exposed to various deadly pathogens annually. Among these, stray dogs are particularly vulnerable, posing a significant risk of zoonotic disease transmission. As the culture of pet ownership becomes more widespread, the close bond between humans and pets presents a critical avenue for the spread of diseases. This growing connection underscores the importance of effective infectious disease management in environments where animals are housed together, such as shelters.

Infectious Disease Management in Animal Shelters is an essential resource for addressing these challenges. This comprehensive guide provides a detailed approach to preventing, managing, and treating disease outbreaks in animal shelters. It emphasizes strategies for illness prevention and disease mitigation, offering practical, evidence-based information on the fundamental principles of disease control. The book also addresses the specific management of important diseases affecting dogs and cats in group living environments, where the risk of infection is particularly high.

The spread of infectious diseases in animal shelters can have devastating consequences not only for the animals themselves but also for public health at large. Effective disease management in these settings is therefore critical, requiring a combination of stringent biosecurity measures, regular health monitoring, and prompt intervention when outbreaks occur.

The book *Infectious Disease Control in Animal Shelters* is divided into nine chapters, each focusing on different aspects of infectious diseases that can affect a variety of animals commonly found in shelters.

Chapter 1 introduces the topic by exploring the types of infectious diseases that are commonly encountered in animal shelters. It explains the conditions that make shelters prone to disease outbreaks and highlights the importance of early detection and prevention.

Chapter 2 shifts the focus to dermatological diseases, which are skin-related issues that frequently affect animals in shelters. This chapter discusses the causes, symptoms, and treatments for skin diseases, emphasizing the need for proper hygiene and care to prevent their spread.

Chapter 3 takes a closer look at small rodents, such as mice and hamsters, which are also found in shelters. It outlines the common diseases that affect these tiny creatures, providing guidance on their care and the steps necessary to minimize disease transmission.

Chapter 4 covers vaccinations and immunology in dogs and cats, two of the most common animals in shelters. It explains how vaccinations work and why they are crucial for preventing the spread of infectious diseases among shelter animals. The chapter also addresses the challenges of ensuring all animals receive appropriate immunizations.

Chapter 5 examines feline upper respiratory diseases, a significant concern in shelters where cats are housed in close quarters. This chapter explains the causes, symptoms, and treatments of this common disease, along with strategies for reducing its prevalence in the shelter environment.

Chapter 6 focuses on avian infectious diseases, highlighting the unique challenges of caring for birds in a shelter setting. It provides detailed information on managing and preventing infections in birds, emphasizing the importance of specialized care for these delicate creatures.

Chapter 7 discusses the infectious diseases that affect domestic rabbits, another animal commonly found in shelters. This chapter outlines the signs of disease, treatment options, and preventative measures that can help protect rabbits from infections.

Chapter 8 explores the infectious diseases of ferrets, providing insights into the specific health issues that can arise in these animals. The chapter offers practical advice on recognizing symptoms and managing outbreaks in shelter environments.

Chapter 9 concludes the book with a focus on guinea pigs, detailing the common infectious diseases that affect them. This chapter includes information on the best practices for caring for guinea pigs in shelters and preventing the spread of disease.

This book serves as a comprehensive guide for anyone involved in the care of animals in shelters. It provides practical, easy-to-understand information on how to identify, treat, and prevent infectious diseases in a variety of animals, ensuring their health and well-being during their stay in the shelter.

CHAPTER 1

INTRODUCTION TO INFECTIOUS DISEASE CONTROL IN ANIMAL SHELTERS

Contents

Introduction.......... 2
1.1. Fundamentals of Disease Control in Shelters.......... 2
1.2. Shelter Missions and Goals.......... 7
1.3. Wellness.......... 20
1.4. The Problem Oriented Approach to Small Animal Medicine.......... 25
1.5. Wellness: Physical Health.......... 30
1.6. Wellness: Behavioral Health.......... 39
1.7. Environmental Wellness.......... 46
References.......... 52

INTRODUCTION

The high animal turnover and density in shelters can promote the spread of pathogens, making infectious diseases a serious concern. There is a greater chance of disease outbreaks in these settings because animals frequently reside there with unknown medical histories. Common infectious diseases found in shelters include skin conditions, respiratory infections, and gastrointestinal infections. These infections can spread quickly through contaminated surfaces, airborne transmission, or direct contact. A shelter animal's immune system may be weakened by the stress of living there, leaving them more vulnerable to illness. An interdisciplinary approach is necessary for the effective management of infectious diseases in shelters. This approach should include strict intake procedures, vaccination protocols, isolation of sick animals, and appropriate sanitation practices. Maintaining a close eye on health and treating sick animals quickly are essential to stopping the spread of illness. To reduce risks, staff must receive training on infection control and biosecurity procedures. A few common elements that contribute to the spread of infections in shelters are overcrowding, inadequate ventilation, and poor cleaning. Receiving a vaccination right away is essential.

1.1. FUNDAMENTALS OF DISEASE CONTROL IN SHELTERS

Maintaining the health and welfare of the animals housed in animal shelters depends on effective disease control. Shelters are especially vulnerable to infectious disease outbreaks because of their high animal density, regular intake of new animals, and stressful conditions. Robust intake protocols, comprising prompt health assessments, vaccinations, and quarantine measures to identify and isolate potentially ill animals before they interact with the general population, are essential for the implementation of effective disease control strategies. To lower the risk of disease transmission, the shelter's common areas, food, and water bowls, and kennels must all be cleaned and disinfected regularly.

Appropriate ventilation systems are also necessary to minimize the concentration of airborne pathogens, which is another way to control the spread of disease. To avoid cross-contamination, staff members must receive biosecurity training on techniques including good hand hygiene, wearing personal protective equipment (PPE), and handling animals safely. An outbreak can be stopped in its tracks by keeping an eye on the wellbeing of the animals during their visit and acting quickly to treat any symptoms of disease. Furthermore, shelters can monitor possible disease trends and react swiftly to new threats by keeping accurate and current health records for every animal. A thorough disease control plan must include both access to medical resources and cooperation with veterinarians. Additionally, overcrowding in shelters can exacerbate stress and impair animals' immune systems, leaving them more susceptible to illness. Coordinated and consistent application of these tactics can greatly lower the likelihood of disease outbreaks in shelters, safeguarding the animals and increasing their prospects of finding long-term homes.

1.1.1. The Production Medicine Model

Controlling infectious diseases, for all its advantages, is just one of the objectives of an effective shelter medical program. Remarkably applicable in this situation is the production medicine model, which was created in the context of commercial animal husbandry businesses. A good livestock veterinarian is aware that their responsibility goes beyond simply curing or even preventing illness in animals. Instead, they offer direction to assist the production system in achieving many objectives, which might include offering a safe and healthy food product, making sure the business is financially sustainable, ensuring good welfare, and upholding regulatory compliance. Reaching one of these objectives cannot come at the expense of another.

In a similar vein, the shelter practitioner needs to approach the task of disease control with knowledge of the organization's mission, objectives, specifications, and top priorities. The real art of shelter medicine is weighing the risks to best achieve the overall goals, particularly if those goals could be mutually exclusive. One example would be striking a balance between allowing exercise, social contact, and contact with adopters and imposing isolation and confinement to control infectious diseases.

There are certain ways in which the balance between risk and reward in a shelter is different from other settings. Techniques are offered to reduce risks and increase the shelter's capacity to fulfill its objectives. Paradoxically, when it comes to controlling infectious diseases, veterinarians can sometimes recommend practices that seem less cautious rather than more cautious to best contribute to the overall success of the shelter. For example, routine quarantine of incoming healthy-looking animals is frequently advised in herd-health contexts to identify animals that might be disease-incubating. However, other objectives of the shelter, like the prudent use of scarce resources and the prompt placement of healthy animals in adoptive homes, are frequently compromised by the longer length of stay (LOS) this practice entails, the increased population density, the expense, and the strain on staff that follows. Other approaches to prevent the spread of disease without the need for a quarantine include taking an accurate history whenever possible, carefully examining and immunizing newcomers, conducting rounds and monitoring every day, maintaining optimal sanitation practices, and using diagnostic tests as soon as necessary.

1.1.2. Shelter Medicine as a Specialty

The growth of shelter medicine as a respected area of veterinary medicine is a result of several factors, such as the growing appreciation for animals and the push to find alternatives to euthanasia in the face of companion animal homelessness; the increased sophistication and resources of animal sheltering organizations, which present previously unheard-of opportunities for the design of high-quality facilities and health-care programs; and the explosion of evidence-based knowledge that informs best practices for shelter animal care.

Veterinarians have long worked with shelters, but it has only lately come to light that this is an extremely complicated field requiring specialized knowledge. Many universities now offer shelter medicine programs, courses, and residencies. Cornell University offered the first official shelter medicine class at a veterinary college in 1999. Shelter medicine lectures are also offered at a number of significant veterinary conferences. The number of members in the Association of Shelter Veterinarians is constantly increasing. More research is being done to find better ways to manage the health and welfare of shelter animals as interest in the field grows.

1.1.3. Roles of Veterinarians in Shelters

Veterinarians assist shelters in a number of ways as consultants, staff members, or volunteers. Authority can have a very wide scope. They might hold positions of authority such as board members or directors of the shelter, or they might only enter the facility to perform per diem medical or surgical services. In the middle, as regular or part-time employees in charge of the healthcare program, are many veterinarians.

Shelter veterinarians are finding more and more opportunities for employment and consulting; these opportunities offer satisfying and demanding options for professional practice. But at the moment, very few veterinarians have a specific training or level of experience in this field. For veterinarians to better serve the populations in shelters there is a pressing need to increase their learning opportunities.

1.1.4. Herd Health Approach to Shelter Medicine

Shelter medicine, put simply, is companion animal herd health medicine. The veterinary professional faces a challenge in designing an all-encompassing program to regulate, manage, and lessen the spread of disease in animal shelters. The current standard of clinical veterinary education either concentrates on providing sophisticated and potentially expensive health care to individual companion animals or on designing cost-effective herd health protocols that emphasize disease prevention and maximize the production of animal products for food. A combination of these two methods is needed for shelter medicine. Often, strict attention to the group's overall wellness is best for the care of each individual shelter animal. Animals that are protected from disease transmission are spared grave illnesses that might not be curable otherwise. More resources are available for people who need a higher level of care when the population is generally in good health.

The emotional attachment and value placed on companion animals, which are not as prevalent in large animal agriculture, is another significant historical distinction between the two approaches to clinical practice. The capacity to provide science and evidence-based management recommendations to shelters is significantly impacted by this bond. Just as slaughter is frequently used to manage disease in large animal herds, euthanasia was once the main tool for controlling population numbers and disease in shelters. The bond between humans and animals is one of the main reasons behind the growing opposition to animal shelters' regular use of euthanasia. While animal advocacy organizations may argue that pets are "disposable," many people see them as members of the family. Shelter animals are subject to the same extraordinary interest in utilizing the most recent medical advancements to enhance their health and well-being.

In a way that has never been done before, shelter medicine aims to bring together individualized animal care with herd health management strategies and principles. It can be extremely difficult for a well-meaning companion animal practitioner who is not experienced in herd health or shelter management to deal with medical issues that arise in shelters. On the other hand, if a large animal herd health professional attempts to implement conventional outbreak control techniques in a shelter—such as depopulation, herd isolation, and testing all newcomers—they will frequently encounter outright resistance.

1.1.4.1. Unique Aspects of the Shelter Environment

It could be argued that shelter medicine does not need all of this attention and that the concept of herd health care for companion animals is not new. While there are some fundamental disease control concepts that apply to shelters, there are also notable distinctions. These concepts have been applied to the management of kennels, catteries, and research laboratories. While the objectives of breeding and research facilities are generally well-defined, animal shelters have particular objectives and difficulties because of their diverse missions. The functioning and priorities of shelter health programs are influenced by variances and fluctuations in funding, resources, philosophy, training, governance, and even community attitudes toward the shelter. Shelter veterinarians are forced to make bold, creative, and inventive decisions because husbandry practices are frequently introduced in shelters that have never been used in any other type of communal housing.

The preventive aspect of shelter medicine is incorporated into a comprehensive health care plan that goes well beyond the basic advice regarding immunizations and dewormings. It can be intimidating to consider the breadth of expertise needed to create a complete shelter wellness program. Almost every program offered by a shelter, including foster care, adoptions, volunteer programs, stray animal management, zoonotic disease control, cruelty investigations, and even the layout of the shelter itself, is impacted by the health aspect of animal sheltering. Put another way, very few, if any, shelter programs are unaffected by issues related to animal health. Shelter veterinarians need to be well-versed in many other areas in addition to infectious diseases, such as animal behavior, nutrition, husbandry, sanitation, stress management, data collection, veterinary forensics, high-volume, high-quality spay/neuter procedures, and much more.

1.1.5. The Growth of Shelter Medicine

The discipline of shelter medicine has evolved quickly, which has had a tremendous impact on disease control in shelters even though scientific advancements have been substantial. Within 10 years of significant events like the first formal course in shelter medicine (taught at Cornell University in 1999), the first residency training program in shelter medicine at UC Davis in 2001, and the founding of the Association of Shelter Veterinarians (ASV) in the same year, the original edition of this book was published. Since then, a lot of veterinary schools have added externships and core and elective courses in shelter medicine to their programs.

Since then, there has been a notable development in both the field of animal sheltering and veterinary medicine: the quick rise of shelter medicine. In terms of disease control, the 2010 release of the ASV Guidelines for Standards of Care in Animal Shelters marked a significant turning point. This document covered a wide range of topics, including population control, facility design, sanitation, and maintaining the physical and mental well-being of animals. It also covered policy, record-keeping, and shelter management. Because all these components are interconnected, the guidelines offer a strong foundation for any program to maintain the health of shelter animals.

This explosion of information and investigation has coincided with the emergence of shelter medicine as a veterinary specialty. In 2014, the American Board of Veterinary Specialties granted shelter medicine provisional recognition as a specialty within the American Board of Veterinary Practitioners (ABVP). The following year, the first diplomats received certification. The number of veterinarians pursuing board certification in shelter medicine will result in increased availability of knowledgeable consultants and research to assist in the implementation of effective disease control programs in shelters.

1.2. SHELTER MISSIONS AND GOALS

For a shelter health program to be designed effectively, it is imperative that the mission of the shelter be understood. A medical program cannot be deemed a full success if it maintains the health of the animals but falls short of the organization's main objectives, which include increasing animal adoption, community spay/neuter rates, or lowering euthanasia rates. It is necessary to comprehend the objectives and resources of that specific shelter, both generally and specifically for that animal or circumstance, to provide advice on anything from outbreak management to individual animal treatment.

There are an estimated 4,000 to 6,000 animal shelters in the United States alone, according to the Community Outreach Department of the American Society for Prevention of Cruelty to Animals (ASPCA). Assuming that all shelters aim to achieve the same things is a mistake. Shelters generally fall into two categories, despite the fact that there is often overlap in the services offered: either they are private, nonprofit shelters or they are municipal shelters whose main responsibility is animal control. While some towns have a large number of private and public shelters, others have none at all.

Adoption and rehabilitation of homeless animals are not the main priorities of all shelters. Municipal shelters may prioritize stray animal capture, public health protection, complaint handling, and law enforcement when allocating resources, while private animal welfare organizations might spend more on vaccinating, deworming, testing for diseases, and treating, and neutering animals in preparation for rehoming. Municipal and private shelters, however, are increasingly inclined to develop programs that prioritize public outreach and the prevention of issues that result in relinquishment, search for alternatives to euthanasia as a disease management strategy, and strive toward a higher adoption rate. Private shelters differ greatly as well; some offer a limited number of animals lifelong sanctuary, those that take all animals that are offered and euthanize those they are unable to place, and many more variations on these approaches.

It is not advisable to draw conclusions about shelter philosophy or resources based on shelter type, title, location, or history, just as it is crucial to refrain from judging clients by their appearance, the breed of their pet, or the car they drive. When management or philosophy changes, priorities might shift and opportunities might present themselves. Even the smallest or most impoverished shelters may prioritize adoption, use progressive spay/neuter, volunteer, participate in foster care or other programs, or look into non-euthanasia methods of illness management. Although these options might not be available right now, shelters might include them in their future plans. As with any patient, shelters should therefore be given all available options and have the best standards explained.

1.2.1. Municipal Shelters

It is a widely held misconception that the majority of municipal shelters run programs that are chronically overcrowded, underfunded, and housed in run-down buildings in undesirable neighborhoods. Although there is such a model, over the past 20 years, as society has grown less accepting of animal abuse and neglect and as the human-animal bond has grown stronger, animal sheltering has experienced a fundamental transformation in many communities. All kinds of shelters are under more pressure from the outside and from within to improve the standard of care they offer. The response to this pressure has been diverse: numerous communities are expanding staff and services, consulting with veterinarians, renovating, retrofitting, and constructing new facilities with the newest innovations, and so on. Many of these changes cannot be implemented effectively without veterinary expertise.

Historically, the duties performed by municipal shelters included collecting stray animals, controlling dangerous animals (such as placing them in quarantine if they may have bitten someone), capturing animals that were on the loose, responding to complaints of nuisances, looking into allegations of animal abuse, managing wildlife, and so on. In addition, they might provide affordable spay/neuter and vaccination services, adoptions, humane education, and a range of volunteer, foster care, and other community programs. The majority of municipal shelters are required to accept all animals, regardless of their ability to find homes or receive the proper care. They also routinely use euthanasia to end the lives of animals that cannot be safely placed for adoption and as a means of controlling both the number of animals in the shelter and the spread of disease.

1.2.2. Private Shelters

Animals in need receive care and temporary housing from private animal shelters, which are run without government funding and are vital to animal welfare. These shelters' funding may occasionally be more constrained than that of public shelters since they frequently depend on grants, donations, and volunteer assistance to run. On the other hand, private shelters frequently enjoy greater operational flexibility, which enables them to introduce creative programs and customize their offerings to the unique requirements of the animals and communities they serve.

A variety of services, such as adoption, vaccinations, spaying, and neutering, and educational initiatives to encourage ethical pet ownership, may be provided by private shelters. They frequently concentrate on the rehabilitation and finding new homes for animals, particularly those who may have been mistreated, abandoned, or neglected. To guarantee that animals are placed in secure and appropriate homes, adopters may be subject to a rigorous screening process during the adoption process at private shelters.

Managing the intake of animals while avoiding overcrowding, which can increase stress and the spread of diseases, is one of the challenges faced by private shelters. Many private shelters use waitlists to control the flow of animals into their care or operate on an appointment-based surrender system to address this issue. Additionally, they frequently collaborate closely with foster networks, which lessen the burden on shelter resources by allowing animals to be temporarily placed in homes until they are adopted.

In addition, private shelters might take part in community outreach and education programs aimed at addressing the underlying issues that lead to animal homelessness, like overcrowding and limited access to veterinary care. Private shelters play a significant role in the long-term decrease of stray and abandoned animals by promoting preventative measures and increasing public awareness.

One of the most recent trends in animal sheltering is the adoption of "no kill" policies by humane societies, which prohibit the euthanasia of adoptable animals due to a shortage of space for housing. This has a significant effect on programs for the care of animals: "no kill" groups or facilities with restricted admissions may limit their intake and detain animals for extended periods, which can present special difficulties for preserving the physical and mental well-being of the animals. Research conducted in American shelters has indicated that the longer an animal stays there, the higher the likelihood of them is getting sick; however, a recent study conducted in U.K. shelters revealed the opposite pattern in terms of feline upper respiratory infections (Coyne et al., 2008; Edinboro et al., 1999; Ward et al., 2004). This demonstrates the influence that differences in management techniques, shelter environments, and even cultural beliefs can have on the health of animals.

Even among private shelters bearing similar titles, there is a great deal of variation. It should be highlighted that very few, if any, descriptive terms can be taken for granted to mean the same thing in every shelter. The phrase "no kill" is only one illustration. While some shelters that use this term actually carry out some euthanasia, others that adhere to guidelines similar to those of "no kill" shelters (e.g., they restrict intake and/or refrain from euthanasia to manage the population) do not employ the phrase.

1.2.3. Other Types of Shelters

Not every shelter falls neatly into the private or municipal categories. Some municipal shelters ask for private donations in addition to accepting contracts from private organizations to provide services that are not required or covered by their agreements with the municipality. Certain breeds, ages, or animals with special needs may be the focus of other foster care and rescue organizations, or they may operate out of private homes. They frequently collaborate closely with shelters to save animals that, with the right veterinary and behavioral care, can be rehabilitated and put up for adoption. There are also a few sanctuaries that will take care of animals for the rest of their lives if they cannot be successfully or safely adopted.

1.2.4. Regulation of Shelters

Shelters are not held accountable to any specific organization, if at all. The ASPCA and the Humane Society of the U.S. (HSUS) are autonomous, independent organizations that do not manage or supervise local SPCAs, humane societies, or other animal rescue or adoption organizations. As such, there is no parent organization to which all shelters belong. The majority of states and the federal government do not regulate shelters. There are very few states with minimum requirements for the care of animals in shelters. Usually, the only regulations that apply to shelters are those that specify the procedures for euthanasia and the length of time that stray animals and bite victims must be held. However, since there is a growing trend toward more regulation and scrutiny of many aspects of shelter practice, shelter veterinarians should familiarize themselves with pertinent local laws.

1.2.5. Requirements for Data Reporting

This isn't usually the case, though a few states do mandate the reporting of specific data regarding the intake and disposal of animals. It's even unclear how many shelters are actually in operation in the U.S. Apart from the dearth of trustworthy information about the number of shelters in this nation, the absence of mandatory reporting mandates complicates the task of precisely ascertaining the number of animals taken in or put to death in shelters, as well as establishing benchmarks for disease prevalence and other crucial aspects of animal health in shelters.

Even though comprehensive data on a national or global level about animal welfare remains hard to obtain, individual shelters and local communities are progressively enhancing their abilities to monitor crucial data on the welfare of animals within their spheres. Thanks to the prevalence of digital, and sometimes web-based, database systems in shelters, the collection and analysis of aggregated data from various shelters could become more feasible in the near future.

1.2.6. Shelter Challenges

When working with a shelter, any veterinary professional must be aware of the difficulties and problems the organization faces to create a comprehensive and successful program that integrates wellness protocols with preventative health care techniques. Regardless of the specific purpose of the shelter, one objective should be to offer a clean, safe, and healthy environment that supports each resident's health maintenance and improvement, regardless of the duration of their stay or eventual outcome.

1.2.7. Shelter Resources

Regardless of their specific purpose or structure, shelters frequently encounter challenges in providing adequate resources for animal control and welfare services. The competition for scarce municipal funding between human and animal services, coupled with insufficient private fundraising efforts, often leaves shelters struggling to meet their program objectives and fulfill the needs of the animals in their care. By seeking the expertise of veterinarians, shelters can make informed decisions on how best to allocate their limited resources, focusing on enhancing animal health while aligning with the shelter's broader goals and mission. Despite financial constraints, shelters can create a healthful environment for the animals by diligently managing animal populations, ensuring good hygiene, and promptly isolating sick animals, reducing stress, and implementing other effective strategies.

Veterinarians should advise on resource allocation in shelters with a broad perspective. When all expenses are taken into account, preventing disease is frequently more economical than treating it, as well as more humane for the animals and better for the general public's health. Adoption fees can partially offset care costs, but the costs of euthanasia and disposal can be significant. This is true even if ethical concerns are disregarded. If more animals are saved and adopted as a result, a small investment in vaccination, diagnostic testing, or sanitation will be amply repaid. Therefore, when preventive measures are prioritized, the optimal strategy for animal health and adoption may also turn out to be a wise financial decision.

Thankfully, many procedures that improve the health of animals in shelters don't cost any more than those that don't work as well. According to immunology and vaccination, vaccinating animals as soon as they are admitted is not more expensive and has a far higher chance of protecting than vaccinating them a few days or even hours later. Best practices can sometimes be more affordable than the alternative. On sanitation lists "spot" or in-home cleaning as the recommended technique for keeping cat cages clean. Compared to more thorough daily disinfection, this requires less time and expensive chemicals, may lessen stress in cats, and may stop the spread of disease. It is urgent to alert management, funding agencies, and the public if there are insufficient resources to implement basic procedures to protect animal health. The welfare of sheltered animals as well as the public and community's animal health may be seriously impacted by the shelter's incapacity to control the spread of disease, which is why it should never be allowed to continue.

When expectations are not met concerning the facilities, staffing, and funding that are available, even shelters with an abundance of resources may face difficulties, potentially compromising the health of both humans and animals. Many ambitious and well-meaning public and private organizations try to adopt more animals than they can adequately care for, which leads to filthy, foul-smelling, packed facilities full of sick and possibly dead animals. This consequently results in increased mortality and euthanasia rates, decreased visits from prospective adopters, negative public relations (particularly in the event of disease outbreaks or the release of sick animals from the shelter), and pain and suffering for the animals. When shelters neglect to provide animals with the proper food, water, shelter, and veterinary care that are required to prevent suffering, they have occasionally been accused of cruelty to animals.

Together, veterinarians, and managers should determine how many animals can be housed in the facility in a humane manner to prevent such dangerous situations. Once this has been determined, resources should be set aside to provide the animals with the necessary care and humane treatment. This serves as the cornerstone for putting many other strategies discussed in this text into practice. When the shelter's basic capacity is not exceeded, it is much easier to prevent disease, reduce stress, and contain outbreaks without depopulating. Reducing the number of animals in the shelter does not necessarily mean that more animals will be put to death or fewer animals will be adopted.

1.2.8. Shelter Administration

It would be incorrect, as tempting as it may be, to attribute all of the issues in shelters to a lack of funding. A deficiency in understanding and communication between the management of the shelter and the veterinarian may be a hindrance to the success of the shelter health program. Even though it is becoming more widely acknowledged that the veterinary department plays a crucial role in the management of the entire shelter, some shelters continue to keep the medical program and general shelter operations apart, failing to understand how decisions about animal health affect every aspect of the shelter's operations.

This division can lead to misconceptions that cause shelter veterinarians to be accused of being callous to the suffering of the animals, of not comprehending the objectives or issues facing the shelter, or of gross incompetence and cruelty when tough medical decisions are made that are unpopular with staff or inconsistent with previous procedures.

The responsibilities and role of the veterinarian should be clearly outlined in each shelter specifically to help prevent some of these problems (Miller, 2007). Establishing decision-making and authority levels, as well as defining which areas fall under the purview of the veterinarian, are crucial. Will the veterinarian, for instance, decide which animals are appropriate for adoption? How are decisions about euthanasia made? Who conducts behavior assessments? Who chooses the food? Who decides how the animals are moved around the facility or how staff members are assigned?

Sometimes, staff members at shelters may insist that veterinarians are not necessary for some aspects of the facility's health care program. This could be especially true if there has been a history of less-than-successful communication with neighborhood veterinarians who were unaware of the limitations and considerations of shelters. It is important to inform staff members about stress reduction, population management, sanitation, and facility design, all need veterinary advice just as much as traditional medical advice regarding vaccines and anthelmintics. The implementation of new healthcare protocols can be challenging for all parties involved if the veterinarian is limited to making medical decisions only and is not involved in strategic planning, administration, training, or management. For shelter veterinarians to be effective, they must be an essential member of the management team, empowered to decide on any issue relating to the health and welfare of the animals or to participate in decision-making processes. If the vet is not on the management team, a clear procedure for regular communication between the veterinary staff and management regarding matters should be established.

Staff members working at shelters may also run into issues when they consult with neighborhood doctors who are unaware of the distinctions between shelter medicine and private practice. As a result, they may criticize procedures that shelter veterinarians recommend, especially when taken out of context. Despite the growing awareness of shelter medicine, some private practitioners might continue to treat shelter animals according to their own standards of care, failing to recognize the differences in recommendations for vaccinations, treatments, spay/neuter procedures, and so on. for this demographic are predicated on a distinct set of risk variables, instruments for assessment, situations, and assets. Similar to how there are various but equally effective standards of care for large animal herds, this different standard of care should not be interpreted as being lower but rather as being shelter-specific.

Shelter medicine is relatively new compared to most other veterinary specialties, and many veterinarians are not even aware that it exists, much less that there are resources available to assist them in resolving the particular problems that arise frequently in shelters. It is normal for there to be updates and modifications to philosophies and practices when a new specialty emerges, as well as occasional arguments among the "experts." Conflicting advice from shelter experts or changes in priorities shouldn't be interpreted as mistakes in judgment. Emerging diseases, new research, and shifting demographics call for veterinarians to be adaptable in reevaluating programs and free to make protocol changes without fear of retaliation. Several commonplace beliefs and practices have come under closer examination in recent years, particularly as new research that pertains to or targets shelters is conducted. Among these unanswered questions are the following: should newly admitted animals always be placed in quarantine? Are footbaths always beneficial? How certain pathogens aerosolize and contribute to disease transmission? How important is footbath cleaning?

1.2.9. Disease Transmission

To create a program that can stop the spread of disease in shelters, a comprehensive understanding of the mechanisms involved in transmission is necessary. Although aerosolization, droplets, and direct contact are important factors in the spread of disease, the most popular way for the disease to spread in shelters is through fomites. It's important to understand not only which species–including humans–are at risk from the pathogen but also its routes of shedding, that is, as in the secretions of the nose, eyes, urine, and feces, and so on. While creating a sanitation program, volunteers, and staff need to be made aware of how much of a role they play in the spread of disease through their hands, clothes, and other items that aren't properly cleaned. If employees are aware of the penalties for breaking the rules or taking short cuts, they are far more likely to follow the tight sanitation regulations.

1.2.9.1. Incubation Period, Shedding, and Carrier States

When it comes to disease control, consideration must be given to the incubation period, duration, shedding pattern, and carrier states. Determining whether an animal came into the shelter sick or contracted the disease there is made easier with knowledge of the incubation period. Setting up suitable quarantines, hygienic practices, and other outbreak control measures requires this information. Unawareness of the parvovirus shedding pattern, for instance, can cause major issues if vulnerable animals come into contact with recovered patients who might still be shedding the virus. Understanding parvovirus shedding patterns is also crucial for interpreting antigen tests accurately and for determining how a recent vaccination may impact the results. Veterinary professionals may find it particularly difficult to control feline URIs in shelters if they are not aware that herpes virus and calicivirus have inapparent carrier states and that herpes recurs about a week after a stressful event.

1.2.10. Shelter Design

Without a doubt, a large number of shelters are located in inadequate spaces. They are frequently found in structures like factories and warehouses that were initially intended for uses other than the care of animals. The purpose of the shelter may have changed significantly since it was first built, or it may have been designed during a period when the population's demographics were different. Adoptee-focused shelters that retain animals longer may discover that they lack the space necessary to accommodate isolation f in an animal becomes ill and needs medical attention. They might not have enough room to meet the animal's physical needs, like areas for play and exercise, areas for grooming, and so on, as well as its emotional needs. Cats, kittens, and teenage dogs with problematic behaviors may now make up a larger portion of the population in shelters that were initially built primarily to handle dogs or litters of puppies. Normal building wear and tear can leave concrete surfaces with cracks and crevices that are difficult to disinfect.

Renovating or retrofitting the shelter to address physical and design flaws is a necessary step towards implementing a successful disease control program. In certain instances, investigating the necessity of building a new facility or making capital improvements may be necessary to best fulfill the shelter's overarching goal of giving the animals a secure and healthy environment. But, inadequate animal care should never be justified by a run-down facility. Even in a building that is far from ideal, there are numerous steps that can be taken to maintain animal health. Important elements of a wellness program that are not building-dependent include vaccinations upon intake, the provision of toys and bedding, and vigilant population monitoring for disease.

1.2.10.1. Special Considerations for Shelter Facilities

Animal shelter architecture differs greatly from that of veterinary clinics, breeding grounds, or labs. A well-constructed shelter ought to possess the flexibility to adjust to different circumstances that it might encounter, such as a pandemic, an unexpected influx of animals taken from hoarding situations or responding to a disaster. There should be multiple smaller areas that can be modified as needed for isolation, quarantine, or other specific uses, rather than a few large, open areas for housing animals. Simple traffic patterns that route people and animals from areas containing young, healthy animals first to areas containing high-risk or sick animals last are ideal for the shelter. All of the animal shelter's housing sections should have proper drainage and be made of sturdy, nonporous materials that can resist frequent washings with hot water, detergents, and disinfectants. Regularly used materials in veterinary hospitals are frequently chosen more for their visual appeal than for their functionality, and it's possible that they won't hold up to the strict sanitation standards used in shelters.

Making sure the facility has enough ventilation is one of the keys to maintaining the health of a confined population. Animal-level ventilation should be measured, and ambient room temperature should be comfortable, species-appropriate, and free of fluctuations. One should never undervalue the benefits of sunshine, fresh air, indoor, and outdoor running, and open windows. To make up for shortcomings in their ventilation systems, a lot of shelters use fans, high-efficiency particulate air (HEPA) filters, and other sometimes ineffective techniques.

The use of noise-reducing materials, both individual and communal cages, runs with guillotine doors to make cleaning enclosures with dangerous dogs easier, housing for species other than dogs and cats, areas for food preparation, laundry, and storage, and euthanasia facilities are some other factors to take into account when designing shelters. When designing or retrofitting a shelter, it is advisable to consult architects and contractors with experience working with shelters due to the unique needs of these structures. More detail is provided on issues about the environment and design of shelters.

1.2.11. Sanitation

Sanitation is one of the main pillars of any shelter health program. Each chapter discusses this idea in detail. The sanitation program ought to be customized for every specific shelter setting, taking into account the staff's degree of training and expertise, the surfaces that need to be cleaned, their state of repair or disrepair, and the prevalence of diseases among the clients of that particular shelter. It is almost always possible to design a program that is reasonably effective, even under less-than-ideal conditions. Using hot water, soap, and degreasers to thoroughly clean all contaminated surfaces and potential fomites should be the first step in eliminating as many pathogens as possible. Any pathogens that are left should then be inactivated using the proper disinfectant. The veterinarian's contribution to the sanitation protocol's design goes far beyond choosing the right disinfectant and jotting down a few guidelines. Regular staff training, periodic review, and updating of protocols are necessary; whenever a disease outbreak arises, a hands-on assessment of sanitation protocols should take precedence.

1.2.12. Stress

Stress plays a well-established role in the spread of disease in both human and veterinary medicine. Unfortunately, many managers and staff at shelters usually overlook it because they are too busy cleaning cages or taking care of other, more pressing issues. One cannot overstate how crucial it is to manage stress. Stress significantly affects an animal's ability to function. Many diseases are known to be either directly or indirectly linked to stress, and it can lead to behaviors that make an animal less likely to be adopted. Apart from the general impacts on immunity and disease susceptibility, the relationship between stress and herpes virus reactivation in cats, which results in symptoms of upper respiratory diseases, is especially significant in animal shelters. In addition, stress can result in lesions and symptoms that are similar to those of actual clinical diseases, such as acral lick nodules, depression, diarrhea, and vomiting. The part stress plays in the spread of illness and strategies to reduce its effects on the people living in shelters.

1.2.13. Treatment, Adoptability, and Euthanasia

Euthanasia decisions are the most emotionally taxing, frustrating, venomous, and polarizing decisions made in the shelter. Selecting when and how to treat illness when it arises is one of the problems shelters have. Many infectious diseases, like ringworm or upper respiratory infections, which are not intrinsically harmful or potentially fatal for individual animals, present serious moral and ethical challenges for animal shelters that lack the resources to treat or otherwise handle them. Because these relatively benign diseases may be highly communicable or zoonotic, they could prove fatal in the shelter. The public and shelter employees may find it very challenging to comprehend that saving the lives of a small number of afflicted animals occasionally requires the expenditure of valuable resources that could be utilized to save the lives of many more animals.

Care of sick animals can expose many others to infection and jeopardize their lives if proper isolation facilities and staffing levels are not met. However, being able to treat at least some animals can boost volunteers' and staff morale and public support, as well as improve their ability to report diseases. Ideally, preventive programs should be implemented so that animals will stay healthy, and thus not require treatment, and isolation facilities or other alternatives (e.g., off-site care) are organized and created in a way that makes treatment both humanely and safely possible. In the interim, adopting policies that both protect the lives of the entire population and benefit the individual animal can be a difficult balancing act.

1.2.14. Monitoring and Measuring Shelter Animal Health

While measuring outcomes like adoption and euthanasia has received a lot of attention lately, metrics reflecting the health or well-being of the animals in the shelter have received less attention. This is unfortunate because, without a plan for systematic measurement, it is hard to judge the relative value of different investments in animal health, communicate challenges or successes to stakeholders and the public, or recognize emerging problems before they become serious. On the other hand, a system that tracks how procedural changes affect disease control can help reduce conflict and improve staff adherence. The extra expense and work involved in this practice will be more easily accepted, for instance, if it can be shown that altering cleaning procedures or vaccination schedules improves animal health.

1.2.14.1. Counting the Number of Cases of Disease

Counting the number of disease cases that arise over time is one of the simplest ways to assess the health of animals in shelters. This calls for a system to identify and document the occurrence of diseases as well as a consistent case definition. In certain instances, this can be achieved by appropriately utilizing shelter software systems, given that a field is present to document a distinct diagnosis associated with a date of diagnosis. The number of animals admitted during the same time period, the total number of "days at risk" (the number of animals present each day who may be susceptible to the disease in question), or ideally both, can be reported concerning the number of cases. It is possible to track the number of cases for all relevant diseases or syndromes, or just a select group of "marker" diseases. For instance, the most prevalent medical issue affecting cats and dogs in many shelters is URIs, also known as kennel coughs. Comparing URI levels over time to more sporadic conditions like parvovirus or feline panleukopenia, differences can be relatively easy to identify. A rise in URI is a warning sign that a more serious disease outbreak could be imminent in the population. On the other hand, a modification in housing, cleaning, or other practices that lowers URI is probably going to lower the chance of other health issues as well.

1.2.14.2. Shelter-Acquired Disease Leading to Euthanasia

Many additional health metrics are available in addition to the counting of disease cases. The number of animals that are brought to the shelter in "healthy, adoptable" condition and are subsequently put to death by illnesses they contracted while in the shelter is arguably one of the most significant for shelters that carry out euthanasia. If the proportion of animals falling into this category rises, there should be significant concern. Commonly used shelter software systems can report this if the intake status and outcome fields are appropriately used to track these data.

1.2.14.3. Sick Animal Care Days

The number of "sick animal care days" is a useful and easily accessible indicator of the health of the population residing in shelters as well as the expense of treatment as opposed to prevention. "The monthly and annual total of days spent caring for sick animals can be obtained by compiling a daily count of all sick animals, either manually or through a computerized report, and adding this over time. Alternatively, this number can be obtained by measuring the length of the illness and adding it to each case over time. This can be done manually or through a computer report, as long as the dates of onset and conclusion are accurately recorded.

Tracking sick leave days for a single illness, like URI, can be very beneficial as it is likely a good indicator of overall shelter health. The true cost of disease can be estimated by multiplying the number of sick days by the average daily cost of care. Investing in preventive measures might be more easily justified in this situation. For example, a one-time investment in cat housing that is expected to drastically lower feline URI may quickly pay for itself in the form of lower staff time and caretaker expenses.

1.2.14.4. Shelter Death Rate

The clearest possible sign of serious health issues in shelters is probably the quantity and proportion of animals that pass away (as opposed to being put to death) there or in foster care. Carefully tracking, monitoring over time, and keeping this number apart from all other "outcome" categories are all necessary. Because the death rate tends to rise slightly during the summer months in conjunction with kitten season, monthly numbers should be compared to the same month in prior years. The average annual death rate, according to author Hurley's small, informal survey of shelters, was 0.75% (range: 0.18 to 1.61%). The higher end of this range was reported by shelters with comparatively high annual intake for death rates. Death rates exceeding 2% to 3%, or any rise in the death rate, should be closely monitored.

Even in situations where a positive policy change (like adding a foster program for neonatal kittens, who are prone to relatively high mortality rates) results in higher death rates occasionally, the impact of higher animal mortality on foster, rescue, and adopter morale needs to be acknowledged and addressed. Every animal death that occurs in a shelter or foster home should be thoroughly looked into and recorded. This includes determining whether the animal was healthy when it was brought in, when health issues were discovered or diagnosed, whether the animal was receiving the proper care for the illness that led to its death, where the animal died (a particular area of the shelter or foster home), how long the animal had been there–both healthy and sick–before it passed away, and the cause of death, if known. Serious concerns should be raised about even a slight increase in the number of healthy animals admitted to shelters that are dying from acquired diseases.

1.2.14.5. Other Indicators of Shelter Animal Health

Additional indicators of health include the quantity of legitimate health-related complaints that shelters receive after adoption; the number of recheck visits for animals covered by post-adoption care programs; the number of claims for pets covered by shelter pet health insurance plans; the number of days from intake to vaccination; the ratio of vaccines administered to animals admitted (if the ratio is lower than the number of vaccines administered, it may indicate that not all animals are receiving vaccinations); the number of daily treatments; and the quantity and cost of medications used. Any changes to these figures should be examined in light of other factors. An increase in the number of treatments, their cost, or the quantity of drugs used, for instance, does not always indicate a negative trend. If, instead, this is the result of more animals entering the population healthy but subsequently becoming ill and needing medical attention, rather than a deliberate plan to expand the variety and caliber of treatments available, this indicates a failure in prevention that needs to be addressed.

These data should be viewed as the essential foundation of a successful shelter health program, even though setting up health monitoring systems in a hectic shelter setting may seem daunting. Investing blindly in health practices aimed at a population is likely to be at best suboptimal, just as diagnosing and treating individual animals cannot be done without conducting a physical examination and gathering additional health measures. In the end, a well-thought-out health measurement system is an economical and humane investment since it provides veterinarians and shelter management with the best ways to sustain animal wellness and promptly pinpoints issues.

1.3. WELLNESS

Put simply, the main objective of any animal shelter, regardless of its resources, mission, or philosophy, must be to ensure that the animals are kept as "healthy and happy" as possible while they are there. Furthermore, maintaining public safety and health must always be the top priority. By setting a good example and modeling proper animal care, shelters also accomplish the goal of public education.

The preservation of health is the definition of wellness. Wellness is made up of either behavioral (or emotional) and physical health. For instance, a dog might be in good physical health and not be afflicted with any physical illnesses, but it might have severe separation anxiety. It is not possible to evaluate this animal as fully healthy, and treating its behavioral issues is necessary to guarantee its wellbeing. Therefore, both physical and behavioral health must be included in a wellness program to maximize animal health in the shelter.

When creating a wellness program for an animal shelter, taking care of the shelter environment is just as important as taking care of the animals. Without the careful application of environmental wellness protocols, infectious diseases and problem behaviors are impossible to control or prevent, even in the best-designed facilities. Environmental wellness is often underemphasized in small animal practice because many owners are used to giving their pets a reasonably healthy environment. On the other hand, regardless of the actual physical layout of the facility, an organized program to address environmental wellness is crucial in the context of an animal shelter. It is crucial to take proactive steps to keep clean, hygienic settings free from overcrowding, where animals are divided into groups according to their health and species and receive regular care from committed, well-trained personnel.

1.3.1. The Critical Importance of Wellness Protocols for Shelters

Cats and dogs kept in animal shelters frequently suffer from infectious diseases, stress, and problematic behaviors. People who are euthanized more frequently than adopted are those whose physical or mental health has been compromised. Animals that come into shelters under a lot of stress and run a serious risk of contracting infectious diseases. Stress from even brief stays in a shelter can harm an animal's behavioral and physical well-being, which will reduce the appeal of the animal to prospective adopters and negatively impact animal welfare.

For obvious reasons, keeping animals in shelters healthy or happy poses significant challenges. Numerous factors can contribute to the spread of infectious diseases in shelters, such as the addition of new animals to the facility, high-density housing, the close proximity of animals with varying ages and susceptibilities, stress induction, and insufficient vaccination or delayed vaccination response. The shelter environment has all of these risk factors and more, so there is always a chance of contracting an infectious disease. Furthermore, a number of illnesses commonly become endemic in establishments housing animal populations.

Pets in foster care may exhibit a wide range of behavioral signs of stress and anxiety when they are confined, such as hyperactivity, activity depression, frustration, aggression toward barriers, and stereotypical behavior like pacing or pawing. To keep confined animals from experiencing excessive stress, shelter environments need to be improved. Because stress has a significant impact on both behavior and the spread of infectious diseases, programs aimed at reducing stress also help to minimize the morbidity of endemic infectious diseases.

Every shelter must implement a wellness program since it is their duty to care for animals in a humane manner. Housing animals in circumstances that could lead to illness and poor welfare is just inhumane. When wellness programs are not implemented and closely observed, these conditions are reasonable to anticipate. Animal shelters must also be able to present healthy animals in a healthy environment if they are to compete with other sources of animals for adoption.

1.3.2. Goals of a Shelter Wellness Program

A shelter wellness program's objectives are to maximize the animals' physical and behavioral well-being, minimize infectious diseases, and stop the spread of zoonotic illnesses. Simply put, the goal of a wellness program should be to maintain both public safety and "healthy and happy" animals. Adoption rates and public relations may both benefit when shelters achieve these objectives.

Programs for shelter wellness should provide a wide range of preventive techniques rather than focusing on the treatment of a specific illness or issue (a holistic approach). Furthermore, wellness programs in an animal shelter need to take into account the general health of the animals as well as the health of individual animals. Herd health and shelter medicine have been compared (Hurley, 2004). In fact, population medicine in the shelter employs a methodical approach to maximize animal health in the collective, very similar to a herd health approach. In an animal shelter, the primary objective is to ensure the welfare of cats and dogs, as opposed to large animals, where the focus is on production through herd health programs. Setting objectives and devising strategies for population monitoring is essential to guaranteeing the health and welfare of the animals in the framework of creating and carrying out a thorough wellness program for the shelter. Medical decisions must take animal welfare and resource availability into account, as well as the context of the population as a whole. In conclusion, evaluation, and subsequent actions are carried out both at the individual and population levels.

It is the ethical duty of every animal shelter to do everything within its power to ensure the well-being of every animal under its care, even in the absence of many laws and standards of care. Neglect is defined as receiving veterinary care that is insufficient or delayed, and certain states have laws against it. A shelter should never participate in any actions or inactions that cause its residents' animals to suffer needlessly or unjustifiably. Euthanasia must be used if there are no other options available within or outside of the shelter to alleviate animal suffering when circumstances develop where animal welfare cannot be controlled, whether because of a physical or behavioral illness or environmental factors like overcrowding. But euthanasia shouldn't take the place of giving shelter animals the care they need and putting in place sincere initiatives to reduce the killing of adoptable animals and feral cats.

The goal, philosophy, and resources of the shelter will all influence the wellness programs offered. Within shelters, these programs may also differ based on the season and the rate of intake. Remember that the secret to developing successful protocols is to maximize each animal's well-being during its stay while preserving the well-being of the population as a whole. Research is required to improve the definition of protocols for controlling and treating behavioral and physical illnesses prevalent in shelter animal populations, as well as to improve evaluations of the welfare and standard of living of cats and dogs.

1.3.3. Quality of Life

Maintaining the quality of life for shelter animals requires every effort. Similar to "happiness," defining quality of life is still challenging. Welfare, well-being, and quality of life are influenced by both psychological and physiological variables. These elements are numerous, intricate, and highly personal. Quality of life "is comprised of an array of affective states, broadly classified as comfort-discomfort and pleasure states," according to McMillan (2000). The general rule is that life quality increases with the amount of positive effects and the decrease of negative ones."

Although there are no objective standards for measuring the quality of life in cats and dogs, medical, and behavioral professionals can and should make subjective evaluations regularly (weekly or even daily, as recommended) based on the best available data. Researchers are focusing more on validating the quality of life measures because they can be used for welfare audits in animal shelters and other settings where populations of animals are housed, define minimum housing standards, and help ensure humane endpoints for healthcare. The five freedoms established by the Farm Animal Welfare Council serve as a standard for evaluating animal welfare and quality of life.

The five freedoms:

1. Freedom from hunger and thirst by ready access to fresh water and a diet to maintain full health and vigor.
2. Freedom from discomfort by providing an appropriate environment, including shelter and a comfortable resting area.

3. Freedom from pain, injury, or disease by prevention or rapid diagnosis and treatment.
4. Freedom to express normal behavior by providing sufficient space, proper facilities, and company of the animal's own kind.
5. Freedom from fear and distress by ensuring conditions and treatment that avoid mental suffering.

1.3.4. Considerations Regarding Infectious Disease Transmission

Health is maintained under normal circumstances even though infectious agents are constantly present in the environment. It is commonly known that the host, the infectious agent, and the environment all interact intricately to determine how an infectious disease develops. Animal health is known to be influenced by a variety of factors, including the species, age, sex, general health, immune status, stress level, and genetic predispositions of the host (Greene, 1998a). Both the virulence and the modes of transmission of infectious agents differ. They frequently survive in the environment due to their resistance to disinfection, and many of them also produce carrier states that further contaminate the environment or expose other animals directly. The possibility that a disease will spread in the shelter environment depends on many factors, including the length and intensity of exposure to an infectious agent, modes of transmission, routes of immunization, carrier states, and mutation rates. Inhalation, ingestion, contact with feces, urine, other body secretions, contact with fomites, or even direct contact with infected animals or carriers can all result in the transmission of the disease.

1.3.5. Principles of Disease Control

It is important to recognize the general principles of infectious disease control:

1. Removal of infected animals (through isolation, foster care or euthanasia).
2. Mass vaccination.
3. Mass treatment.
4. Good husbandry practices (animals and environment).
5. Education of personnel 6. Quarantine of new arrivals.

In the context of an animal shelter, it is imperative that these broad guidelines be followed whenever feasible. When disease is present, these principles form the basis of all disease control efforts when combined with attentive monitoring and early disease recognition. However, it should be mentioned that while mass vaccination, treatment, proper husbandry, and staff education can be accomplished in animal shelters, putting an actual quarantine in place is difficult. In most cases, maintaining strict biosecurity for animals that are most vulnerable to infectious diseases (such as young puppies and kittens) may be the best course of action, unless specific conditions like serious disease outbreaks call for a full quarantine or even a temporary closure of admission. However, creating, and putting into practice thorough wellness protocols for the environment and animals in shelters continues to be the most crucial approach to disease control.

1.4. THE PROBLEM ORIENTED APPROACH TO SMALL ANIMAL MEDICINE

It is commonly acknowledged that the problem-oriented approach to medicine is the best method for diagnosing and treating small animals. Any abnormality that needs to be managed medically or surgically, or that compromises an animal's quality of life, is considered a problem in small animal veterinary medicine (Lorenz, 1993). Problems in the context of an animal shelter can also be defined as circumstances (like aggression) that compromise public health and safety.

An animal's problems are methodically identified and addressed using the problem-oriented approach. The four steps of this approach are (1) database collection, (2) problem identification, (3) plan formulation, and (4) assessment and follow-up. These represent the foundation of the clinical reasoning process. With this method, the doctor can approach each patient logically and ensure a complete and accurate assessment, allowing the right course of action to be taken. Effective systems that enable the evaluation of individual animals while taking the population into account are crucial in the setting of an animal shelter, where populations of animals are kept. A special fusion of population health and individual patient care is known as "shelter medicine."

1.4.1. Database Collection

Every patient should have an initial or guaranteed minimum database obtained. While opinions on database size are frequently divided, it is universally agreed upon that every patient should, if at all possible, have a comprehensive history and physical examination included in the database. A shelter's resources and philosophy must be taken into account when determining what information should be included in the guaranteed minimum database for each patient.

The author suggests including PCV/TS (packed cell volume and total plasma solids), urine specific gravity, and a dip stick in the minimum database of elderly animals that will be available for adoption about senior animals in the shelter. These procedures are highly economical and provide a broad screening of numerous body systems. These ought to be seen as additions to the senior animal's physical examination. The veterinary clinician should also think about fine-needle aspiration for in-house cytologic evaluation of all cutaneous and subcutaneous masses whenever time and resources permit. This easy and affordable procedure of assessing lumps and bumps may reveal potentially cancerous growths that would otherwise go undiagnosed or reassure prospective adopters that any growths are not excessively concerning.

1.4.2. History

The history takes precedence over the physical examination when it comes to solving medical puzzles. The history makes the clinician aware of any possible issues. Of course, getting a precise history may not always be feasible in a shelter. Some animals that have little to no information about them will be brought in by good Samaritans or animal control officers. Certain animal shelters offer a designated area, such as drop-off cages or runs, wherein animals can be returned after regular business hours. In this instance, every attempt should be made to gather a history by way of questionnaires that surrenderers can complete at the time the animal is given up. It is greatly preferred that staff members be present to accept the animal directly and get a history at the time of relinquishment. Nevertheless, relinquishing owners may or may not give complete or accurate information out of concern that the animal may be put to death if they are forthright about its issues.

However, if one is available, history can be very helpful and could save a lot of money, time, and stress for both the staff and the animal. Procedures for taking basic patient data, such as behavioral and physical characteristics as well as the reasons for relinquishment should be in place for intake. One cannot stress how crucial it is to gather historical data. Historical data can frequently be utilized to hasten the pet's disposition.

1.4.3. Physical Examination

At the time of admission to the shelter, or as soon as possible thereafter, every animal (that is judged safe to handle) should undergo a physical examination. After the animal has been in the shelter for two to five days, in addition to the physical examination, a behavioral examination and assessment should be carried out. Assessing and processing seemingly socially owner-relinquished animals with no history of aggression in their prior home as soon as possible would be an alternative plan for high-volume shelters that are unable to hold animals for additional days while they acclimate to their new surroundings. This would allow for immediate placement of the animals in adoptions, foster care, or with rescue organizations. The guaranteed minimum database's most crucial component is its physical inspection. Adhering to a physical examination form guarantees a thorough and methodical assessment of every bodily system. Similarly, using a standardized behavior examination form will guarantee that assessments are carried out in-depth.

A precise physical description of the animal and a close look for identification are especially crucial during the physical examination at the shelter, as these details could help with pet-owner reunification. A meticulous check for the presence of a tattoo or microchip should be part of the examination. Every animal should have a routine microchip scan done on it when it is brought in, before it is put up for adoption, or before it is put to death.

Microchips are activated by low-power radio frequencies that are released by scanners, and upon activation, they transmit a unique identification number that can be associated with an owner via a database. In the United States, there is currently no standard radio frequency for microchips; instead, three distinct frequencies are marketed. For this reason, when scanning animals, it is imperative to use a global scanner that can read the frequencies of all available microchips. Furthermore, the scanner needs to have fully charged batteries to reliably identify microchips. Additionally, a consistent technique that is described below must be used to scan the entire animal to obtain an accurate reading. To enhance the chances of successful detection and ideal chip orientation, gently shake the scanner back and forth. In other words, the scanner should be very close to the animal, either lightly touching the skin or held just over the skin less than an inch away from contact with the patient. Most scanners should be held parallel and in near or close contact with the animal during the scanning process.

No more than 0.5 feet per second should be the maximum speed for the scanner. This is so that they can read all possible chip frequencies using global scanners, which have to switch between different modes. The typical implant site, which is halfway between the shoulder blades, is where scanning should start. If the microchip cannot be found, scanning should proceed down the sides, back, shoulders, neck, and elbows cranially and stifles caudally. The scanning areas should be covered by the scanner moving in a transverse (side to side) S-shaped pattern. The scanner head should be turned 90° if no microchip is found, and the scan should then be repeated on both sides in an S-shaped pattern running longitudinally.

Tattoos are frequently found in the inguinal region, abdomen, and inner pinnae. Visual examination of the pinnae should be done for cats (even those that might be feral or too feisty to handle) to spot a cropped ear tip, which could be a sign that the animal is a sterilized member of a supervised cat colony.

Finding conditions that call for particular housing considerations is a crucial component of the intake exam for animals in shelters. Typical instances include animals that appear close to term in pregnancy, animals suspected of carrying infectious diseases for which isolation would be necessary, and other animals that might have unique medical requirements. To properly house animals in enclosures that are extra secure and made to minimize animal handling, such as those with guillotine doors or other separators, animals that are judged unsafe to handle upon entry should be identified.

1.4.4. Plan Formulation, Assessment, and Follow-up

After issues are recognized, a strategy can be developed to deal with them within the parameters of the shelter's philosophy, mission, and resources as well as the legal holding periods (which give pet owners the chance to locate and reclaim their animals) as set forth by regional and state legislation. Three components should be considered in the plan: (1) treatment, (2) any necessary diagnostic testing, and (3) staff or adopter education.

The assessment and follow-up phase is the last in the medical problem-solving process. In the shelter, animals may be prioritized for adoption, foster care, rescue organizations, isolation, or euthanasia. To ensure the welfare of each individual animal as well as that of the population and the shelter staff, it is essential for animals that spend extended periods in shelters to undergo routine reassessments to detect any new medical or behavioral issues that may arise.

Certain animal shelters may decide to accept animals that already have behavioral or medical issues. It is essential to implement a humane plan for diagnosis, treatment/management, monitoring, and housing when such animals classified as "special needs" are kept in the shelter. Animals with special needs shouldn't be housed in the shelter unless they can receive sufficient medical attention, including pain management. When determining if animals with special needs can be humanely cared for in the shelter, the following goals and considerations should be addressed:

1. Will the care provided to the animal result in a cure or adequate management of the disease or problem behavior?
2. Will the animal be adoptable?
3. What steps can be taken to minimize the holding time required for treatment?
4. What measures must be implemented to prevent transmission of disease to other animals or people?
5. Can the shelter afford the cost of and time for care? How will this impact resource available for other animals?

6. Can adequate care realistically is delivered in the shelter or in foster care?
7. What factors will be used to assess if the treatment plan is working or should be modified?
8. If the animal is adopted, what can be done to decrease/ eliminate the return of the animal for its special needs?
9. If the pet is not adopted, what welfare assessment will be used to measure the quality of life in the shelter?
10. Do humane long-term care options exist in the shelter?

For the purpose of monitoring all animals, a consistent system of behavioral and physical health surveillance ought to be established. Medically trained personnel or volunteers should, at the very least, make rounds twice a day to check on each animal and its surroundings for any indications of health issues. For infectious diseases to be effectively controlled, early detection and prompt intervention are essential.

1.4.5. Medical Record Keeping and Data Collection

Medical records are necessary to guarantee prompt and high-quality care. Federal drug laws, state, and local practice acts, national veterinary medical associations' guidelines, and record-keeping protocols should all be followed. Each animal should have a medical record that contains the following information: the date of intake; an animal ID number; signalment; physical description; findings from the physical and historical examinations; the results of the microchip scanning; the animal's weight and condition; the names and dosages of all prescribed and administered medications, including those for vaccines, parasite control products, other treatments, and anesthetic agents; the outcomes of any diagnostic tests carried out; the surgical procedure(s) carried out; any abnormalities that are discovered; and any other relevant information about the animal's condition.

Although standardized examination and operative/surgical reports can be used, they should also accommodate additions and medical updates as needed and suitable. During a systematic body system review, standardized examination forms that record both normal and abnormal findings are especially useful. A determination of the animal's adoptability as well as the behavioral history, examination, and assessment results should all be included in the medical record.

The population gains from careful record keeping and data collection in addition to the obvious need for medical record keeping in the context of individual animals. Within the field of population medicine, for instance, the wellness program's objectives could include lowering the incidence and prevalence of infectious diseases both within the shelter and after adoption, as well as lowering the incidence of problem behaviors there, the rate at which animals are returned to the shelter due to behavioral issues, raising the adoption rate, and so on. It is feasible to gauge progress toward these objectives by identifying and monitoring quantifiable elements, which are frequently referred to as performance targets in large animal medicine. The term "welfare targets" may be a more suitable term in shelter medicine. By assessing individual welfare targets, it might be possible to gauge the effect of protocol modifications on population health once baseline data (like disease rates) have been established. It will be simpler to recognize both positive and negative trends in animal health if there is a system for routine reporting.

1.4.6. Policy and Protocol Development

Written policies and procedures outlining the management of medical and behavioral issues ought to be in place at shelters (Hurley, 2004). Research and facts, as well as the goals and values of each particular organization and the resources (staff, facilities, and veterinary care) at its disposal, should all be taken into consideration when developing policies and procedures. The best group to create policies and procedures is the committee that oversees shelter health, which should include the manager or director, medical personnel, and other important figures. A general policy regarding the disposition of animals affected by the condition, a description of the methods that will be used to diagnose or recognize the condition, and definitions or descriptions of the disease or problem behavior in question should be the objectives of the committee. Protocols should also cover information on notifying the right parties, housing, decontamination, treatment, and documentation for each case. The methodical triage and care of the animals in the shelter should be governed by these documented protocols.

1.5. WELLNESS: PHYSICAL HEALTH

Through the wellness program, the fundamental physical health needs of dogs and cats should be methodically addressed. The vaccination, parasite control, spay/neuter, identification, appropriate diet and exercise, prevention of periodontal disease, and breed-specific care should all be covered by the protocols.

1.5.1. Vaccination

Due to the high risk of exposure and the potentially fatal effects of illness, vaccination against a few specific diseases is imperative in shelter settings. The

AAFP (American Association of Feline Practitioners) and the AAHA (American Animal Hospital Association), respectively, have provided comprehensive guidelines for the vaccination of dogs and cats in animal shelters (Carmichael et al., 2006; Richards et al., 2006). It is recommended that all cats and dogs who enter shelters receive a few core vaccinations. Core vaccines target illnesses that are prevalent in shelters, have high rates of morbidity and mortality, and for which vaccination has been shown to offer comparatively good disease protection. *Feline parvovirus* (FPV or panleukopenia), *feline calicivirus* (FCV), and FHV-1 (*feline herpes* virus type 1 or *feline rhinotracheitis* virus) are the three main vaccinations given to shelter cats. While vaccination against FeLV is not deemed essential for cats in shelters, it might be necessary if cats are kept in groups. According to AAFP guidelines, this vaccine should be injected into the left hind leg when first administered. A booster shot should be given within two to four weeks after the initial vaccination. *Chlamydophila felis* (*C. psittaci*) and *Bordetella bronchiseptica* are not recommended as core vaccines, but they may be of benefit when clinical signs of these diseases are present in the shelter and diagnosis is confirmed by laboratory evaluation. Their efficacy is moderate, and reactions are more common than with most other feline vaccines; ongoing use should be periodically reassessed. Canine parvovirus (CPV), canine distemper virus (CDV), canine adenovirus (CAV-2), hepatitis, parainfluenza (CPiV), and *Bordetella bronchiseptica* are among the core vaccinations for shelter dogs.

When a licensed veterinarian is available to administer the vaccine (or shelter staff in compliance with state laws), rabies vaccination is advised for both cats and dogs before adoption. The vaccine should be injected into the right hind leg. As an alternative, the veterinarian of the new owner might give the animal a rabies vaccination as soon as possible after adoption. This substitute could potentially motivate recently acquired pet owners to form a bond with a private veterinarian. It is important to inform new owners that most jurisdictions require dogs to be vaccinated against rabies, and that obtaining a dog license may require proof of vaccination. In addition, rabies vaccination is required for animals kept in shelters for an extended period of time. A licensed veterinarian should administer a rabies vaccination to animals under quarantine for rabies bites in compliance with the guidelines outlined in the most recent Compendium of Animal Rabies Prevention and Control. Finally, it is strongly advised that shelter employees receive a rabies preexposure vaccination.

Due to their lack of proven effectiveness and/or low risk of disease transmission, several vaccines are generally not advised for use in animal shelters. These comprise the canine adenovirus type 1, giardia, leptospirosis, canine corona virus, *Crotalus atrox* toxoid (rattlesnake), and Porphyromonas vaccine for dogs. For cats in shelters, vaccinations against giardia and FIP are typically not advised. FeLV vaccination is regarded as a noncore vaccine and is not usually advised, except foster homes or shelters where cats are kept in groups. In the same way, vaccination against FIV is not usually advised in the shelter. Another confusing aspect of the FIV vaccination is that cats who receive the shot experience false positive test results. To make their status more clear, cats who have received the FIV vaccination should be permanently identified (e.g., by wearing a microchip). Costs are decreased when unnecessary vaccinations are avoided, which is crucial when shelters have limited funding. Additionally, it helps to reduce the likelihood of vaccination-related side effects. The risk/benefit ratio strongly favors vaccination in the shelter environment, and serious adverse reactions from core vaccines are incredibly rare.

Without a doubt, following the right vaccination schedule significantly lowers disease in the shelter and enhances the health of the animals. Having said that, it's critical to recognize the limitations of vaccinations in terms of disease prevention. Vaccines are not "magic bullets" that can completely prevent disease, even though they are a crucial part of an animal shelter's overall wellness program. Shelter staff members ought to be informed about the fundamentals of vaccination. They must realize that vaccines are medical devices that elicit immune responses in animals and equip them to fend off infections from agents that cause disease; they do not cure illness or grant immediate immunity. Staff members also need to be aware that vaccinations don't always guarantee sterilizing immunity or complete disease prevention. Often, they only offer limited protection, reducing the severity of illnesses in the future but not curing them. In fact, vaccination failure can happen if animals come to the shelter carrying disease already, and even the best vaccinations take some time to start working. Specifically, vaccinations are only partially effective in preventing canine and feline upper respiratory diseases, but they can effectively prevent canine distemper and canine and feline parvovirus when administered appropriately. Nevertheless, because of the window of susceptibility to these diseases and waning maternal antibodies, there may occasionally be isolated cases of FPV, CDV, and CPV in shelters, particularly in young puppies and kittens. It is imperative to acknowledge that vaccine failure may occur in certain individuals, irrespective of the protocol employed and that vaccines are not accessible for all diseases observed in shelters.

An individual's reaction to vaccination is influenced by a number of factors, such as age, the existence of maternal antibodies, the presence of another illness at the same time, fever, nutritional status, and stress level (Greene, 1998b). Furthermore, vaccine response is influenced by the time elapsed between vaccination and exposure, the infectious dose administered to the animal, and the specific disease agent involved. And last, the efficacy of vaccines depends on how they are handled. Vaccines are unstable once reconstituted and extremely sensitive to temperature changes because they contain live or modified live agents. Vaccines should always be kept well chilled, refrigerated right away, and never left out of the refrigerator after a shipment arrives. On the other hand, since they can become inactivated by both extreme heat and cold, they should never be frozen. Vaccines should only be taken out of the refrigerator when needed, and they should be given within 30 minutes of reconstitution for optimal effects. Each patient's medical record should contain information about the vaccine type (including the manufacturer and serial number), the date of administration, and the name of the person who gave it.

Since modified live vaccines elicit a more robust and rapid immune response and are more effective than killed products at overcoming maternal antibody interference, they should be used in the shelter in the majority of cases (Carmichael et al., 2006; Richards et al., 2006). Actually, for animals older than four months, a single dose of a modified live virus (MLV) vaccine can provide protection; in contrast, killed vaccines usually need a booster shot to provide immunity. Although parenteral vaccinations are generally preferred, intranasal vaccinations may have benefits for use in respiratory diseases in dogs and cats since they have been demonstrated to quickly develop local immunity at the site of exposure. The vaccination's schedule is very important. The ideal time to vaccinate an animal is at least one week before it is brought into the shelter. Since this is typically not feasible, the next best course of action is to vaccinate as soon as possible before or upon entry. This can offer significant protection for the majority of cats and dogs who are admitted to shelters. Actually, immunity usually starts to take effect a few hours after vaccination. If a mother's antibody or another factor contributes to vaccine failure, modified live vaccinations against the canine and feline parvovirus can provide complete protective immunity in as little as three and five days, respectively (Brun et al., 1979; Carmichael et al., 1983). Protection against CDV from modified live and recombinant vaccines may be conferred even more rapidly (Larson and Schultz, 2006; Schroeder et al., 1967). Intranasal vaccines against respiratory infections including FHV, FCV, and Bordetella may protect in two to four days (Cocker et al., 1986; Gore et al., 2005).

All dogs and cats should be assumed to be unvaccinated upon arrival at the shelter unless a medical record proving the animal's current vaccination by a licensed veterinarian is readily available and can be independently verified. Upon arrival, all cats and dogs four weeks of age and older should receive a vaccination. The vaccine's capacity to offer prompt protection will be severely hampered by even a one- or two-day delay (Paul et al., 2006). All animals, including those with medical conditions and those who are injured, should receive vaccinations. The majority of the time, vaccinations is successful in building immunity, and the high risk of infection-related exposure and development in the shelter outweighs any potential negative reactions or side effects. Additionally, to maximize response, a booster shot can be given in two weeks. Practicality dictates that an animal exposed to infectious disease and stress in the shelter will probably result in decompensation and death or euthanasia if the animal is so severely crippled that vaccination is deemed unsafe.

It has been demonstrated that animals suffering from fever or hyperthermia with a temperature higher than 103.6°F will respond less well to vaccinations (Greene, 1998b). Before receiving vaccines, cats, and dogs with rectal temperatures higher than 103°F should be cooled. The vaccination of pregnant animals with modified live vaccines carries some risk to developing fetuses, but the risks of vaccination must be balanced against the risk of disease exposure. Pregnant animals are often vaccinated upon arrival at shelters and spayed before being adopted. Lastly, while it is ideal to vaccinate all cats and dogs (that are safe to handle) upon entry to maximize disease prevention and control, not all shelters may be able to afford to do this. In this instance, animals that are expected to be in the shelter for an extended period of time or those that are deemed adoptable upon entry should receive vaccinations right away, while animals that are anticipated to be put to death following legal holding periods should not receive vaccinations. As soon as it is possible to determine which animals will remain unvaccinated, vaccinated animals should be kept apart from the latter.

To reduce the window of infection susceptibility and guarantee that a vaccine is received as soon as possible after maternal antibodies have decreased sufficiently to allow vaccine response, kittens, and puppies younger than four months of age should receive a series of vaccinations. Vaccinations for kittens and puppies should be given every two weeks until they reach 16 weeks of age, or until the eruption of their permanent incisors. Less than two weeks between vaccinations is not advised as this could actually lessen the immune response from the prior shot (Greene, 1998a). The author suggests against putting puppies or kittens in the shelter if they are younger than six or eight weeks old because, even with rigorous vaccination programs and environmental control, they always end up very sick from infectious diseases. Whenever feasible, within 48 hours of their arrival, underage puppies and kittens should be taken out of the shelter and placed in foster care.

Booster shots for modified live vaccines are typically not needed before one year, similar to owned pets, but they may be given every two weeks if finances allow. Revaccination in long-term shelters should adhere to state and local rabies ordinances as well as the guidelines established for pets (i.e., booster at one year, then every three years for feline viral rhinotracheitis, calicivirus, and panleukopenia, or FVRCP), and distemper, hepatitis, parainfluenza, and parvovirus (DHPP).

1.5.2. Parasite Control and Prevention

Programs aimed at promoting shelter wellness must include both prevention and control of parasites. In dogs and cats, parasites can be both internal and external. Particularly common intestinal parasites with the potential to spread to humans are roundworms and hookworms. To manage these possible zoonoses, the Companion Animal Parasite Council and the Centers for Disease Control and Prevention strongly suggest routinely administering broad-spectrum anthelmintics. One of the safest, most economical, and most effective anthelmintics for treating roundworms and hookworms is pyrantel pamoate. The author suggests giving all adoptable dogs and cats 10 mg/kg of pyrantel pamoate upon admission, followed by monthly doses after two weeks of treatment. Furthermore, until they are four months old, puppies, and kittens should receive treatment every two weeks. Fecal flotation, direct fecal smear, and stained fecal cytology should be done on dogs and cats experiencing diarrhea, and treatment should be based on the findings. The administration of broad-spectrum anthelmintics should be seriously considered, even in the event of negative results. Dogs and cats that enter shelters frequently have ectoparasites, such as fleas, ticks, ear mites, lice, sarcoptes, and cheyetiella. These parasites need to be regularly diagnosed and controlled. It is strongly advised that adoptable dogs be tested for canine heartworm disease in areas where the condition is common. Antigen testing is the gold standard for diagnosing dogs, but some shelters may not be able to afford it. Although testing for microfilaria is far more affordable, up to 20% of dogs with heartworm infections might not be microfilaremic (Nelson et al., 2005).

Many shelters conduct microfilaria tests (direct smear and/or concentration test), which are inexpensive per test, in areas where heartworm disease is prevalent. If this test yields negative results, money is spent on an antigen test to determine the dog's actual status. Cats shouldn't have heartworm testing because it can be challenging to interpret the results. It is advised that cats and dogs that are sheltered in endemic areas be protected against heartworms. Dogs can be given a heartworm preventive dose of 6–12 mcg/kg orally (PO) once a month at a very cost-effective rate by diluting 1% ivermectin solution in propylene glycol for use in shelters. Ivermectin prophylactic doses given continuously every month have been shown to be highly effective against young adult heartworms and both early and late precardiac larvae; as a result, dogs who have never received preventive medication before or whose exposure history is unknown may benefit more from this medication (Nelson et al., 2005).

1.5.3. Spay/Neuter

Ensuring that cats and dogs are spayed or neutered before adoption is a crucial part of a shelter wellness program. Almost all animal shelters demand that adopted pets be sterilized; nevertheless, despite the implementation of spay/neuter contracts, coupons, other incentives, and time-consuming follow-up, the average national compliance rate is only 50% to 60% (Moulton, 1990). The American Veterinary Medical Association recommends that all pets, including young puppies and kittens, be neutered before adoption to ensure compliance (American Veterinary Medical Association, 1999). The safety of spaying and neutering puppies as early as six to eight weeks of age has been confirmed by a number of controlled prospective studies and sizable retrospective cohort studies (Howe et al., 2000; Slater et al., 2001; Spain et al., 2004a, 2004b). Animals adopted from shelters must be sterilized according to laws in many states.

Reproductive stress from estrous cycling in queens and bitches, as well as sex drive in tomcats and dogs, can cause appetite reduction, an increase in urine spraying and marking, intermale fighting, and a significant increase in social and emotional stress in shelters where pets up for adoption may be kept for extended periods of time. In shelters that house cats and dogs for longer than two to four weeks, spaying, and neutering animals up for adoption is imperative. These methods greatly reduce stress and eliminate heat behavior and pregnancy, as well as spraying, marking, and fighting. This makes it easier to live in groups and take part in monitored playgroups for emotional and physical development. Furthermore, the medical advantages of spay/neuter have been extensively documented. These benefits include notable decreases in the likelihood of tumors, prostatitis, benign prostatic hyperplasia, and ovarian cancer in males, as well as significant reductions in the risk of mammary cancer, pyometra, and ovarian cancer in females (Johnston et al., 2001).

1.5.4. Identification (Collar/Tags, Microchip)

In particular, where animals are housed in groups, the identification of animals in the shelter through the use of a microchip, collar, and tag, or neck band is crucial for ongoing monitoring and preventive healthcare. Many microchips have been released into the U.S. market during the last 20 years market without thinking about setting any kind of national radio frequency standards. Regrettably, this has led to circumstances in which specific microchips cannot be detected by existing scanners. At the moment, there are initiatives underway in the U.S. to standardize micro chipping, such as the broad distribution of international scanners to guarantee the accurate identification of every implanted microchip. Since the 134 kHz (ISO) microchip frequency is accepted as the international standard for microchips in the rest of the world, the AVMA advises adopting it as the American standard once global scanners are generally accessible.

While collars and tags serve as useful, readily identifiable means of identification, permanent identification in the form of a microchip could prove to be an even more advantageous method of facilitating pet-owner reunions, as collars and tags are prone to being misplaced. Another significant objective or welfare target that animal shelters should work toward is improving pet recovery after adoption. As a result, fitting collars and tags, implanting microchips, and conducting microchip scans are additional proactive measures that shelters can take to set an excellent example of pet care for the general public.

1.5.5. Proper Nutrition and Exercise

Fitness and healthy eating have a big impact on wellbeing. Regular exercise and a healthy diet are not only linked to improved immune function and behavioral health and well-being, but they are also crucial for maintaining a healthy body weight and condition. To maintain an appropriate nutritional plan, a regular diet of tasty, high-quality commercial food appropriate for the child's life stage should be provided, and appetite should be watched. Animals going more than a day or two without eating should have their health and stress levels assessed, and if necessary, appropriate action should be taken in response to the results. Furthermore, fresh water needs to be accessible at all times. In conclusion, it is important to weigh animals both upon arrival and at regular intervals during their stay in the shelter. During the first month of shelter care, body weight should ideally be recorded weekly. After that, it should be done once a month or more frequently as needed. This is crucial for cats in particular, as stress or URIs during the first few weeks of confinement may be linked to noticeable or even dramatic weight loss. However, there is a chance that you will gain too much weight.

1.5.6. Grooming

The animals in the shelter also need to receive proper grooming, which includes bathing, brushing, and cleaning their ears and nails to get rid of matted hair. Long-haired dogs and cats, which may be painful and prone to skin infections, often arrive at animal shelters with severely matted coats and/or overgrown nails. A routine grooming program must be put in place for animals kept in shelters for extended periods of time to prevent the development of uncomfortable matting and overgrown nails. Dogs are frequently kept in damp runs, which can make them more susceptible to skin conditions like pododermatitis and strong body odor. Animals need to be kept clean and dry at all times.

The author has seen numerous animals suffering from chronic pyoderma and being heavily matted in long-term shelter facilities. This constitutes an intolerable standard of care and ought not to be approved. Regular grooming and keeping the surroundings dry and clean can often prevent such situations. Obviously, certain animals need more upkeep than others when it comes to their conformation, hair type, and coat type. Regular grooming also offers a great way to keep an eye on body condition and health while looking for lumps and skin issues. Furthermore, certain animals take pleasure in the interaction and focus.

1.5.7. Periodontal Disease Prevention

Another aspect of wellness is dental health, which goes far beyond bad breath. Plaque and tartar accumulation has been linked to several grave health issues, including organ failure, chronic, intermittent bacteremia, and mouth pain. One of the most prevalent medical conditions in dogs is periodontal disease, which is thought to affect 80% of patients over the age of five. It is particularly prevalent in dogs of small breeds (Debowes, 1998).

Prevention of periodontal disease may not be the top priority for wellness in an animal shelter, but it should still be taken into account when providing individual care. If an animal has a painful dental disease and will be kept for adoption or extended periods of time, a prompt treatment plan needs to be determined and put into action.

When it comes to easy and effective ways to prevent, chewing activity-promoting products are known to be beneficial because they maximize salivary flow stimulation and self-cleaning. Additionally, chewing is a common behavior for puppies and dogs, and it can be a coping mechanism for dogs who are stressed, confined, bored, or lonely. Dogs of all ages should have access to a variety of chew toys that are suitable for their size and age due to these factors as well as to help maintain dental hygiene.

1.5.8. Breed Specific Care

Wellness protocols may also be determined by the unique requirements of particular dog breeds. For instance, because of the way their airways are shaped, care must be taken when dealing with brachycephalic dogs to prevent heat exhaustion, to which they are particularly susceptible. This could have an impact on the people's choice of housing and holding locations as well as their workout regimens. Additionally, brachycephalic dogs and cats are more likely than other breeds to experience severe URIs due to poor airway conformation. Because of these factors, brachycephalic animals should be given priority when it comes to being removed to foster care or rescue, and they should be kept in well-ventilated spaces away from sick animals. Based on the author's observations, it is advisable to avoid administering even intranasal vaccinations to these breeds due to the potential for serious respiratory disease symptoms. Similar to this, some other breeds need particular attention in the shelter due to genetic predispositions related to their health or behavior. Another dog that needs special consideration when it comes to housing in a kennel setting is the pit bull. This is because many of these dogs have a tendency to learn or display dog-dog aggression, and we don't want that to get worse.

1.6. WELLNESS: BEHAVIORAL HEALTH

Cats and dogs are not normally housed in animal shelters; they are not natural settings. They are designed to function as transitional homes for animals awaiting return or adoption, as well as in certain situations, as temporary homes for animals awaiting death. In the last 10 years, animal shelters have seen an increase in the need to provide longer-term stays for adoptable pets. This is particularly prevalent in shelters with limited admission that offer an "adoption guarantee," meaning that animals are not put to death unless it is medically or behaviorally necessary. An animal may remain in the shelter for weeks, months, or even years if no adopter chooses to adopt it.

In animal shelters, stressors for both dogs and cats start the moment they are admitted. Even in the best of circumstances, animal shelters are stressful places by definition because new arrivals are kept in close quarters and subjected to a range of infectious disease agents in addition to intensely novel and new stimuli.

Long-term confinement can cause anxiety, social isolation, poor mental stimulation, and insufficient exercise in cats and dogs, all of which can negatively impact their behavioral and physical well-being and reduce their adoption. In certain shelters, this can lead to the animal's euthanasia; in others, it might prolong their stay if the animal's poor physical or behavioral health prevents an adopter from taking an interest in it. The animal's physical and/or emotional health may deteriorate even more with time.

Prevention is key when it comes to treating behavioral health in the shelter. Proactive methods to reduce stress are the first step in a behavioral wellness program, and they apply from the time animals arrive at the shelter until their departure. A comprehensive behavioral history and examination are crucial, as previously mentioned, and they will establish a crucial baseline for further action and monitoring.

1.6.1. The Role of Stress

Stress has secondary effects due to elevated cortisol and catecholamine secretion. Chronic activation of these hormones causes a number of well-documented negative effects, such as insulin resistance, peptic ulcer formation, mental depression, dehydration, and increased susceptibility to infection (Greco, 1991; Moberg, 1985). Prolonged stress can also change metabolism to the point where an animal loses weight, cannot grow normally, and exhibits harmful abnormal behavior. Additionally, there is a close relationship between immunity and stress responses. Stress weakens immunity, which reduces resistance to infection (Griffin, 1989). In actuality, stress can cause some viral pathogens to shed, such as the reactivation of feline herpesvirus (rhinotracheitis) latent infections in cats (Gaskell and Povey, 1977). Reducing stress can significantly improve animal welfare, lower infection and disease transmission rates, and increase adoptability in the context of an animal shelter.

Any stimulus or factor that causes stress is referred to as a stressor. There are many opportunities for stressors to be introduced and stress to be produced when cats and dogs are housed in animal shelters. Stressors can include disease, confinement, transportation, crowding, isolation, alterations in temperature, light, and/or ventilation, odd odors, noises, other animals, dietary modifications, handling, restraint, erratic daily routines, the presence of unfamiliar human contact, and the absence of familiar human contact. In actuality, a cat or dog may become anxious and experience stress over anything they are unfamiliar with. The degree, duration, novelty, predictability, chronicity, and intensity of the stressor, in addition to the person's perspective, all affect how they react to it (McMillan 2002; Moberg 1985). An animal's personality, socialization history, genetic make-up, and level of experience all affect how it interprets a stressor.

When given the chance, animals use coping mechanisms to decrease the harmful effects of a stressor (Carlstead et al., 2002; Strawn 1993). Behavioral coping strategies can take many forms, such as hiding, looking for social support, and engaging in mental stimulation. Regarding their capacity for adjustment, individual cats and dogs differ significantly from one another. Individuals who can effectively manage their stress levels will experience fewer negative effects on their bodies and minds and will adapt to living in an animal shelter more easily. Having said that, it is uncommon for a single animal to genuinely flourish in a long-term shelter setting.

Stress in animals kept in shelters often stems from their inability to participate in proactive behavioral responses that would otherwise function as a coping mechanism. The stress response that follows is most severe when stress is thought to be unavoidable or uncontrollable (Carlstead et al., 1993; McMillan, 2002). This is a crucial factor to take into account when creating housing and husbandry procedures for dogs and cats in shelters.

1.6.2. Behavioral Needs of Cats and Dogs

For cats and dogs to be healthy, certain behavioral requirements are also vitally important in addition to their basic physical needs (like adequate food and shelter). Being social creatures, most dogs and cats do not thrive in isolation, so having the chance to socialize with other people satisfies a fundamental behavioral need. Both dogs and cats need to be able to designate distinct spaces in their homes for eating, sleeping, and urinating. They need daily care regimens or patterns that are predictable, including regular times for light and dark. The capacity to locate a hiding spot, to sleep through the night without interruption, and to be free from ongoing harassment by people, other animals, or environmental stressors are other crucial behavioral needs. Dogs and cats need to be able to play and exercise whenever they want, as well as mental stimulation. Ultimately, dogs need to chew, and cats need to scratch. Scratching is a common behavior for cats that helps to stretch their claws and acts as a visual and olfactory marker. Chewing is a normal behavior for dogs as it helps to maintain their teeth, explore their surroundings, and can be entertaining.

The way an animal behaves, both normal and abnormal, can reveal how well it is adjusting to its surroundings. Behavioral expressions can take the form of defensive, disruptive, inhibited, or withdrawal behaviors, as well as stereotypical behaviors (Hubrecht, 1993; Overall, 1997). Withdrawal or inhibited behavior is the term used to describe activity depression or the lack of regular activities (like eating, sleeping, removing waste, stretching, introducing oneself to others, etc.). Defensive behavior is characterized by distinctive vocal and/or postural reactions and is frequently brought on by fear. Disruptive behavior includes making a hiding place or destroying the contents of the cage.

Stereotypical behaviors associated with anxiety include pacing, pawing, and circling. Stress-related behavioral indicators can take the form of passive or aggressive actions. Scent marking (urine, feces, various skin glands), vocalization (growling, hissing), visual cues (facial expression, body posture, ears, and tail), and overt aggression are examples of subtle or obvious signals of anxiety, fear, aggression, and submission. Inability to sleep or rest, pretending to sleep, poor eating, hiding constantly, lack of grooming, activity depression (reduced play and exploratory behavior), and social disengagement are examples of passive symptoms of stress (Griffin 2006; Rochlitz et al., 1998; Wemelsfelder, 2005). These indicators are aggravated by high-density housing. Cats and dogs exhibit a wide range of typical behaviors, such as a healthy appetite and activity level, sociability, grooming, appropriate play behavior, and sound sleep when their housing meets their behavioral needs and they are well-adjusted.

A behavioral wellness program should include adequate housing, regular daily schedules, mental stimulation, social companionship, and training based on positive reinforcement. Maintaining appropriate behavioral welfare in cats and dogs requires an awareness of the significance of reducing stress and the capacity to identify and address it. Employees should receive stress indicator recognition training. Animals exhibiting behavioral indicators of stress and/or social conflicts, such as persistent hiding, agonistic behavior with conspecifics, activity withdrawal, or other markers as previously described, require active daily monitoring to identify and address their needs. Employees should keep a daily log of their observations to guarantee prompt identification of signs of stress and the implementation of suitable measures to mitigate the animals' suffering and improve their adaptability to the shelter setting. Even though this is subjective, staff members should make an effort to gauge the level of stress and observe any patterns: is the animal adjusting to its new surroundings, getting less or more stressed, or measuring the incidence and prevalence of stress in the population?

1.6.3. Housing

The behavioral needs of the animals are met by appropriate housing, which reduces stress. Provisions for housing individual animals, litters, or compatible pairs for intake assessment, triage, isolation, and quarantine should be included in the design of short-term housing. Safe, well-ventilated, and easy to clean and sanitize housing is essential for both caregivers and animals. Even temporary housing should meet the bare minimum of an animal's behavioral requirements by giving them enough room to sit or lie down fully extended, stand, and take several steps, and have separate places for eating, sleeping, and urinating. Comfortable surfaces and a safe hiding place for cats or another visual barrier for dogs are essential components of resting areas. Long-term housing, or confinement in a shelter for longer than two weeks, should be designed to provide a space that is both physically and mentally stimulating, and ideally, one that is aesthetically pleasing to the public, particularly in adoption areas. Enriched single or group housing is recommended for the long-term housing of cats, as an alternative to cage housing should be provided.

At the very least, cats kept in cages should be released daily to allow them to exercise and explore in a safe, enriching environment, even though this isn't always possible in busy shelters. Since cats are social animals, group housing will usually be beneficial for long-term housing. This is especially true if there is enough room, easy access to feeding and waste areas, a sufficient number of cozy hiding and resting spots, and attentive staff activity monitoring. However, not every cat will do well in a group home; some cats, due to their unique physical or behavioral requirements, will need enriched single housing. These could be mothers who are nursing young kittens, cats with special medical needs, or cats who have not been socialized with other cats. Since it is generally easier to monitor each cat's health and well-being in smaller groups and because larger groupings make it more difficult and complex to establish successful social groupings, the author advises housing cats in small groups of up to four to eight people for social group housing. Amazing individual variations. Adding a new cat to an established colony is actually like inviting someone you don't know into your house and sharing your living room and bathroom. Because of this, groups should be kept small and new cats should be introduced gradually and under close supervision. A range of elevated resting spots and hiding boxes should be included in enriched single and group housing areas to expand the living area's size and complexity and divide it into distinct functional areas that permit a range of behavioral options. Opportunities for hiding, playing, scratching, climbing, resting, feeding, and urinating should all be present in the physical environment. According to the author's observations, most cats can be grouped together in small, friendly groups provided that the enclosure is not too full. Housing options include small rooms or several runs inside one larger room.

For colonies with a maximum of eight cats, the author suggests a minimum enclosure size of roughly 10 to 12 feet by 16 to 18 feet, or three to four feet by four to six feet long and high for groups of two to four cats. Runs ought to be at least six feet high when they are used. It is not always possible to house twice as many cats in an enclosure when the enclosure's size is doubled. Staff members in group housing must be closely supervised and educated to spot even the smallest indicators of stress. When more cats are added, there is a period of stress that can be lessened by offering more hiding places, feeding areas, and elimination areas, as mentioned. These improve people's capacity for adjusting to and coping with their new environment. Enriched single housing may be the best option for bully cats and others who, after a few days, do not adapt to group housing or do not demonstrate significant progress toward adaptation. To minimize the stress that could result from social conflicts, enriched single or pair housing may be the best option for shelter cats that are not regularly housed for longer than a few weeks.

It can be particularly difficult to maintain a suitable kennel environment for dogs in long-term housing that satisfies their behavioral needs. Although it is generally preferred to have access from indoors to outdoors, dogs should be able to see outside so they can observe their surroundings. Dogs need room in their runs to walk around freely, and there should be a nice, comfy bed available for them to rest in a safe place. There should be visual barriers available to protect against other kennel mates. Boredom and stereotypical movements can be lessened by periodically rearranging the structures and adding three-dimensional space (such as platforms, steps, or ramps).

Furthermore, dogs need to be given daily opportunities to exercise outside of their runs. Playgroups under supervision provide chances for socializing and physical activity. Play areas with features like kiddie pools, tunnels, and platforms may improve interactions. Dogs should be taken for walks outside if there are no outdoor play enclosures available. While retractable leashes or long lines may offer more opportunities for exercise during walks, care should be taken to ensure that volunteers and staff are properly trained to prevent animal escapes or injuries to humans. Cats may also be suited for outdoor enclosures in some circumstances. Advantages include good ventilation, lots of natural light exposure, and mental stimulation. Fencing specifically made for cat enclosures or galvanized wire chain link panels with a 1-inch mesh (including a top panel) may be used. Lastly, "real life" rooms–that is, rooms with a setting more akin to a home–away from the kennel or cattery might also be helpful, particularly for animals who stay at the shelter for longer than a few weeks.

Both the environment's quality and the animal's capacity for adaptation determine how well cats and dogs adapt to new situations. While the majority eventually adjust to new surroundings, some people never do and live under constant stress, which eventually leads to a decline in both mental and physical health. Behavior issues can arise in cat colonies that are moderately populated, carefully introduced and have enriched environments. Although many cats are incompatible as housemates, manipulating the social environment by regrouping cats may help solve these issues. Cats can be taken out and transferred to another colony if they are the only one or two causing social instability in the colony. Frequently, the issue lies with the social grouping rather than the individual.

In time, a careful observer will see that cats usually divide themselves into amicable groups. Observers should note feline personality types (e.g., bold versus shy) and watch for behavioral signs of stress as well as affiliative behaviors (e.g., those behaviors that promote social contact). The best colony environments are those that produce the most "normal" behaviors, including grooming, sleeping, playing, stretching, rubbing, and exploring. Colonies should be carefully monitored to ensure that all colony members are content in the social environment.

Older cats and dogs and those who have not been well socialized often find unfamiliar surroundings particularly stressful. It is best to stay away from housing feral cats whenever feasible. Placement in foster care is generally beneficial for elderly cats in animal shelters, and neuter and return programs are beneficial for feral cats. The author advises shelters to continue offering a range of housing options to accommodate each animal's unique behavioral requirements. It is imperative for animal welfare that staff members receive adequate training in identifying and preventing stress. Cat shelters that use colony housing are advised that while well-managed colony housing can improve welfare, mishandled colony housing has the potential to worsen stress levels. In particular, an all-in/all-out strategy should be used to prevent the constant introduction of new cats, which causes stress, or overcrowding of cats in colony housing must be avoided to ensure welfare. Similar precautions must be taken to ensure that dogs in playgroups or pairs are closely supervised to identify signs of stress, stop fights, and protect the safety of the staff as well as the welfare of the animals.

1.6.4. Behavioral Care and Environmental Enrichment

A staff of animal care professionals who enjoy working with animals and who are able and willing to spend enough time interacting with them on a daily basis to ensure social contact and tractability may be the most effective form of environmental enrichment. Dogs and cats typically react strongly to their human caregivers and become accustomed to daily routines. To help caregivers get to know each animal's personality, it is best to assign them to care for the same animals regularly. This familiarity helps staff identify behavioral or physical issues and facilitates more successful adoption matches. In addition to feeding and cleaning, caregivers should set aside time each day to spend with "their" animals. While some dogs and cats prefer to play with toys, others might prefer to be handled and petted (e.g., cats chasing dangling feathers or dogs fetching a ball). Dogs need to engage in regular aerobic exercise, the duration and intensity of which are determined by the needs of the individual.

For cats in particular, having scratching boards available is essential. In addition to encouraging scratching, empty cardboard boxes and chapter bags are cheap and disposable. They also encourage play and exploration. To encourage play, cats, and dogs should have disposable or subitizable toys. Dry food or cat treats can be hidden in empty soda bottles, cardboard boxes with holes, or commercial food puzzle toys so the pet has to work to get the food pieces out. In a similar vein, you can encourage dogs to chew by giving them commercial chew toys filled with treats. Including special toys or treats in the daily schedule gives the animals something to look forward to, which may help to reduce their feelings of boredom and frustration. Dogs can enjoy olfactory stimulation, so giving them a scented room deodorizer (like lavender) regularly could be a good way to give them positive stimulation in the kennel.

For both dogs and cats, clicker training with food or play rewards can offer extra stimulation, exercise, and social interaction. Teaching dogs manners not only makes them more appealing to adopters but also sets another example of proper pet care in addition to offering mental stimulation. Lastly, since these products have been shown to help reduce anxiety, using a commercially available feline facial pheromone spray (Feliway, Veterinary Product Laboratories, Phoenix, AZ) or a dog-appeasing pheromone (DAP ® Veterinary Product Laboratories, Phoenix, AZ) may be beneficial for cats and dogs, respectively.

One of the most important requirements for an animal's mental health and wellbeing is a sense of control over their surroundings (McMillan, 2002). Different people have different preferences for different environmental factors, activity levels, and social interactions with humans and other animals. Cats and dogs require variety and choice. Every one of these options will be provided by the best environmental enrichment.

1.7. ENVIRONMENTAL WELLNESS

Animal health and well-being are significantly impacted by the shelter environment; as a result, it is necessary to develop systematic wellness protocols that address the shelter environment. In addition to other facility operations, staff training, animal segregation, cleaning, and sanitation, and proper population density maintenance should all be covered by protocols.

1.7.1. Population Density

One of the biggest known stressors for animals kept in homes is overcrowding. In a given group, overcrowding increases the number of susceptible animals as well as asymptomatic carriers, increasing the risk of disease transmission through contaminated food and direct contact. In addition to impairing animal husbandry and increasing the intensity of various stressors in the shelter environment–such as noise, air pollutants, and infectious agents–overcrowding also raises the risk of significant disease outbreaks among the population.

The optimal number of animals to house in a particular facility cannot be determined by a straightforward formula. Animal overcrowding affects not only how much room there is for them, but also how well the shelter can take care of their mental and physical needs. Staff members at the shelter suffer from overcrowding because it makes it difficult for them to give the animals the care they need and to treat the public with good customer service. Adoption rates can be adversely affected by overcrowding because prospective adopters frequently perceive an overcrowded shelter as intimidating and unwelcoming, which exacerbates the shelter's crowding issue. Shelters are required to house no more animals than they are able to care for reasonably. Unexpected intake can occasionally lead to brief instances of overcrowding, but a strong wellness program requires that procedures be in place to reduce overcrowding and maintain a moderately populated environment for the staff and animals' health and safety.

Euthanasia, intake restrictions, and/or shorter stays for animals at the shelter are ways to alleviate overcrowding. To minimize euthanasia for space in open admission shelters and to maximize intake in limited admission shelters, programs that increase and speed adoption, redemption, and transfer (to rescue or foster care) are helpful.

Although euthanasia decisions are never easy, every shelter has some responsibility in this area. It might be necessary to carry out euthanasia while taking the population into account to reduce overcrowding or stop disease outbreaks from getting out of control. Individual animals that are ill physically or behaviorally or who endanger the public's health or safety may also require euthanasia. When the decision is made to put an animal to sleep, the process must be carried out right away. In many cases, this will benefit the population overall as well as the welfare of individual animals and public safety. For instance, aggressive dogs that are not suitable for a new home shouldn't be kept longer than is permitted by law. Rather, they ought to be mercifully put to death as soon as possible to spare the dog unnecessary stress and worry as well as danger to the public and shelter employees. These processes are always challenging, regardless of the underlying reasons for the death of an animal. However, inaction frequently has a detrimental effect on population health and increases the stress levels of individual shelter animals.

During daily routine surveillance rounds, it is important to reflect on the reasons behind each animal's stay in the shelter and explore ways to improve or expedite their chances of success (Hurley, 2004). Overcrowding may prevent shelters from having enough time to properly care for and market resident animals, which may reduce the length of time animals stay in the shelter and eventually enable them to serve a greater number of animals. Furthermore, money saved on housing expenses could be put toward initiatives that promote adoption, aid in prevention, and reduce intake (like spay/neuter programs).

1.7.2. Cleaning and Sanitation

An environment that is both clean and hygienic is necessary for wellness programs to be successful. This not only encourages public support and staff pride, but it also promotes pet health. Protocols should be in place for periodic deep cleaning and disinfection in addition to regular daily cleaning and sanitation procedures. Additionally, protocols should outline what to do in the event of a disease outbreak.

1.7.3. Segregation of Animals

Animal welfare, staff safety, the prevention of infectious diseases, and adherence to animal control protocols all depend on the segregation of animals entering shelters. Animals arriving should be sorted according to species (e.g., cats, dogs, etc.); age range [e.g., preweaning (till eight weeks of age), pediatrics (6–16 weeks of age), juvenile (4–12 months of age), and adults (till one year of age); sex or reproductive status (till in heat, pregnant, nursing); and physical and behavioral health status (till in heat, dangerous, feral). Therefore, adequate segregation requires a range of holding, adoption, and isolation areas.

Healthy animals should be kept apart for observation in quarantine areas. By using these spaces, it is possible to monitor seemingly healthy animals for the emergence of symptoms of illness and to give them time to react to vaccination in a highly bio secure setting with reduced exposure risks. When quarantine spaces are scarce, the animals that are most vulnerable when they come into the shelter–typically young puppies and kittens–should be given priority. Although quarantines are generally advised as the cornerstone of any population health disease control program, their applicability in shelter health programs has been questioned due to restrictions that make it impossible to follow traditional quarantine protocols. The truth is that when an all-in/all-out system is used, most shelters are unable to properly implement a quarantine. Most shelters add new animals to the quarantine group every day, which essentially negates the goal and just lengthens the animals' stay and promotes the spread of infectious diseases. When utilized for these purposes, quarantines can be very beneficial. When quarantine is used, the length of time recommended depends partly on the availability of space and resources, the prevalence of infectious disease in the population, and/or the geographic region and the incubation period of the disease in question. For example, strict quarantine or enhanced biosecurity measures for young puppies may be more important in shelters in the southern U.S, where there are large numbers of puppies and canine parvoviral infections are extremely common.

A 14-day quarantine should be enough to make sure that no common infectious diseases, like feline panleukopenia or canine parvovirus, are being incubated by the animals and could be brought into the shelter setting. On the other hand, length of stay has been connected to a higher risk of respiratory disease development in dogs (Edinboro et al., 2004) and cats (Scarlett, 2006), so it might be preferable to have a shorter quarantine period. Furthermore, research shows that after receiving a modified live product vaccination for 72 hours, both cats and dogs develop protection against parvovirus, unless recent exposure (disease incubation) or interference from maternal antibodies prevents response (Brun et al., 1979; Joubert et al., 1983), which further disproves the need for an extended period of quarantine or holding.

Adult animals at intake that are less bio secure than those in quarantine may be placed in general holding areas. The length of legal holding periods mandated by state and local laws that give owners an opportunity to retrieve lost animals may also have an impact on the holding time. For owner-returned pets, there is no legal holding period required; however, a medical hold for assessment and triage is nearly always necessary.

Sick animals are kept apart from the general population in isolated areas. For effective disease control, sick animals must be isolated right away. Targeting isolation based on illness, age, and species is important. For instance, dogs, and cats with respiratory illnesses should have access to separate isolation areas. Cats and dogs with non-respiratory infections should also have access to separate isolation areas. In quarantine and isolation areas, strict adherence to biosecurity protocols is crucial. This includes monitoring traffic patterns and wearing protective clothes, such as gowns and shoe covers. Animals in these sections should, whenever possible, be taken care of by designated staff. The groups that are healthiest and most susceptible to disease should have the least amount of traffic, followed by the isolation areas that house sick animals. Signage and observation windows can be used to lessen the amount of traffic that enters isolation and quarantine zones. It is crucial for staff members to maintain good hygiene, and frequent hand washing cannot be stressed enough. Foster care may be a reasonable and medically sound alternative for the isolation or quarantine of some people in situations where space or facilities are not available.

1.7.4. Other Facility Operations

Housing design and facility operations determine whether a wellness program is successful or unsuccessful in almost every way. Success requires the adoption of stringent management procedures, in-depth training, and staff supervision under the guidance of an experienced professional. While creating a wellness program for the shelter environment, there are many other crucial aspects of facility operations to take into account in addition to appropriate housing design, segregation, and sanitation protocols. Inadequate ventilation or poorly thought-out air pressure gradients that recirculate or induce air exchange between rooms can be harmful to an animal's health. High humidity and inadequate ventilation encourage the buildup of pathogens as well as dust and other pollutants that can irritate the respiratory system. Strategies for preserving good air quality need to be carefully considered (Hurley, 2005). These include using dust-free litter (or just dumping dusty litter), maintaining filters regularly, controlling dust and dander with routine or as-needed vacuuming, deep cleaning on occasion, and improving ventilation.

The ideal situation, which is common in settings involving laboratory animals, is for the HVAC system to enable 100% fresh (e.g., air that has not been recycled) in each room to ensure that the air that enters one is exhausted out of the structure and isn't recycled into another. 10 to 15 air changes per hour is the standard recommendation for an animal room; however, depending on the species housed and the expected housing density, more or less airflow may be acceptable or necessary (ILAR, 1996). It's possible that ventilation systems won't reach the level of cages or other animal enclosures even if they produce 10 to 15 room air changes per hour. Better ventilation can be achieved through smart housing design. For cats, this can mean using flow-through cages, runs (particularly the indoor-outdoor variety), or colony housing. To avoid air exchange between the different functional areas of the shelter, it has always been recommended to have separate ventilation systems for each of them.

Though few diseases in shelters are actually spread by aerosolization; instead, fomites are the main means of transmission, some experts have recently questioned this recommendation. Installing and maintaining this kind of ventilation system is highly expensive, even though it seems sensible to take this advice into account whenever it is feasible. This recommendation might not be required to maintain animal health if the shelter maintains comprehensive wellness protocols and the air quality stays good. The author suggests speaking with an HVAC expert to assess the shelter's requirements and optimize the system's potential while more research on the topic is conducted.

It is important to regulate humidity and temperature to ensure the comfort of animals. Avoiding drafts is advised. It is advised that cats and dogs be kept between 64°F and 84°F, with a typical temperature setting in the low to mid-70s (ILAR, 1996). The temperature should, however, be adjusted to the needs of the animals. In contrast to healthy animals, sick animals, puppies, kittens, and animals recovering from surgery are more vulnerable to colder temperatures. Given that cages nearer the floor tend to be a few degrees colder than those higher up, the animal's location should be taken into account. Furthermore, it's critical to take into account the specific shelter circumstances, finances, and location. For example, during hot weather, power companies advise keeping the temperature between 78°F and 79°F because it may result in an electricity bill savings of at least 10%. According to laboratory standards, the ideal humidity range for cats and dogs is between 30% and 70%. Higher humidity (70%) may be advantageous in URI treatment areas, whereas less humidity (40% to 50%) in other areas may help with disease control by curtailing airborne transmission. Although the range considered acceptable is large, a given room should have a fairly constant humidity (e.g., it should not have large fluctuations). It is recognized that hosing or even mopping a room will cause humidity spikes, but they should be short-lived with a well-ventilated room.

Staff should be trained to make sure that lights are turned on during the day and off at night. Timer systems can also be used to maintain control. Regular cycles of light and dark are particularly important. One benefit of exposure to natural sunlight is that it can reduce the spread of disease. It is impossible to overstate the significance of consistent feeding, cleaning, playing, and exercise schedules. Scheduled events, even the most stressful ones, become less stressful. Unpredictable aversive stimuli can cause persistent fear and anxiety. On the other hand, recurring stressful situations permit a time of peace and comfort in between stressful reactions (McMillan, 2002). Positive experiences in daily routines are also rewarding to animals; they may look forward to mealtimes and playtimes. Another crucial factor in shelter design is noise control, which can be achieved through soundproofing and housing design. Examining the dog's motivations may help to resolve the issue, reduce stress for the barker, and lessen the dog's negative effects on the surroundings. Keeping a building well-maintained and controlling pests are crucial factors in maintaining a healthy atmosphere. Systematic maintenance schedules will be carried out if written standard operating procedures and daily, weekly, monthly, and quarterly checklists are developed and followed.

1.7.5. Staff Training

It takes regular staff training to put wellness programs into action. All facets of animal care, including intake, adoption, redemption, transfer, and euthanasia, require training for staff members. Every aspect of the shelter's wellness program will be largely determined by the knowledge, attitude, and skill of the staff. It is imperative that staff members receive training on gathering crucial historical data during intake, handling all facets of animal care such as cleaning, feeding, diagnosing illnesses, and providing behavioral enrichment, and providing adoption counseling to facilitate a successful match. Written records and guidelines for following up with their veterinarian should be provided to adopters. Post-adoption counseling and follow-up should also be provided. It is impossible to overstate the value of consistent staff development and evaluation, as well as strong staff management and leadership. The best-run shelters have knowledgeable, kind, and professionally trained staff members who collaborate to deliver superior animal care and community service (American Veterinary Medical Association, 1999).

REFERENCES

1. American Veterinary Medical Association. (1999). *Position on Spay/ Neuter of Dogs and Cats at Early Age (prepubertal)*. Schaumburg, IL: AVMA.
2. Brun, A., Chappuis, G., et al. (1979). Immunisation against panleukopenia: Early development of immunity. *Comparative Immunology, Microbiology and Infectious Diseases, 1*(4), 335–339.
3. Carlstead, K., Brown, J. L., & Strawn, W. (1993). Behavioral and physiological correlates of stress in laboratory cats. *Applied Animal Behaviour Science, 38*, 143.
4. Carmichael, L. E., Joubert, J. C., et al. (1983). A modified live canine parvovirus vaccine with novel plaque characteristics: Viral attenuation and dog response. *Cornell Veterinarian, 73*(1), 13–29.
5. Cocker, F. M., Newby, T. J., et al. (1986). Responses of cats to nasal vaccination with a live, modified feline herpesvirus type 1. *Research in Veterinary Science, 41*(3), 323–330.
6. Debowes, L. J. (1998). The effects of dental disease on systemic disease. In S. E. Holmstrom (Ed.), *The Veterinary Clinics of North America: Canine Dentistry* (Vol. *28*, pp. 1057–1062). Philadelphia, PA: WB Saunders.
7. Edinboro, C. H., Janowitz, L. K., et al. (1999). A clinical trial of intranasal and subcutaneous vaccines to prevent upper respiratory infection in cats at an animal shelter. *Feline Practice, 27*(6), 7–13.
8. Edinboro, C. H., Ward, M. P., et al. (2004). A placebo-controlled trial of two intranasal vaccines to prevent tracheobronchitis (kennel cough) in dogs entering a humane shelter. *Preventive Veterinary Medicine, 62*(2), 89–99.
9. Edwards, D. S., Coyne, K., et al. (2008). Risk factors for time to diagnosis of feline upper respiratory tract disease in UK animal adoption shelters. *Preventive Veterinary Medicine, 87*(3–4), 327–339.
10. Gaskell, R. M., & Povey, R. C. (1977). Experimental induction of feline viral rhinotracheitis in FVR-recovered cats. *Veterinary Record, 100*, 128.
11. Griffin, B. (2006). Recognition and management of stress in housed cats. In J. R. August (Ed.), *Consultations in Feline Internal Medicine V* (pp. 717–734). Philadelphia, PA: WB Saunders.
12. Howe, L. M., Slater, M. R., et al. (2000). Long-term outcome of gonadectomy performed at early age or traditional age in cats. *Journal of the American Veterinary Medical Association, 217*, 1661–1665.
13. Hurley, K. F. (2004). Implementing a population health plan in an

animal shelter: Goal setting, data collection and monitoring, and policy development. In L. Miller & S. Zawistowski (Eds.), *Shelter Medicine for Veterinarians and Staff* (pp. 211–234). Ames, IA: Blackwell Publishing.

14. Larson, L. J., & Schultz, R. D. (2006). Effect of vaccination with recombinant canine distemper virus vaccine immediately before exposure under shelter-like conditions. *Veterinary Therapeutics, 7*(2), 113–118.
15. Lorenz, M. D. (1993). The problem-oriented approach. In M. D. Lorenz (Ed.), *Small Animal Medical Diagnosis* (pp. 1–12). Philadelphia, PA: Lippincott, Williams and Wilkins.
16. McMillan, F. D. (2002). Development of a mental wellness program for animals. *Journal of the American Veterinary Medical Association, 220*, 965.
17. Miller, L. (2007). A blend of science and art: What every shelter should know about shelter medicine. *Animal Sheltering*, January/February, 49–51.
18. Overall, K. L. (1997). Recognizing and managing problem behavior in breeding catteries. In J. R. August (Ed.), *Consultations in Feline Internal Medicine 3*. Philadelphia, PA: WB Saunders.
19. Rochlitz, I., Podberscek, A. L., & Broom, D. M. (1998). Welfare of cats in a quarantine cattery. *Veterinary Record, 143*, 35.
20. Schroeder, J. P., Bordt, D. W., et al. (1967). Studies of canine distemper immunization of puppies in a canine distemper-contaminated environment. *Veterinary Medicine and Small Animal Clinician, 62*(8), 782–787.
21. Wemelsfelder, F. (2005). Animal boredom: Understanding the tedium of confined lives. In F. D. McMillan (Ed.), *Mental Health and Wellbeing in Animals* (pp. 79–91). Ames, IA: Blackwell Publishing.

CHAPTER 2

DERMATOLOGICAL DISEASE IN ANIMAL SHELTERS

Contents

Introduction 56
2.1. Dermatophytosis 56
2.2. Etiology 57
2.3. Pathogenesis 59
2.4. Disease 59
2.5. Transmission 61
2.6. Clinical Presentation and Differential Diagnoses 62
2.7. Diagnosis 64
2.8. Initial Risk Evaluation and Response 82
2.9. Treatment 84
2.10. Outbreak Management and Response 95
References 97

INTRODUCTION

Animal shelters often face dermatological diseases because animals' skin conditions can worsen due to stress and cramped quarters. From bacterial and fungal infections like pyoderma and ringworm to parasitic infestations like fleas and mites, these illnesses cover a wide spectrum. The high population density of animals in shelters, many of whom may arrive with weakened immune systems or pre-existing conditions, frequently contributes to the prevalence of these conditions. In addition to weakening the animals' defenses, stress, and malnutrition increase their susceptibility to infections. Contagious skin diseases can also spread quickly in a shelter setting because of the shared bedding and spaces. Strict hygiene guidelines, quick medical attention, and occasionally isolating sick animals are necessary for effective management. In these situations, prevention, and management of dermatological problems depend heavily on routine veterinary examinations and appropriate diet. To enhance the general health and adoptability of shelter animals, the aim is to decrease the prevalence of skin diseases.

2.1. DERMATOPHYTOSIS

A superficial fungal infection of the skin, hair, and, less frequently, the claw or hoof is called dermatophytosis. The most important pathogens of veterinary importance are:

1. *Microsporum canis* (affects cats, dogs, and to a lesser extent large animals).
2. *Trichophyton mentagrophytes*, *T. verrucosum*, and *T. erinacei* (affects hedgehogs).
3. *M. gypseum* (soil organism that causes inflammatory lesions).

The genera Microsporum and Trichophyton are being reclassified into the genus *Arthroderma*. In otherwise healthy animals, dermatophytosis is a self-healing disease that goes away on its own without medical intervention. Given that it causes easily treatable skin lesions in humans, it is regarded as zoonotic. Direct contact with an infected animal is the method of transmission, though illness is not always the result of exposure alone. Spores alone are insufficient for efficient transmission from the environment; microtrauma is required (e.g., clipping of hair coat with contaminated clippers, tack). Since it is not a disease that needs to be reported in pets, it is challenging to estimate the prevalence because many studies that do so fail to distinguish between true infection and fomite carriage. As a whole, the disease is rare, with a true disease prevalence of less than 4% of all skin disorders. It is not the most typical cat skin condition. Risk factors include young dogs, animals that roam freely, hunting dogs, and warm environments. Cats that test positive for FIV/FELV are not more likely to become infected. Animals experiencing physiological stress or living in crowded conditions—such as in cases of animal hoarding—are more likely to contract diseases. Treatment for infections in sick animals, animals experiencing physiological stress, and animals with underlying diseases or factors can be more challenging. Enough infectious spores, microtrauma to the skin, and moisture on the skin are necessary for lesion development. The degree of infection is a good indicator of the host's general health; more virulent strains do not cause severe infections; rather, host cell-mediated immunity fails to mount an effective defense against the infection. Under ideal circumstances, dermatophyte arthrospores can germinate and begin invading skin and hair shafts in as little as 6–8 hours.

2.2. ETIOLOGY

A collection of fungi belonging to the genera *Microsporum*, *Trichophyton*, and Epidermophyton make up the dermatophytes. These organisms have evolved to be able to break down keratinous waste. These genera contain pathogens that affect the claw, hair shafts, and stratum corneum. Three classes–anthropophilic, zoophilic, and geophilic–are used to categorize the species. The anthropophilic species hardly ever affects animals because they have evolved on humans. Although zoophilic species are suited for living with animals, they have the potential to spread to people. Although they occasionally harm animals, the soil-adapted geophilic species are not as frequently responsible for human illness.

Microsporum canis (*M. canis*), *M. gypseum*, and *Trichophyton* spp. are the most frequently isolated pathogens from cats and dogs, despite the fact that more than 30 species of dermatophytes affect dogs, cats, and birds. *M. canis* is the most significant pathogen affecting cats; it is responsible for at least 90% of feline dermatophytosis cases. Other species, like *Microsporum persicolor* or *Trichophyton* spp., can also infect cats., although these illnesses are not common. Dogs are more likely to have *M. gypseum* isolated from them, but they can also have *M. canis* or, less frequently, *Trichophyton mentagrophytes*. Clinically significant *Trichophyton* infections are uncommon, but they are more common in dogs than in cats, though they can also affect the latter. Trichophyton species. Infections are also more prevalent in hedgehogs and rats. As *M. canis* is the pathogen of greatest significance in shelters, this chapter will primarily focus on its identification and control as its primary management goals.

2.2.1. Key Factors Affecting Susceptibility to Infections

Dermatophyte infection is strongly correlated with seasonality in many areas. Animal infections are typically more prevalent in warm, humid climates across the globe. Infections may be more common in the warm months than the winter in more temperate climates. Infection in dogs is rare in areas with colder winters. A higher rate of infection in shelters could also be associated with the breeding season, that is to say "Spring kitten season." For instance, October, and May typically see the biggest influx of infected cats and kittens in the northern Midwest of the United States (Newbury et al., 2005; Verbrugge et al., 2006). Kittens are more prone to infection and frequently show up at shelters underweight, anxious, and infected with both internal and external parasites. More often than not, shelters in the warm, humid southern parts of the United States and those in the north that take in dogs from farther south report cases of dermatophytosis in dogs. Rarely are infections in shelter dogs from the Northeast and northern Midwest noted as being problematic.

Although there are numerous variables that affect infection, age is typically the most significant one. Animals under a year old are most frequently afflicted with this illness. Adult animals are more likely to have limited or focal infections or lesions, whereas young animals are also more likely to experience severe, generalized infections (i.e., skin lesions covering much of the body). Cats who live close to other cats and gregarious, social cats tend to be more likely to contract an infection, most likely due to the higher risk of exposure through contact. Persian and Himalayan cats, who have longer hair, might be more vulnerable since they are less adept at maintaining their own grooming, which makes it harder for them to mechanically remove the spores.

Whether a cat is housed alone, in a group, or is allowed to roam freely, overcrowding in shelters should be taken very seriously as a risk factor. Animals from places like barns or hoarding situations, where there are a lot of animals, are also more likely to get sick (Polak et al., 2014). No correlation with intake status, such as stray or owner surrendered, has been found (Verbrugge et al., 2006).

Dogs that are free, hunt, or live in crowded conditions are more likely to become infected (Cafarchia et al., 2004). One well-known dog breed that is prone to infection is the Yorkshire terrier; the cause is unknown but could have to do with the length of their hair or the fact that dogs that need frequent grooming are more likely to contract infectious skin diseases (Cafarchia et al., 2004, Cerundolo, 2004).

Other animals kept in shelters, such as rabbits, ferrets, mice, rats, guinea pigs, hedgehogs, and so on, may also contract dermatophytosis. Because the disease has a similar etiology, the clinical signs will also be similar. Treatment protocols that work for kittens are generally safe to use in these other species. Before using systemic antifungals in any species other than cats and dogs, the reader is advised to confirm that the species in question is susceptible to the medication and that there may be potential toxicity.

2.3. PATHOGENESIS

Exposure can happen through coming into contact with a contaminated environment, a fomite, or another infected animal. It's not always the case that spore exposure causes infection. Spores need to breach host defenses and reach the skin's surface in order for an infection to take hold. The first line of defense is sebum, natural skin flora, grooming, and the skin immune system. To penetrate the skin and spread infection, dermatophyte spores need to cause some kind of microtrauma or trauma. High humidity-induced skin maceration is a frequently disregarded cause of "microtrauma." Another frequent cause of skin microtrauma that can make animals more susceptible to infection is external parasitism. Typically, infection results from exposure to infected spores plus one or more predisposing factors, such as being young or elderly, having a debilitating illness, having a compromised immune system, being in poor physical condition, having matted hair, and/or having predisposing environmental factors linked to poor husbandry. These elements are frequently present in combination. An infectious spore that reaches the hair coat in an animal may fall off, be groomed off, or simply be mechanically carried and never cause an infection in environments where the animal is not under stress, illness, or poor physical condition. Exposure may more easily result in infection in shelter animals, where stress, parasitism, and coexisting illnesses may hinder normal grooming behavior and negatively affect the immune system.

2.4. DISEASE

2.4.1. Incubation Period

The incubation period is typically two to four weeks from the time of exposure until the development of clinically noticeable lesions. Infections that are induced experimentally have been observed to manifest as infected hairs in less than a week (DeBoer and Moriello, 1994). Highly skilled observers may be able to identify lesions early. An infection of the hair follicle occurs before clinical lesions appear. When mapping an outbreak or figuring out the possible source of infections (in-shelter spread vs local resource). It is not advised to just quarantine potentially exposed animals and wait for clinical lesions to appear because of the protracted incubation period and inconsistent clinical signs.

2.4.2. Immunity after Recovery

Research on the immune system of cats has revealed that, although they respond to infections both humorally and cell-mediatedly, it is the cell-mediated response that is linked to recovery. In comparison, many other infections do not heal (DeBoer and Moriello, 1993; Sparkes et al., 1996). Prolonged infection is linked to a lack of a cell-mediated response. Moreover, immunity is relative. The majority of recovered infected animals are resistant to re-infection, although infection is still possible in response to a high spore challenge. It is important to shield recovered animals from harm to avoid the spread of infection. It is unknown how long immunity lasts after an infection has healed.

2.4.3. Carrier State

In a shelter environment, the researchers have not observed any instances of "carrier state" animals that are unaffected or of animals that have infections that do not go away despite receiving the right care. Research on cat fungal flora has revealed that *M. canis* is NOT a typical component of cat fungal flora. A cat with visible clinical lesions, a cat with less noticeable lesions, or a cat that is mechanically carrying spores on its hair coat are the three scenarios that a positive fungal culture from a cat suggests. These spores can spread to humans and other animals, so action is needed in all three scenarios. In contrast to other diseases found in shelters where "latent" infections can become active, the idea of "asymptomatic carriers" concerning dermatophyte infections is distinct. Both "carrier" and "asymptomatic carrier" are terms that the researchers strongly advise against using.

Dermatophytosis is a disease that can be cured with treatment. Considerations should be made regarding medication and treatment options for animals with chronic or recurrent infections, as well as environmental contamination. Because there are still spores in the housing environment, the animal frequently maintains a positive culture. The researchers have not come across any situations in a shelter setting where cats with chronic or "incurable" infections could not be healed after the proper care was given, underlying illnesses were treated and resolved concurrently, and the environment was appropriately decontaminated.

Whether an animal is genuinely infected or if the spores are just settled like dust on their hair coat, it can still test positive for dermatophytes. Differentiation is crucial because these two scenarios offer radically different risks and challenges. Dust mop cats–cats that mechanically carry spores on their hair coat–may contract an infection if the spores are not eliminated, but frequently, cats will groom themselves to get rid of the spores.

What kind of risk these "dust mop" cats pose to other animals and the environment of the shelter is unknown. Dermatophytes are not regarded as typical canine or feline flora, despite the fact that animal hair coats frequently contain spores that are representative of those found in their surroundings. There is a chance that an infection will spread or that spores will be released into the environment when there are a lot of spores being carried on the hair coat. The danger decreases with low spore carrier counts. To differentiate between these two situations, quantitative fungal culture can be helpful; please refer to the risk assessment section.

2.5. TRANSMISSION

Animals with the infection can readily release naturally infectious spores and hairs into the surrounding environment. Since infectious spores are tiny–roughly the size of dust–they are easily caught in or spread to other animals' hair coats when they are in or out of contaminated areas. A susceptible host can become infected with infectious spores through direct contact, contaminated fomites (such as hands, bedding, blankets, toys, lab coats, and environmental surfaces), or even external parasites. Staff members may mechanically transfer dermatophytosis from one cat to another while administering medication, feeding, or performing other nursing care in specific situations, such as when medicating cats. In addition to causing microtrauma to the skin, these activities also raise the risk of infection development.

Shelters and other animal-housing facilities frequently report concerns about airborne transmission. The possibility of spores passing through heating and cooling ducts and contaminating the environment has drawn a lot of attention. Although infectious spores can travel great distances and spread by air, most of the time airborne transmission of dermatophyte spores is negligible when husbandry and disinfection procedures are followed to the letter. Spores did not spread by air in the authors' work at a shelter equipped with a heating and cooling system like one found in a house. Infected cats were kept in cages at this facility, and the floors were mopped once or twice a week with detergent and swept every day.

Fungal culture plates were frequently left open and exposed to the air throughout the facility when it was fully occupied by infected cats. Furthermore, to determine whether spores were dispersing through forced air heating vents, fungal culture plates were placed directly over the vents with the culture side facing down. Even in cases where infectious spores could be isolated from regions where contamination was anticipated (e.g., cat cages that are actively infected). Nonetheless, the furnace filter's fungal cultures were consistently positive. These findings suggest that rather than being blown throughout the building, infected hairs and spores were drawn into the air vents and trapped in the furnace filter. This illustration also implies that, in the majority of shelter scenarios, dermatophyte transmission can be successfully managed without the need for independent air circulation–so long as appropriate care is concurrently given to reduce infectious dose and environmental contamination. Author Moriello has, however, observed airborne transmission in a cattery where the owner used fans to cool the cats and move air around in the isolation room. The possibility of airbo.

It is possible to minimize transmission in shelters by closely monitoring fomite and environmental control. Frequent mechanical cleaning and disinfection significantly reduces the risk of environmental contamination. The worst places for environmental contamination are shelters and foster homes, places where hair has been allowed to grow, and houses with a high population of sick kittens (Mancianti et al., 2003).

If the area is not properly cleaned and disinfected, along with taking particular measures to remove any infectious material mechanically, spores will continue to exist in the environment and may remain contagious for several weeks to months.

A higher infectious dose and more frequent exposure raise the probability of transmission. When animals are kept together, either in cages or in environments free of cages, as is the case in shelters and rescue homes, the spread of dermatophytosis is especially problematic. A higher risk of introduction and/or increased exposure to a critical mass of infectious material may exist in higher turnover shelters due to the continuous influx of new, potentially infectious cats; however, any time an infectious animal is admitted to a group setting, particularly when infection remains undetected, infection may spread to other members of the population.

2.6. CLINICAL PRESENTATION AND DIFFERENTIAL DIAGNOSES

In veterinary dermatology, dermatophytosis is one of the most pleomorphic skin diseases seen. Circular patches of alopecia with inflammation and scaling are commonly referred to as classic lesions of dermatophytosis. The term "ringworm" originated from the original theory that the "classic" round, hairless lesions were caused by a worm moving in a ring within the skin. While many dermatophytosis cases will show those typical clinical signs, many others won't. Dermatophytosis can present itself in many different ways besides the traditional one. Clinical signs in dogs and cats can include any combination of the following: "feline symmetrical alopecia," mild to severe scaling and crusting, erythema, easily broken hairs, excessive shedding, minimal to marked pruritus, follicular plugging, hyperpigmentation of the skin, otitis externa, inflammation of the ear margin, pododermatitis, papules, pustules, and presentations severe enough to mimic immune-mediated skin diseases (Figure 2.1).

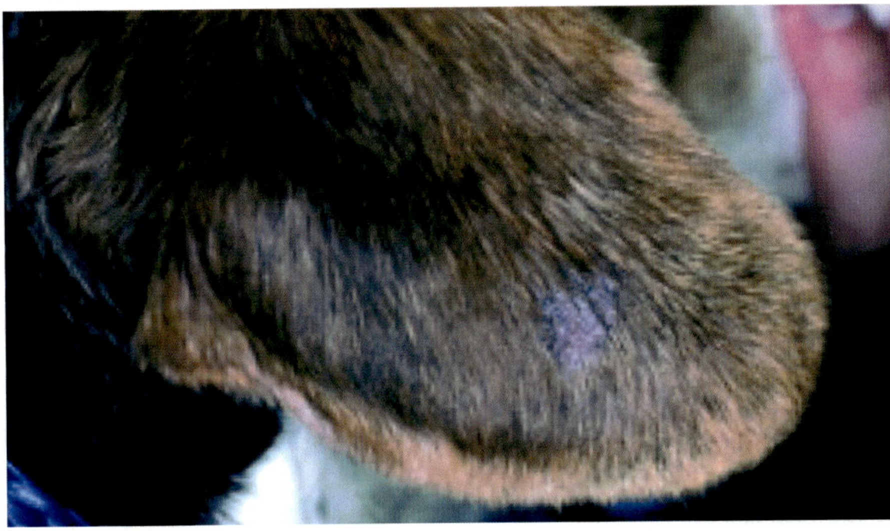

Figure 2.1. *Puppy with an area of hair loss caudal to the ear.*

Source: https://images.ctfassets.net/4dmg3l1sxd6g/7eCqbIhiVmcgpRdY1SwSp8/2b7b5350d158f064b00fa3be561c3107/untitled-1-2341-article.jpg_-_en.

Any size or form is possible for lesions. The extent of crusting, hair loss, and inflammation varies greatly. Most lesions include some level of inflammation, either with or without hair loss. Numerous different non-contagious, non-infectious skin conditions can mimic or conceal skin lesions.

Without a thorough, in-depth examination of the entire body, subtle lesions can frequently go unnoticed. In other situations, lesions that appear "classic" might not even be the result of dermatophytosis. Dogs that have skin lesions that resemble "classic" ringworm lesions are much more likely to have bacterial pyoderma or demodicosis than dermatophytosis, even though cats with "classic" dermatophytosis appearances are frequently actually infected with dermatophytes. Dog dermatophytosis is more likely to cause skin lesions in puppies than in adults, though Demodex or bacterial pyoderma may also be to blame. Although examination of lesions is an essential component of identification, diagnosis, and outbreak response, dermatophytosis cannot be diagnosed based only on the presence of "classic" skin lesions due to its multifaceted presentation (Figure 2.2).

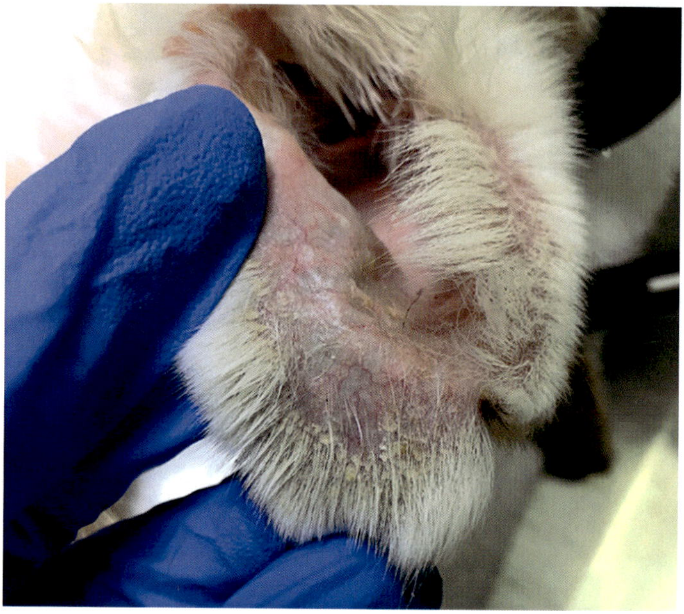

Figure 2.2. *Ear margin crusting due to dermatophytosis. This was caused by a Trichophyton spp. infection.*

Source: https://images.ctfassets.net/4dmg3l1sxd6g/3srBW6QzMXRlsHobhX9Te2/8bf9fdf0ddb80176824f14469d1b908c/top-5-pinnae_fig7.png_-_en.

2.7. DIAGNOSIS

In animal shelters and other animal welfare organizations, diagnosing dermatophytosis in animals can be difficult. Frequently impacted are choices regarding which animals can be placed in foster care or offered for adoption. Due to the potential for infection in both humans and animals, early detection and quick action are crucial. A comprehensive explanation of the suggested systematic screening procedure, including particulars on how each diagnostic tool can be applied successfully in animal welfare organizations.

2.7.1. Basic Diagnostic Tools

The basis for the appropriate application of all other diagnostics is a careful and comprehensive dermatologic examination. The diagnostic tools must be utilized in conjunction with a comprehensive screening procedure that includes a trained observer's meticulous physical examination of the skin and hair coat. It is important to make every effort to find out the animal's history, including its previous housing circumstances and health issues. To identify animals that pose a higher risk, information regarding possible exposure can be highly beneficial. The most popular and useful diagnostic techniques are the use of a Wood's lamp, direct microscopic examination of suspicious hairs, and fungal culture, which come after the history and physical examination. Skin biopsies are rarely performed, but when one is performed on an unusual skin lesion, it can surprisingly reveal dermatophytosis as the cause–especially if the condition hasn't been ruled out beforehand.

2.7.2. Diagnostic Protocols for Preventive Management and Control: Systematic Screening Protocol

The recommended systematic screening protocol described here consists of coordinated and consistent use of the five basic diagnostic tools:

1. History.
2. Visual (physical) examination.
3. Wood's lamp examination.
4. Direct examination of fluorescing hairs.
5. Fungal culture.

Developing an efficient protocol for dermatophyte screening, prevention, and treatment in a shelter necessitates a comprehensive comprehension of all five elements of this process. Certain personnel may occasionally receive specialized training to concentrate on certain protocol steps (e.g., reading fungal culture plates). Using these five fundamental diagnostic tools makes early identification easier and aids in directing choices about housing and treatment. This methodical screening process can be applied during admission, at any significant control point, or in the event of an outbreak.

To stop the spread of disease, screening is essential at the time of admission to the shelter. When resources permit, screening at admission offers the chance to quickly begin treatment after identifying sick, potentially contagious animals and taking preventative measures. The systematic process of screening for dermatophytosis only takes a few extra minutes during the admissions process, provided that the designated area is adequately equipped for the screening examination and the intake staff has received the necessary training. To detect any lesions or infections that may have been overlooked or developed since intake, the examination steps of the screening process should, at the very least, be repeated at critical infectious disease control points (such as bringing animals into public spaces, group housing, or foster or adoptive homes). Having a systematic screening procedure in place is also crucial for an outbreak response strategy. To help inform decisions about treatment and management, additional diagnostics are most effective when used in concert with data acquired during the screening process.

2.7.2.1. Step One: History

At the time of admission, or even earlier, all pertinent historical data for every single animal and every examined group should be reviewed. Living with other infected animals in the same household should make an animal extremely suspicious, or high-risk. Although not all members of a litter will exhibit clinical signs, littermates or cagemates of confirmed infected animals should be treated as infected due to the close exposure and direct contact.

2.7.2.2. Step Two: Visual (Physical) Examination

Bright light examinations are necessary to look for skin lesions on animals. When working with the animal, the examiner should always wear disposable exam gloves. A bright, focused examination light is very helpful in locating small lesions. If this isn't available, use a flashlight with a strong beam. A flashlight's focused beam can sometimes reveal more lesions than overhead lighting because it can illuminate a larger area. The muzzle, lips, periocular area, ear margins, the skin inside the ear cone, digits, and nail-bed area, abdomen, medial aspects of limbs, and the tail are common areas where lesions are missed, so special attention to these areas is required. Employees need to be taught how to distinguish potential dermatophyte lesions from scars or other quiet, non-inflammatory skin lesions. Even if another clear cause has been found, all inflammatory lesions need to be reported. For instance, the inner pinnae and ear canal in cats may become inflamed due to ear mites. The obvious cause of inflammation or hair loss in and around the ears may seem to be the infestation or the self-trauma that goes along with it. The presence of ear mites does not, however, rule out concurrent dermatophytosis. The author has found that because the ear-mite infestation causes microtrauma that promotes infection, it may actually raise the risk of dermatophytosis. There may be another way for spores to spread in a shelter because fleas have been shown to be culture positive. Furthermore, keep in mind that cats, especially kittens, may have multiple illnesses, even in cases where the obvious cause of an animal's pruritus (such as a louse infestation) is apparent. Furthermore, this meticulous visual examination of animals is a useful means of detecting additional medical issues.

2.7.2.3. Step Three: Wood's Lamp Examination

The next step is to look for fluorescing hairs on animals by using a Wood's lamp. An ultraviolet light with a wavelength of 365 nm is a Wood's lamp.

Wood's lamps serve as screening instruments to detect contaminated hairs for microscopic analysis. *M. canis* is the only significant veterinary pathogen capable of producing apple-green fluorescence. A positive test results in a strong indication of infection and permits the selection of hairs for microscopic analysis. Although a negative examination does not rule out dermatophytosis, it does indicate the absence of infection in an uninfected animal. When an animal has lesions and a negative Wood's examination, the animal is much less likely to be infected. It is best to use a plug-in Wood's lamp instead of a battery-operated UV lamp because the latter seldom provide enough light at the right wavelength for diagnosis. Examining with smaller hand-held models is simpler than using large wands or lamps that have built-in magnification. A hand-held magnifying glass is useful if more magnification is required.

Lamps don't need to warm up, but it frequently takes a few minutes for the examiner's eyes to adjust to the light before they can see glowing hairs. To prevent missing any affected areas, the entire body should be examined slowly and carefully in a dark room with the lamp held close to the skin and hair coat. When using the Wood's lamp, care should be taken to examine the commonly missed lesion sites described previously as well as the common sites of infection.

The fluorescence linked to dermatophytosis is not limited to the tips of the hairs; it is also visible at the base of the hairs, close to the hair follicle. A metabolite is deposited on the hair shaft by the fungus in the hair follicle as it grows, which is why infected hairs fluoresce. It will appear as though the entire hair shaft is glowing in many instances. The hair's metabolite coating prevents the hairs from adhering to one another. In certain instances, the afflicted hairs may break, leaving the stubble glowing only near the skin's surface at its base.

The most typical fluorescence from infected hairs is a vivid apple-green color (*See* Figure 2.3). The author's study involved daily examinations of cats, and in the early stages, the fluorescence was blue-white before eventually turning apple green. Most of the time, other colors are artifacts. Sebum that is either inflammatory or not will glow a duller yellow color on the skin, which is not a sign of dermatophyte infection. When applied or crusted on the fur, doxycycline or terramycin give off a yellow glow that should not be mistaken for an infection. If a dermatophyte infection is the cause of the fluorescing substance, it cannot be ignored. In case of doubt, carefully remove any hairs and check for fluorescence at the roots.

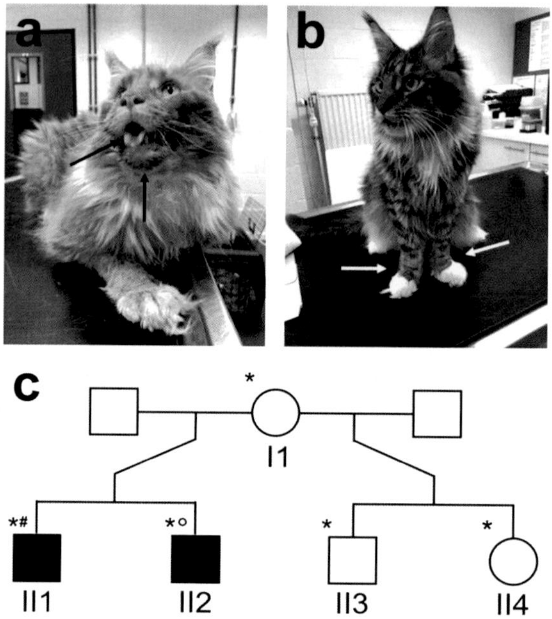

Figure 2.3. *Cat studies.*

Source: Beckers E, Cornelis I, Bhatti SFM, Smets P, Shelton GD, Guo LT, Peelman L, Broeckx BJG. A Nonsense Variant in the DMD Gene Causes X-Linked Muscular Dystrophy in the Maine Coon Cat. Animals. 2022; 12(21):2928. https://doi.org/10.3390/ani12212928.

In cases where thick crusting is evident, it is crucial to carefully scrape off the crusts and search for any infected hairs that might be hidden by the crusting. Broken hairs beneath the crusts are frequently Wood's positives even though the crusting is not. It is frequently stated that using a Wood's lamp, 50% of cases of dermatophytosis caused by *M. canis* will fluoresce. The metabolite that gives *M. canis* its fluorescent glow is only produced by this species, and not all strains of *M. canis* will produce it. Nevertheless, in the author's experience diagnosing hundreds of afflicted cats at animal shelters, glowing hairs were found on the great majority of the cats with lesions upon close inspection using Wood's lamp. Fluorescing hairs in cats that initially appear non-lesional on visual examination often indicate pre-clinical infection or help spot subtle lesions that were overlooked during the initial examination.

2.7.2.4. Step Four: Direct Examination of Fluorescing Hairs

When a Wood's lamp examination is positive, a quick and easy way to get an instant diagnosis of *M. canis* dermatophytosis is through direct microscopic examination of fluorescing hairs. Although it requires some training and practice, it is not difficult to learn how to interpret the microscopic examination. Every step in the initial training process for assessing direct examinations should be verified by a fungal culture until a satisfactory accuracy rate is attained. When done correctly, Wood's lamp and direct hair examinations can save a lot of money and time when compared to performing a fungal culture. One thing to be careful of is that, even if the diagnosis is confirmed by the Wood's lamp and direct examination of the hairs, a fungal culture should always be carried out if the case involves any legal issues.

The "gold standard" for diagnosing dermatophytosis is fungus cultures. It is also advisable to take pictures of the animal in its entirety as well as any microscopic examinations if there are any legal concerns. By aiming a digital camera through the microscope lens, this can be done with ease. The only tools needed are a Wood's lamp, a microscope, and a curved-tip hemostat. Glass microscope slides, coverslips, and mineral oil for the suspension solution are required supplies. Mineral oil can be substituted with clearing agents like potassium hydroxide (KOH). Clearing agents make hairs and spores more refractile by causing the background material to swell. The KOH solution should be handled carefully since it can be caustic to the skin of the animals and the lenses of the microscope. Mineral oil is a good, safe substitute that is easily accessible for use in shelter environments.

Procedure:

1. A microscope slide that has been placed aside where it won't be easily knocked to the floor but is still close to the examination area should have a drop of mineral oil (or KOH) on it.
2. To guarantee that the hair root is obtained for analysis, a few luminous hairs should be removed in the direction that they grow. To accomplish this task, a minimum of one helper is needed to support the animal and the Wood's lamp while the other person locates and cautiously removes the glowing hairs. Before the hairs are put in the suspension solution on the slide, they can be closely inspected using the Wood's lamp.
3. The coverslip should be placed on the microscope slide, and the hairs should be placed on the drop of mineral oil or KOH. Once more, you can verify that there are glowing hairs on the slide by using the Wood's lamp.
4. After turning off the microscope light and dimming the lights in the room, the slide should be placed on the microscope stage and examined under a 4x magnification. Holding the Wood's lamp close to the microscope stage and pointing it at the slide is the proper

position. This technique draws attention to glowing hairs so that they can be seen through the eyepiece and lens. Once the fluorescing hair is visible through the eyepiece, the Wood's lamp can be turned off, and the room lights and microscope can be turned on.

5. At low power, infected hairs appear more filamentous and are wider and more irregular than healthy hairs. Cuffs of refractile spores and hyphae are visible at higher magnification. To confirm the presence of ectothrix spores, a 40 × magnification lens may be required, but a 10 × lens is typically sufficient to confirm infected hairs. Immersion in oil is not required. Once more, the examiner is searching for hair shafts that appear swollen, broken, filled in, or frayed in comparison to normal hair shaft structure, as well as stacks or rows of tiny beads on either side of the hair shaft.

2.7.2.5. Step Five: Fungal Cultures

Performing a fungal culture on all animals entering the shelter, regardless of whether lesions were present, was thought to be the best or most prudent option for detecting the disease at the time this textbook's first edition was published. The most cautious method for shelters was to perform a fungal culture, which is the gold standard for diagnosing dermatophytosis in individual animals. This method provided a safety net to identify any infectious animals that might have escaped the notice of visual and Wood's lamp exams.

Unfortunately, most shelters are unable to adopt this practice due to its extreme complexity and resource requirements. A shelter's dermatophytosis was successfully managed by thorough fungal culture screening of all cats upon intake (Newbury et al., 2005). However, the same shelter and many others adjusted to culturing only Wood's positive cats and those with suspicious inflammatory lesions while keeping comparable levels of control over the infection (Table 2.1).

Table 2.1. *Clinicopathological Features of Cutaneous and Extracutaneous Invasive Dermatophyte Lesions in Cats*

Cat no.	Breed	Age-Years	Sex	Lesion Type	Lesion Location	Biopsy	ITS2 PCR	28S rDNA PCR	IHC	Culture	Woods Lamp
1	Persian	7	FN	C, U, S	Neck, dorsal	EB, Co	– ve	ND	+ ve	ND	ND
2	Persian	10	Ma	C, U, M EC, S	Right pinna Hard palate	IB	– ve	– ve	+ ve	ND	ND
3 *	ESH	7	MN	EC, S	Left SCLN, ventral thorax, abdomen	IB	*Trichosporon inkin*	ND	+ ve	ND	ND
3 *	ESH	8	MN	EC, S	Digit, left hind limb	IB	– ve	*M. canis*	+ve	ND	ND
4	DSH	6	F	C, U, M	Interscapular, Limbs	IB	*M. canis*	*M. canis*	ND	ND	+ve
5	ESH	6	F	C, U, M	Trunk (hip to tail)	EB, IC	*M. canis*	ND	ND	ND	ND
6	BLH	2	MN	C, S	Digit, left hind limb (P2)	EB, Co	*M. canis*	ND	+ve	ND	ND
7	BSH	0.58	MN	C, S	Head, dorsal	EB, Co	*M. canis*	ND	ND	ND	ND
8	Persian	9	MN	EC, S	Sub-lumbar mass	IB	– ve	ND	+ ve	ND	ND
9	ELH	10	F	C, S	Digit, RHL (P2)	EB, Co	– ve	ND	+ ve	ND	ND

10	ESH	10	F	C, M EC, S	Dorsolateral head, Oral cavity, Intracranial	IB	*M. canis*	ND	ND	ND	ND
11 *	BSHX	.8	FN	C, U, M	Dorsal head, trunk, flanks	EB, Co	*M. canis*	ND	ND	− ve	+ ve
11 *	BSHX	2	FN	C, S	Pinna, left	EB, Co	*M. canis*	ND	ND	ND	ND
12	BSH	1	MN	C, S	Left hind limb	EB, IC	− ve	ND	+ ve	ND	ND
13	ESH	10	MN	C, M	Trunk, ventral abdomen, left lateral thorax	EB, Co	− ve	ND	+ ve	ND	ND
14	Persian	1.5	FN	C, U, M	Unknown	EB, IC	*M. canis*	ND	ND	ND	ND
15	ESH	2	MN	C, U, M	Dorsal trunk, interscapular	EB, Co	*M. canis*	ND	ND	ND	ND
16	ESH	7	MN	C, U, M	Trunk, ventral abdomen	EB, Co	*M. canis*	ND	ND	ND	ND
17	BSH	8	MN	C, U, S	Trunk, dorsal	EB, Co	*M. canis*	ND	ND	ND	ND
18	BSH	11	MN	C, U, S	Trunk, ventral abdomen	EB, Co	*M. canis*	ND	ND	−ve	ND
19	Persian	10	MN	C, S	Neck, dorsal	EB, Co	*M. canis*	ND	ND	ND	ND
20	ESH	7	FN	EC, S	Intra-abdominal/ vaginal	EB, IC	*M. canis*	*M. canis*	ND	+ve	ND
21	DSH	4	MN	C, S	Trunk, flank	EB, Co	− ve	ND	ND	ND	ND

22 *	BSH	1	FN	C, M	Trunk, dorsal	EB, IC	*M. canis*	ND	ND	ND	ND
23	BSH	10	MN	C, U, M	Trunk, left axilla	EB, IC	– ve	ND	ND	ND	ND
24	Persian	6	MN	C, U, S	Trunk, dorsal	EB, Co	*M. canis*	ND	ND	ND	ND
25	ESH	4	MN	C, U, M	Trunk, dorsal	EB, Co	– ve	*M. canis*	ND	ND	ND
26	ELH	3	Ma	C, U	Unknown	EB, IC	*M. canis*	*M. canis*	ND	+ve	ND
27	DSH	1.5	F	C, U, M	Thorax, bilateral	EB, IC	*M. canis*	ND	ND	ND	ND
28	ESH	5	FN	C, M	Trunk, axilla	EB, Co	*M. canis*	ND	ND	ND	ND
29	Persian	1	MN	C, U, S	Base of tail	EB, IC	*M. canis*	ND	ND	ND	ND
30	BSH	4	MN	C, U, M	Pinna, tail, hind limb	IB	ND	*M. canis*	ND	+ve	+ve

** Previous history of dermatophytosis; + ve, positive; – ve, negative; BSH, British shorthair; BLH, British longhair; BSHX, British shorthair cross; C, cutaneous; Co, complete; DSH, domestic shorthair; EB, excisional biopsy; EC, extracutaneous; ESH, exotic shorthair; F, female; FN, female neutered; IB, incisional biopsy; IHC, immunohistochemistry; Ma, male; MN, male neutered; ND, not done/not reported; M, multiple; S, singular; SCLN, superficial cervical lymph node; U, ulcerated.*

Source: Hobi S, Tam WYJ, Tse M, Nekouei O, Chai Y, Hill FI, Cheung E, Botes W, Saulnier-Troff F, McDermott CT et al. Microsporum canis Causes Cutaneous and Extracutaneous Feline Dermatophytic Pseudomycetomas: Molecular Identification and Clinicopathological Characteristics. Journal of Fungi. *2024; 10(8):576. https://doi.org/10.3390/jof10080576.*

The majority of suspect and infectious cats can be identified by a thorough intake evaluation that makes use of the first four diagnostic tools previously mentioned: the history, visual physical examination, Wood's lamp, and a direct examination of fluorescing hairs. This significantly lowers the infectious risk associated with dermatophytosis. The process is aided by selective fungal culture screening, which is primarily required to diagnose infections in animals with suspicious lesions that do not fluoresce. It also helps to rule out infections in positive or suspect animals. Since the majority of infected cats will have fluorescing lesions, if the screening process is successful, non-fluorescing lesion culture results will typically be negative. To sum up, the most effective way to definitively diagnose dermatophytosis in animals with fluorescing lesions or rule out the condition in animals with inflammatory lesions that do not fluoresce is through the use of fungal cultures. Most often, culture is done as a safety measure; it lowers risk by helping to rule out infection and facilitating the cats' quicker transition through the shelter.

Screening for all animals is most important for shelters that commonly:

1. House animals in group or community settings.
2. House animals in rooms that are difficult to disinfect.
3. Send animals to private temporary homes for foster care.

It should be standard procedure to screen cats for visual and Wood's exams before placing them in foster homes or group housing.

2.7.2.5.1. Sample Collection and Preparation Technique

Every animal sample needs to be taken with a brand-new toothbrush. Fresh toothbrushes are mycologically sterile when kept in their original packaging. Bulk individually wrapped toothbrushes are inexpensively available from a number of sources. When brushing a cat, make sure to brush them vigorously over their entire body until the brush bristles show signs of hair growth. If lesions are found, the lesions should be brushed last to prevent spores from being distributed throughout the hair coat. Unless there are no lesions and the dog is only being sampled because there has been a known or suspected exposure, only lesions should be sampled in dogs. Contaminant organisms frequently overgrow fungus cultures from a dog's whole hair coat quickly; these cultures should be closely monitored, and dogs should be cultured again if necessary.

The fungal culture plate should then be gently stabbed with a toothbrush bristle onto its surface to initiate the inoculation process. To support the pathogen scoring system that is explained below, the entire plate needs to be covered, and a consistent method needs to be used every time. To ensure that the bristles are embedded in the plate and transfer the spores, it is crucial to start in the center of the plate, work your way outward, and cover the entire surface. However, avoid using too much pressure as this could lift the medium off the plate.

If it is not feasible or possible to inoculate toothbrushes immediately after collection, they can be double-bagged into a sealable Ziplock bag after being placed in plastic sandwich bags. If cultures are inoculated at a later time, this should be done in a clean area that is easily disinfected. It is important to identify toothbrush samples carefully; pre-printed labels with blank spaces for ID numbers and dates can make this process easier. It is also important to indicate on the bag if a sample is very questionable. To avoid cross-contamination, high-suspect samples should be vaccinated last when inoculating multiple samples. When handling samples, gloves should always be worn. Used toothbrushes should be thrown away in sealed trash.

Labeling culture plates can be challenging. After being chilled, plates are frequently damp; therefore, before writing on them, wipe them dry. If using Petri dishes, identification, and dates should be marked on the bottom of the plate and not the cover to ensure that there are no mix-ups in the laboratory. Culture plates must always be labeled with the name and unique animal ID number of the animal, along with the dates of sampling and vaccination.

Dermatophyte Test Medium (DTM) from six commercial companies, including flat trays, jars, and "micromedium plates" (In-Tray), was tested in a series of laboratory tests conducted by the author to ascertain whether one type of fungal culture plate was superior to another concerning time to sporulation. Some media were advertised as "rapid sporulation." Tests were conducted on multiple strains of *M. canis*, *M. gypseum*, and *Trichophyton* spp. using multiple replicates of each DTM brand. Furthermore, tests were conducted to determine how temperature and light exposure affected the growth period (Verbrugge et al., 2007).

Regarding growth time, there was no difference in performance between the brands, except for one fungal culture medium that underperformed. Identification of growth and pathogens became more challenging the smaller the plate's surface area; the color indicator lost its usefulness once the surface turned red. Furthermore, the medium plates with reduced surface area frequently dried out before the pathogen growing. The time to grow was insensitive to light. Temperature was the primary factor influencing growth. Growing cultures between 27 and 30□ (80 and 86°F) was optimal for sporulation. These results suggest that DTM plates with a large, easily inoculable surface area be used for performing cultures. Because it is challenging to inoculate the surface and sample colonies, using jars should be avoided. Moreover, yeast, and bacteria frequently congregate on the surface.

2.7.2.5.2. Animal Holding Time for Culture Results

It takes time for fungal cultures to grow and incubate before pathogens can be identified or the culture can be declared negative. Waiting for the results of fungal cultures has little effect on the length of stay (LOS) or level of crowding when the legal or required holding periods are comparable to the time required for culture incubation and identification, or when animals are expected to stay for extended periods of time. Holding cultured cats, however, frequently results in longer holding periods. The health of the populace as a whole may suffer as a result. It is necessary to strike a balance between the danger of possibly introducing an infectious animal into the system and the risks of stress, infection, and shelter crowding brought on by keeping animals for long periods of time. An animal that has developed clinical signs of respiratory disease while awaiting culture results cannot successfully have dermatophytosis ruled out. Unless there is a cause for greater suspicion, the risks of holding animals while awaiting culture results outweigh the chance of failing to detect a dermatophyte infection. While employing a comprehensive.

The waiting period will be significantly reduced if fungal cultures are conducted internally or in a lab that will regularly report on the progress of the cultures rather than waiting until the cultures are finished to report results. Since performing fungal cultures in-house is a skill that is easily learned, the author suggests doing so. Additionally, internal fungal culture management allows for daily culture monitoring and greatly lowers screening program costs. Community practitioners should also think about creating internal culture systems as a service for their clients, such as rescue organizations and shelters. Upon retrospective analysis of the data, it was found that between days 10 and 14, over 95% of the positive fungal cultures were identified. In these cultures, suspicious growth was almost always observed by day seven after vaccination. Animals with serious infections frequently had their cultures finished in seven days. It is unlikely that cultures that are not suspicious by day 7 or that are positive by day 10 are *M. canis*. When fungal cultures are kept warm and assessed in-house, decisions can typically be made by the tenth day of growth, though this is not always the case. Certain species of *Trichophyton* grow more slowly than others.

2.7.2.5.3. Fungal Culture Processing, Pathogen Identification and Reporting Culture Results

Cultures are typically reported as "culture negative" or "culture positive" when a pathogen has been identified. According to the author's observations, there is not enough information here to create and carry out a management program that will reduce the risk of infection while optimizing the chances of saving the lives of sick or suspect cats. Just as it is with bacterial cultures, the number of colony forming units (CFU) on the plate is crucial for decision-making when interpreting culture results. The author suggests using a semi-quantitative reporting system for fungal culture results, which was created for treatment studies using CFU on cats with dermatophytosis.

This is a synopsis of how fungal culture results are reported using the CFU scoring system, how the results are interpreted based on physical findings (lesions or not), and how treatment decisions are made. The Pathogen Score, or P score, is what the author refers to as a way to make the language of CFU more "user-friendly" for staff. DTM culture Petri dish plates or other sizable flat-bottomed plates of a similar kind should be used with this system.

The following describes an in-house system for inoculating, incubating, and reading fungal culture plates.

1. Plates need to be correctly identified using a methodical approach. The animal's ID or name, the date of sample collection (DC), and the date of immunization (DI) should all be written on the plate. Additional data, like age, sex, hair length, and source, can be added to the ID label and added to a spreadsheet to find infection trends in a specific community or organization.
2. In situations where plates are not immediately inoculated during the sampling process, it is crucial to record both the DC and the DI. An unexpected outbreak in a facility that had not been impacted before might be explained if there is a long period of time between DC and DI and the culture is positive in a cat that was not suspected.
3. Samples are inoculated onto DTM, preferably on a Petri dish style plate.
4. Fungal culture plate examinations every day are essential for the early and quick identification of suspects and the timely diagnosis of pathogens. In contrast to what many people think, this doesn't take a lot of time. It is possible to evaluate one hundred or more plates in less than 7 to 10 minutes. Any small white colonies that have a growing red ring around them ought to raise suspicions. In many cases, the colonies might be too young to identify with certainty, but once a plate is labeled as "suspect," further observation is warranted. For prompt intervention, a communication system that reports positives promptly is necessary.
5. After being held for 21 days, plates should be "officially" read on days 7, 10, 14, and 21. Day 10 is a particularly helpful time point because, by this point, the majority of positive cultures exhibit questionable growth.

6. Detecting potential suspect colonies is relatively straightforward. During the utilization of DTM plates, the colonies to focus on are those that are white or have a buff hue, encircled by a red ring as they expand. Although the red color shift is not definitive for a dermatophyte, it serves as a marker, signaling colonies that warrant microscopic verification and are considered suspect. Colonies with color pigmentation are not harmful and can be disregarded. Similarly, pale or entirely white colonies without a red ring can be disregarded.
7. If a suspect colony is identified and it is too small or too soon to sample, it should be circled with a marker so it can be observed.
8. The animal should be re-cultured if plates quickly overgrow with contaminants within the first 7–10 days, as this could lead to a false negative or an incomprehensible fungal culture. Plates that do not meet these criteria are not good for screening. This mostly affects kittens and stray cats. As soon as overgrowth is noticed in these cats, especially if a questionable lesion was noticed, they should be re-cultured. Furthermore, the usefulness of the red color change is gone once the plate's entire surface is red.
9. Data can be entered into a spreadsheet with ease using a shorthand system for recording culture results. A shorthand system is also simple for non-technical staff to comprehend.
   - i. NG" is for no growth; it is not uncommon for cats to have no growth on their fungal culture plate due to their fastidious grooming skills.
   - ii. "C" is for contaminant growth.
   - iii. "HC" is for heavy growth of contaminants and signals the staff to re-culture the cat. This may not be cost-effective in every shelter but minimally, cats with skin lesions or kittens with HC should be re-cultured.
   - iv. "S" is for suspect growth; many culture-positive plates will have S growth in the first week. Identified pathogens should be listed by name. If a pathogen (e.g., *M. canis*) is identified, the number of CFU should be counted and recorded.
10. Unless a pathogen is discovered earlier, the plate turns entirely red, or the plate is overgrown, plates should be stored for 21 days. Most *M. canis* positive cats can be identified 7–14 days after culture when they are kept in a warm room or incubator (about 80–86□ or 27–30□) (Stuntebeck et al., 2017). The reason for the 21-day holding of plates is *Trichophyton* spp. Dermatophytes grow more slowly. Practically speaking, cats can be classified as "culture negative" if,

after 10–14 days, they test negative for *M. canis*. Since reference laboratories consider 21 days to be the standard culture incubation time, it is still advised to hold cultures–not animals–for the entire 21 days. People who are infected with less common and less contagious species may still benefit from identification and treatment.

2.7.2.5.4. The Pathogen Scoring System: Using Colony-Forming Units to Aid in Management and Treatment

The calculation of CFU is a crucial tool for treatment planning. It may not be very helpful to simply report a fungal culture as positive or negative as part of a management plan because there is no way to distinguish between genuinely infected cats and fomite carriers, also known as "dust mops." Using a Pathogen Score (also known as a "P-score") to account for the quantity of CFU detected on the plate aids in guiding diagnosis and treatment choices. The number of CFU per plate, if a culture has been appropriately obtained and inoculated, typically indicates the degree of contamination when assessing environmental cultures or the severity of infection for animals in culture. The system is very simple:

- P1 score = 1–4 CFU/plate.
- P2 score = 5–9 CFU/plate.
- P3 score = 10 or more CFU/plate.

The significance of a P1, P2, or P3 cat can be easily understood by shelter staff, even if they are not trained or involved in the laboratory side of fungal cultures. This data can expedite the identification of cats that test positive for the virus and distinguish fomite carriers from those who are genuinely infected, along with the presence or absence of lesions at the time of the culture. The P-score should ideally be included when the laboratory notifies the shelter that an animal has been identified as positive. The animal needs to have its lesions checked again right away. Because most commercial laboratories only report one finalized result after three weeks and do not currently report CFUs or P-scores, in-house culture evaluation is even more crucial for shelters.

1. ***P-1 Cats***: P-1 cats are divided into two groups. Cats that carry the fomite virus are the first. Upon careful reexamination, these cats show no lesions and no Wood's lamp negative results. Without any intervening treatment, non-lesional P1 and P2 cats are typically negative on re-cultivation one to two weeks after an initial screening culture. These cats probably comb the spores out of their fur. It's also conceivable that most of the spores were eliminated during the initial toothbrushing procedure used for sampling. As a precaution, these cats ought to be re-cultured and given a single topical dose

of lime sulfur. The author refers to these creatures as "Dip and Go" cats. P-1 cats that were incubating infections at the time of admission fall into the second category. These cats may have had very subtle lesions that were missed during the initial examination. Re-examination may reveal lesions that are still extremely subtle. A thorough examination should focus on the face, lips, chin, ear margin, hairs in the bell of the ear, and the ear canal. Depending on the strain of *M. canis*, Wood's lamp examinations may or may not yield a positive result; infections with *Trichophyton* do not fluoresce. When reexamination yields positive culture results, animals with lesions that are identified should be considered genuinely infected.

2. ***P-2 Cats***: P-2 cats may be carriers of fomite but most often belong to the "infected group." The main distinction with these cats is that lesions if they are discovered, are typically more obvious and are frequently reported by shelter employees nearly simultaneously with the fungal culture report. The index of suspicion for a missed lesion should be higher when reexamining a cat with a "positive P-2" culture than it should be for a P-1. Similar to P-1 cats, cats without lesions after a very thorough second examination can be dipped once in lime sulfur and put up for adoption normally; cats with lesions, however, need to be treated as if they are genuinely infected.

3. ***P-3 Cats***: Regardless of whether lesions are found or not, P-3 cats should be taken out of the general population right away. Generally speaking, these cats have already been classified as suspects after comprehensive intake exams. On reexamination, lesions are frequently discovered in hard-to-find places in P-3 cats that were originally reported to be non-lesional. P-3 cats shouldn't be kept with other cats because they pose a risk to the public. It's possible that a tiny percentage of cats with P3 cultures were "dust mops." These cats frequently lacked proper grooming or had long or thick coats, and they frequently came from contaminated environments in the past. Examining the treatment data revealed that these cats typically became culture-negative very quickly, frequently within the first week of treatment.

4. ***Use of the P-Scoring System:*** Using the P-scoring system to track treatment is also very helpful. Once a week, animals receiving treatment should be cultured, and P-scores should be shared. The P-score decreases as the animals recover. Reporting the outcomes of environmental cultures also benefits from the use of the P score system and CFU counts. Environmental cultures assist in locating

polluted areas so that cleaning efforts can be concentrated and targeted. For personnel tasked with cleaning and disinfecting the area, it is very helpful to report the area as culture positive with some indication of the level of contamination. Staff and volunteers at the shelter have shown better compliance and "buy-in" as a result of the P-score system.

2.7.3. PCR Testing

When testing for dermatophytosis, real-time polymerase chain reaction (PCR) testing is available and may offer a quicker turnaround time than fungal culture (results from a commercial laboratory could be available in one to three days). Since the test has a high sensitivity, false negative results are unlikely, much like with PCR tests in general. To rule out dermatophyte infection in cats that have been exposed, PCR testing can be useful in this regard. Additionally, the costs and risks of diagnostic testing may be comparatively lower than those of holding cats for extended periods of time while awaiting the results of their cultures.

PCR testing does, however, have disadvantages when it comes to managing dermatophytosis in shelters. This test's exceptional sensitivity makes it less accurate at identifying cats who are not actually infected (lower specificity). False-positive results, or those arising from contaminated samples, may cause cats who were not actually infected to undergo unnecessary treatment and a prolonged length of stay in the hospital. Furthermore, the PCR results do not differentiate between viable organisms and those that are not, nor do they offer the clinical recommendations linked to the quantitative CFU scoring system mentioned previously for diagnosis or treatment efficacy monitoring. Quantitative polymerase chain reaction, or qPCR, is a technique that can be used to differentiate between a pathogen that is incidentally present and a clinically significant infection. For example, it can be used to distinguish between a parvovirus or canine distemper infection and a true infection. However, in a field study of shelter cats, this method did not help differentiate between cats that were culture-positive and cats that were culture-negative, nor was it helpful in tracking the course of treatment.

In many cases, a thorough history, visual examination, Woods lamp testing, and direct examination as previously mentioned will be adequate to rule out or appropriately rule in cats that need a fungal culture and additional testing. Nonetheless, there might be circumstances where the entire group is deemed potentially high-risk, such as in a hoarding case or when a group of cats is transferred from a shelter where ringworm is endemic. Under such circumstances, PCR might be a useful addition to swiftly identify and handle the intake of low-risk cats into the shelter system. A culture should be performed on any animal that tests positive for PCR; this can be started at the same time as the PCR samples are submitted. This will speed up the process of verifying more negative cases.

2.8. INITIAL RISK EVALUATION AND RESPONSE

Animals must undergo an initial risk assessment and response during examination and screening as part of a systematic screening procedure. The first four stages of the screening procedure previously outlined gather the data that will be taken into account for this preliminary response. During intake, or following an initial assessment if screening is taking place outside of intake (e.g., during an outbreak), the initial response should be given. Before incubation or confirmation of the fungal culture results, the first response strategy should be created. Personnel, volunteers, and members of the public should be made aware of the identity of positive, suspect, and high-risk animals.

When an animal receives a positive diagnosis, management should begin with a swift and decisive response. For animals classified as high-risk due to Wood's positive lesions, an initial response plan should at the very least involve separation or isolation from the general population. Furthermore, animals exhibiting distinctive inflammatory lesions ought to be regarded as questionable until additional findings from alternative diagnostic tests are obtained. While isolation is the best option, suspect animals must still be kept apart using appropriate procedures when isolation is not feasible. The animals that test positive for wood must be properly identified, along with a warning that they pose a risk to the public, employees, and volunteers for zoonotic illnesses. Signage on their cages, notes in their medical records and on the cage card, and identification tags attached to them in case they are moved around the shelter are all appropriate forms of identification. Even though they probably pose less risk, animals with suspicious lesions that are not Wood's lamp positive should still be considered potentially contagious if their suspicion is strong enough to justify detaining them until culture results are obtained. In the event that a positive result is confirmed, treatment would start if suspicions are high.

2.8.1. Isolation and Separation

As soon as possible, suspect animals (including those that test positive for PCR) and those that have positive Wood's lamp should be kept apart from the general public. A physically distinct housing ward, distinct equipment, and cleaning and care personnel assigned to that area exclusively are all ideal components of isolation. Separate air circulation may have been included in the design of more recent facilities with defined isolation areas. This can be very expensive, and it's probably not worth the investment in cases of dermatophytosis. Rather, there will probably be more profit if rigorous measures are taken to reduce the spread of fomite and environmental contamination.

Animal attendants assigned to work in isolation areas ought to be mandated to wear protective clothing, such as gowns, disposable gloves, shoe covers, and caps, due to the possibility of zoonotic disease. Nothing in the isolation ward should be taken out of its designated use or moved to another location. Staff members who work in multiple areas should, if at all possible, work in isolation wards only after working in other areas, as assigning separate staffing for an isolation area can be problematic in shelters. Before going to another area, staff members must, at the very least, change all of their clothes completely. Ideally, employees in isolation areas shouldn't handle sick, vulnerable, or high-risk animals–especially young ones. At the end of the day, all shelter employees should change into new clothes before leaving. If employees follow these instructions, there is very little chance that they will bring contaminated spores home.

It is important to designate "clean" and "dirty" areas within the isolation zone. These don't have to be physically distinct rooms or wards, but they should be kept apart in both appearance and functionality (e.g., by colored tape or painted lines). "Dirty" areas should be designated as those that are right next to residential buildings. To prevent spore contamination of the environment, bedding, litter, and food that comes into contact with the soil should be carefully bagged within the designated "dirty" area. Animals that are infected should stay in the "dirty" area.

Protective clothing worn in the unclean area should stay there unless it is bagged and taken out right away to be laundered. It is important to set aside a spotless space for activities such as food preparation and supply storage. A separate, contained area within the isolation area for cats with both dermatophytosis and upper respiratory infections may be helpful when designing an isolation area for the treatment of cats with dermatophytosis. Temporary separation of suspect or positive animals in a designated section may be adequate to help control the spread of disease when physical isolation is not feasible. Effective temporary separation requires specific warnings that the animals housed there might be contagious to people and other animals, the implementation of precautionary care protocols, close staff supervision, and strict adherence to control protocols. There should be a designated "dirty" zone or area around these housing units.

Keeping suspicious or high-risk animals close to or mixed in with other animals, or in places where other animals are frequently kept, puts the group as a whole and people who come into contact with them at risk. It might also introduce difficult-to-get rid of infectious spores into the environment. During this waiting period, treating suspect animals with topical lime sulfur lowers the infectious risk by minimizing the amount of spores released into the environment. It is important to stress that animals with positive fungal cultures should only be kept in this area temporarily while a final plan is being developed. Only enough time should pass in this kind of isolation area for suspect animals to validate the findings of diagnostic testing. In general, partial physical separation during treatment shouldn't be regarded as an adequate defense against the spread of illness. Seclusion is highly advised.

2.9. TREATMENT

The length of treatment depends on the severity of clinical signs and the treatment plan, which are based on the dog or cat's overall global health. There isn't any proof that certain dermatophyte strains are more "virulent" by nature based on clinical expertise with dogs and cats that were infected in shelters. Three clinical presentations of practical value:

1. Simple infections,
2. Complicated infections, and
3. Culture-positive, lesion-free animals.

Animals in otherwise good health can get simple infections, and they heal quickly from these infections. Complicated infections can range from mild to severe and diffuse, but they always occur in animals that are ill concurrently or have other conditions that make treatment difficult, such as long hair, difficulty bathing, poor client compliance, or poor body condition from prior treatments.

2.9.1. Minimizing Disease Transmission and Environmental Contamination

2.9.1.1. Clipping the Hair Coat

It's not always required to clip the hair coat when dermatophytosis occurs. Sedation is usually necessary for extensive clipping to reduce patient discomfort and anxiety. It's crucial to keep in mind that electric clippers have the potential to get hot, and thermal contact injuries are frequent but frequently go undetected for several days or weeks after cutting. Clipping of the hair coat can also spread the infection to other uninfected sites. The following is recommended:

1. Only use children's blunt-tipped metal scissors.
2. Clip mats from the hair coat to facilitate bathing and topical therapy.
3. Scissor clip margins of focal lesions if this can be done safely and easily.

4. DO comb the hair coat before using topical therapy to remove broken and shed hairs.

2.9.1.2. Confinement

Restricting an infected pet primarily serves to lessen the area that needs to be cleaned during the course of treatment, when shedding of contaminated hairs is likely to occur. Cleaning is required to stop fomite contamination from producing positive fungal culture or PCR test results. However, direct contact–rather than exposure to the environment–is the main means of transmission. When considering confinement, caution must be used and the least amount of time should be spent. Although behavior issues and socialization issues in young or recently adopted animals can persist a lifetime, this disease is treatable and curable. When recommending this course of action, veterinarians must take animal welfare and quality of life into account.

2.9.1.3. Cleaning

Empirical research has demonstrated that decontaminating contaminated environments is a simple process. Decontamination involves mechanically removing debris, cleaning the surface with a detergent until it is observably clean, rinsing to get rid of detergent residue, and draining extra water. It has regularly cleaned items without any further intervention. Decontamination was found to be simple to achieve with standard cleaning procedures in a recent study involving 70 homes (Moriello, 2019). Killing any infectious agent that washing hasn't mechanically removed is the goal of any further disinfection step. Anything that is marketed as effective against *Trichophyton* species in bathroom disinfectants is my preferred disinfectant. Dilution errors and irritant reactions are minimized when using an over-the-counter, ready-to-use product.

2.9.2. Topical Antifungal Treatments

When treating dermatophytosis, topical antifungal therapy is just as crucial as systemic therapy. Infectious spores are eliminated from the hair shafts by topical therapy. This is significant because, even after the infection has been completely removed from the hair follicle, infectious material will still be present on the hair shaft. The most frequent cause of "failure to cure" or "resistant ringworm" in second opinion cases is improper disinfection of the hair coat. In topical therapy, the entire body is treated twice a week starting at the time of diagnosis and continuing until the mycological cure. Although shampoos containing miconazole, ketoconazole, climbazole, and chlorhexidine are alternatives, lime sulfur and enilconazole are recommended due to their residual activity (Moriello et al., 2017). Family pets that are exposed but not infected can be treated with these shampoo-based products. Furthermore, it is currently advised to apply extra specialized topical therapy to focal lesions in challenging to treat areas like the face and ears. According to Gyanfosu et al (2018), 2% miconazole cream is a safe treatment option for lesions on the face and periocular area. The treatment for fungal keratitis is this cream (Figure 2.4).

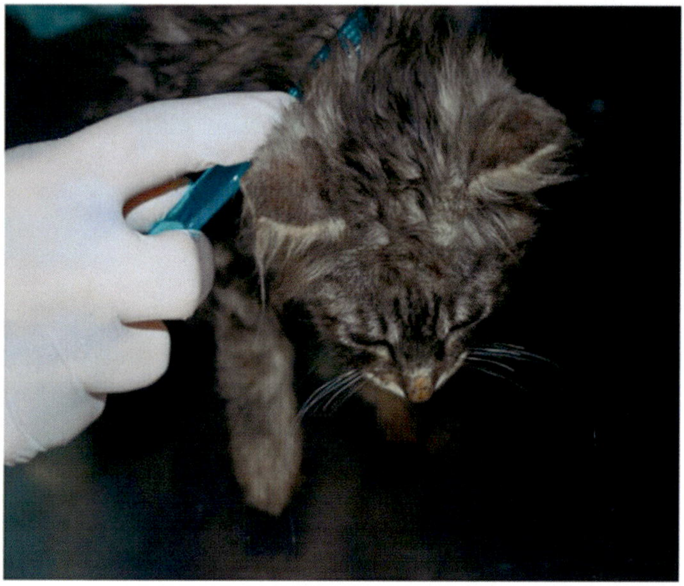

Figure 2.4. *A kitten with M. canis dermatophytosis.*

Source: Pasquetti M, Min ARM, Scacchetti S, Dogliero A, Peano A. Infection by Microsporum canis *in Paediatric Patients: A Veterinary Perspective.* Veterinary Sciences. *2017; 4(3):46. https://doi.org/10.3390/vetsci4030046.*

2.9.2.1. Lime Sulfur

Lime sulfur is a topical treatment that effectively reduces spore shedding that causes environmental contamination and eradicates infection in both *in vivo* and *in vitro* settings. Lime sulfur is the best option to use as part of a shelter treatment protocol because it is readily available, affordable, safe, and effective when applied correctly and diluted.

The effectiveness of commercial lime sulfur products, including those that contain odor-masking agents, is comparable. Lime sulfur has a bad smell that goes away quickly, so using it shouldn't be discouraged even though it might stain some materials. The application of topical lime sulfur is strongly advised anytime an animal shelter is going to treat animals for dermatophytosis because they should prioritize treatments that minimize animal care days and the potential for contagion or environmental contamination.

2.9.2.1.1. Application of Lime Sulfur Dip

The recommended dilution of lime sulfur is 8 oz (236 ml) per gallon (3.78) of warm water, which is the higher dilution of the two that is most frequently specified on commercially available products. It is important to add the warm water to the mixing container after adding 8 oz (236 ml) of the concentrated lime sulfur. Even though the word "dip" is frequently used, the animal does not have to submerge itself in the solution. The solution ought to be applied topically to the animal's coat instead. A laundry sink that is configured to empty into a bucket can be converted into a straightforward, affordable, and transportable "dip" sink.

Before receiving treatment, the cat shouldn't be bathed or pre-wet. A garden rose sprayer that holds a gallon or half-gallon can be used to easily apply the appropriately diluted lime sulfur solution. To ensure that the spray covers the cat like a mild shower, the sprayer's nozzle should be held very close to their skin. The solution will frequently initially roll off the hair coat and bead up. The entire hair, down to the hair's base, should be covered in the solution. For the treatment to work, the cat needs to be soaked all the way to the skin. To gently sponge the dip on the nose, inside the ears, and other areas of the face, use cloths. Because the lesions found there are typically the hardest to treat, these areas are the most crucial. Cats seem to handle this application method much better than soaking in towels or getting wet from watering cans, which is surprising. It is also a very economical application method. Even though aggressive cats can be sprayed through a wire carrier, it is still advisable to try to keep the spray close to the skin since cats seem to tolerate the treatment best when the nozzle is close enough to avoid spraying liquid in their direction or making a hissing sound. The solution should not be rinsed off. Both rinsing and pre-wetting will cause dilution and could lead to less efficacious treatment.

Even on kittens as young as four weeks old and nursing queens, lime sulfur can be applied topically without risk; however, hypothermia in young kittens should be taken seriously. Teats of nursing queens should be thoroughly cleaned of the solution following dipping. While they are wet, animals need to be kept warm. For kittens, puppies, or other animals who might struggle with thermoregulation, this is especially crucial. Unfavorable side effects are rare. In the author's experience, treating hundreds of cats, e-collars were not required following treatment. Oral ulcers were only observed in cats receiving lime sulfur treatment when they also had a serious upper respiratory infection, and these were consistent with an upper respiratory infection in cats rather than an irritant reaction.

2.9.2.1.2. Clipping the Hair Coat Before Topical Treatment

Although it's frequently advised to clip a cat's hair before applying topical treatment, the author hasn't found that clipping is required for cats with short or medium hair. Nonetheless, the entire body of a long-haired cat with a high pathogen score or one that seems messy, matted, unwilling, or incapable of grooming needs to be clipped using a 7 or 10 blade (not a surgical blade). For cats whose coats become clumped or matted during the dipping process, clipping can also be beneficial. It is advised to exercise caution because cutting is likely to cause environmental contamination. Hair needs to be carefully contained. After cutting an infected cat, the surrounding area needs to be meticulously cleaned and disinfected. Clipping should be done with clippers made specifically for that purpose, in a supervised and designated area that won't be used to house other animals. Care should be taken when cutting to prevent abrasive wounds on the skin. To prevent unintentionally inflicting thermal injuries, it is advised to switch between two clippers or to take breaks so that the clippers can cool. Using overheated clippers to trim cats can cause deep thermal burns. Sometimes an eschar forms several days after clipping, making these burns invisible until then.

2.9.3. Systemic Antifungal Treatments

The infection inside the hair follicle is eliminated by systemic antifungal treatment. Oral itraconazole 5 mg/kg once daily on a week-on, week-off schedule is the preferred treatment for cats. This regimen should be followed for three cycles. Recurring cycles are required for cats with complex infections or those who have not achieved mycological cure. Until mycological cure, you presently advise giving dogs 5 mg/kg of itraconazole or 30 to 40 mg/kg of terbinafine orally once daily. Because of its poor bioavailability in studies involving both dogs and cats, itraconazole that has been compounded or reformulated should not be used.

In terms of safety, itraconazole, and terbinafine were found to be both safe and well-tolerated in the evidence-based evaluation of clinical trials and treatment studies conducted on dogs and cats. There were no recorded fatalities or significant side effects. Soft stools or vomiting were the most often reported adverse effects. The use of itraconazole to treat dermatophytosis has not been linked to any serious side effects. For extended periods of time (months), animals with intermediate or deep mycoses were given high doses (> 10 mg/kg) of medication to prevent vasculitis or other severe side effects. The first oral antifungal was griseofulvin, but due to terbinafine and itraconazole's greater effectiveness, griseofulvin is no longer advised. Because of its poor efficacy against dermatophytosis, fluconazole should not be used. Although ketoconazole works well for dogs, cats should not use it because of intolerance. Lufenuron is ineffective in treating dermatophytosis.

2.9.3.1. *Itraconazole*

At the moment, the medication most frequently used to treat dermatophytosis in cats and dogs is ittraconazole. When combined with efficacious topical therapy, it is possible to treat dermatophytosis with ittraconazole, a triazole antifungal that accumulates and persists in hair and epidermis, either on a daily or pulse-therapy basis, or as a short-term treatment protocol.

Since the medication is licensed for use in Europe, the most popular pulse-therapy protocol in practice is a week-on/week-off regimen (5–10 mg/kg) until the patient is cured. Before starting pulse therapy, a 14-day daily dose loading period is needed to reach the proper therapeutic levels. Because the medication is stored in body fat, pulse therapy is not recommended for sick animals or animals with low levels of body fat. It is recommended to combine topical therapy with 21–28 days of daily therapy for animals who are stressed, thin, or debilitated. Pulse therapy is generally not advised for cats in shelter settings because the published shelter protocol only used 21 days of oral treatment.

The medication itraconazole comes in 100 mg capsules that can be manually divided into smaller doses and repackaged in empty gelatin capsules or combined with a small amount of food; the oral solution is also an option. Although toxicity issues are uncommon, animals may occasionally develop an allergic reaction. Liver enzyme tests should be performed and the medication should be discontinued if this persists. The author should be advised that while itraconazole powder can be bought in bulk for a reasonable price abroad, there are anecdotal reports that this treatment is not always successful. Additionally, there have been numerous reports of shelters experiencing unanticipatedly long cure times with compounded itraconazole; this issue is resolved when the brand-name product is modified. To guarantee its absorption, itraconazole must be carefully formulated in the right vehicles; using materials other than those that are official and approved is not advised as it could lead to treatment failure.

2.9.3.2. *Terbinafine*

Oral terbinafine (Lamisil, Novartis) is another reasonable course of treatment. Despite appearing to have no advantages over itraconazole, the medication is nearly as effective (Moriello et al., 2013). Although it happens infrequently, terbinafine can occasionally be used to treat cats with *M. canis* infections that have been shown to be resistant to azole medications. Different oral dosages (10–40 mg/kg, once daily) have been used; however, cats treated at the higher end of the dosage range show notably faster recovery. It is advised to keep an eye on liver enzyme levels; although no clinical toxicity is always observed, this medication may cause cats' alanine aminotransferase (ALT) levels to rise. The medication may be helpful for pulse therapy because it has been demonstrated to have residual activity in the skin comparable to that of itraconazole.

2.9.3.3. Fluconazole

As an itraconazole substitute, fluconazole has drawn some interest. There is little data regarding the use of fluconazole in shelter cats, and its application has not been as widespread as that of itraconazole. There have been reports of treating two groups of cats with fluconazole: one group received 5 mg/kg, and the other received 10 mg/kg (Moriello and Newbury, 2009). Cats given 10 mg/kg of treatment recovered more quickly than those given a lower dosage. It is advised to take 10 mg/kg orally once day, if applicable, until research shows that smaller dosages are equally effective.

2.9.3.4. Griseofulvin

Since griseofulvin has a small safety margin and may take longer to cure dermatophytosis, it is not advised for use in shelter cats. In one study, experimentally infected cats were cured after a mean of 70 days of griseofulvin therapy, as opposed to 56 days of itraconazole therapy (Moriello and DeBoer, 1995). It should not be administered to pregnant animals due to its high teratogenicity. In cats, it may result in bone marrow suppression; this adverse effect is severe, erratic, and unrelated to the dosage, breed, or duration of therapy. White blood cell (WBC) counts are advised before therapy and every two weeks after it is started to ensure safe use, which renders the medication cost-inefficient.

Before use, cats should be tested for feline immunodeficiency virus (FIV) and feline leukemia virus (FeLV), as there may be a connection between infection and these negative effects. Although the author does not advise using griseofulvin in cats, when treating large dogs, it might be less expensive per dose than itraconazole. However, taking into account all treatment expenses and recovery times, itraconazole might still be the more economical course of action. The formulation determines the dose (microsize vs. extremely tiny). An elevated absorption occurs after a fatty meal. Common side effects include inappetence, diarrhea, and vomiting. Getting this drug is getting harder and harder.

2.9.3.5. Ketoconazole

There are few anecdotal reports on the clinical effectiveness of ketoconazole in treating dermatophytosis, so its actual effectiveness is unknown. *M. canis* strains that are resistant to ketoconazole exist. It is not advised to use the medication in cats because it is not well tolerated and has a limited therapeutic margin. Ketoconazole is best used as a systemic antifungal for infections in dogs, especially *Trichophyton* infections. Aside from its decreased cost in recent years, ketoconazole is not more beneficial than other medications when treating common cases of dermatophytosis.

2.9.3.6. Lufenuron

Ben-Ziony and Arzi (2000) reported on a case that sparked conjecture regarding the potential efficacy of lufenuron (Program, Novartis) in treating feline or canine dermatophytosis. Although there have been anecdotal reports of "cures" with lufenuron, controlled studies using natural exposure models and feline experimental infection models have consistently found that lufenuron did not prevent the onset of dermatophytosis in cats or have a faster curative effect once the infections were established (Deboer et al., 2003; Moriello et al., 2004a) and did not work in concert with terbinafine. It is NOT advised to use lufenuron to treat dermatophytosis.

2.9.4. Verification of Cure

Repeated fungal cultures are required to confirm that dermatophytosis has been successfully treated and that a mycologic cure has been obtained some of the clinical symptoms, such as the erythema, pruritus, and scaling, have resolved may be observed two weeks after treatment, and hair growth in cats with cropped hair will soon follow. As long as treatment protocols are followed effectively, a mycologic cure may occur before hair regrowth. However, cats may appear clinically cured before mycologic cure is achieved when times to cure are longer or lesions are mild; if these cats are released into the general public or adopted out too soon, they may continue to spread infection. This is especially likely to happen if a treatment that doesn't work is used. The most common way to determine a mycologic cure is to obtain three consecutive negative cultures at weekly intervals; cultures need to be kept for 21 days. After a week of treatment, treatment monitoring ought to commence.

No cats became positive again following two consecutive negative weekly fungal cultures in a shelter treatment trial that followed the suggested combination treatment protocol (Newbury et al., 2007). Cats were kept in the facility for treatment and were given topical medication until the fungal cultures were completed. The waiting period for adoption was shortened by using two negative cultures instead of three when defining mycologic cure, and this approach had no negative effects. However, it has only been shown to be dependable when the systemic itraconazole and lime sulfur treatment, as previously mentioned, were used in conjunction. To be fairly certain of a cure, using any other treatment still requires three negative cultures.

P-score values are predicted to decline quickly as the course of treatment continues. Wood notices that the hairs' fluorescence is getting progressively weaker. In fact, observing treated animals using a Wood's lamp examination can be a useful way to confirm that topical medication is being administered correctly. According to the author's observations, brightly fluorescing, sparingly treated hairs in the nose and eyes are typically the cause of longer recovery times when a topical solution is not applied to the face with enough care–possibly out of concern for the animal's safety. When treatment personnel saw these hairs, they paid greater attention to the areas where topical treatment was being applied. This resulted in a significant change in the amount of fluorescence observed upon re-examination.

The treatment trial that was previously discussed resulted in an average of 18.4 ± 9.5 SEM (standard error of the mean) days of treatment required for cure (range: 10–49 days). More serious infections in cats meant longer treatment times. The extra 21 days needed to complete the fungal cultures and show a mycologic cure are not included in the time to cure data. When considering an animal for treatment, one should generally estimate that it will require at least one month of treatment and housing. For kittens, the amount of time it takes to recover and reintroduce the animal to the population for adoption may be particularly crucial. Occasionally, after receiving treatment, kittens may become young adults before the mycologic cure is confirmed. Weekly animal culture sessions during treatment help shorten the duration of the course of treatment by minimizing the amount of time that passes between a cat's actual mycologic cure and the taking of the culture.

2.9.5. Housing of Animals Undergoing Treatment

Free roaming of suspect or infected animals throughout the facility will increase the risk of zoonotic diseases, environmental contamination, and possibly lengthen recovery times. Animals with dermatophytosis should be housed individually, in small family groups, or in bonded pairs while receiving treatment in an isolation area as previously mentioned. In addition to allowing for better observation of general health and behavior, housing animals singly or in small groups of two or three kittens helps prevent cross-infection and potential cross-contamination with spores that could complicate the fungal culture process. Releasing animals for adoption should only occur after all co-housed animals have achieved mycologic cure; therefore, limiting the number of animals per housing unit can also avert release delays. Housing size and quality should be appropriate for an extended stay, as animals will likely be housed and treated for a minimum of one month without being allowed to roam freely. It is important to provide toys and frequent opportunities for positive human interactions as well as behavioral and environmental enrichment. Training is required for volunteers or staff to adhere to isolation procedures for both treatment and enrichment. Animals receiving treatment may have bedding, but it should be changed every time the animal receives topical treatment, is determined to be contaminated, and is kept contained when being brought dirty to the laundry area. Every cat in the aforementioned treatment trial had bedding in their cages. Every day, the bedding was taken out and cleaned in the heated dryer and regular washing machine provided by the shelter. Bedding was laundered and dried independently from all other shelter laundry, even though separate laundry facilities were not utilized.

2.9.5.1. Environmental Decontamination: Cleaning, Disinfection, and Preventative Planning

Naturally infectious hairs and spores that are resistant to many disinfectants frequently used in shelters can contaminate environments. The foundation of decontamination must always be mechanical cleaning and spore removal. However, as a supplement to removal, a number of efficacious disinfectant products are offered. Bleach, or sodium hypochlorite, works well at ratios of 1:10 or 1:32; however, in a shelter setting where both humans and animals are present, a lower dilution with a longer contact time (at least 10 minutes) is recommended. Both accelerated hydrogen peroxide and 2% (but not 1%) potassium peroxymonosulfate work well at a dilution of 1:10. Enilconazole (Clinafarm) is most frequently used in poultry farms; it is typically found in agriculture supply stores. It works well at a dilution of 1:100 and is available as a spray or a fogger. Chlorhexidine, 1% potassium peroxymonosulfate (Trifectant), calcium hypochlorite, and the quaternary ammonium products are all ineffective. *Trichophyton* spores are more effectively inactivated by heat (50□ or 122°F) than by repeated freezing and chilling.

Given that toothbrush cultures from severely infected cats left in a car on a sunny day all unexpectedly came back negative, the same is probably true for *M. canis*. Numerous studies have examined how UV light affects different aspects of dermatophyte growth; however, because these studies used mycelial (culture plate) rather than naturally infectious material, data from these studies cannot be applied practically.

Thorough sweeping or "Swiffering" of the floor, mechanical scrubbing with a good detergent, and daily mechanical removal of infectious material are the keys to minimizing contamination and successfully decontaminating the environment. (Proctor and Gamble created the Swiffers brand of cleaning products, which includes disposable cleaning cloths. Only after a thorough cleaning should the environment be disinfected; the goal is to eliminate any spores that were not mechanically removed. Although chemicals have been shown to inactivate spores *in vitro*, it is difficult or impossible to achieve sufficient contact time on many surfaces, so the use of chemical disinfectants is advised as an adjunct sanitation procedure. Spores can remain in difficult-to-fill cracks and crevices. According to the author's experience, if there are no objects in the room that cannot be thoroughly cleaned, thoroughly cleaning a space three times in a row and disinfecting it with bleach that has been prepared and applied correctly each time frequently results in negative environmental cultures.

Effectively decontaminating areas with soft, upholstered, or carpeted furniture and other non-washable items is generally very difficult. Cat trees from community rooms or other indoor housing units have been disinfected by using a lime-sulfur solution in conjunction with direct sunlight (cat trees were placed outside because light filtered through windows will not be effective). Two negative cultures were obtained by culturing cat trees. In those situations, vacuuming, and steam cleaning might be useful. However, challenging or dubious objects should be thoroughly inspected for contamination after cleaning, as they might still pose a risk of exposure and infection to both people and animals. It is frequently more economical and safe to dispose of these contaminated items.

Environmental cultures of furnishings and rooms are a very useful tool for focusing cleaning efforts and confirming that decontamination was accomplished. When confirming that hard-to-clean spaces like foster homes or group rooms have been sufficiently decontaminated and are safe enough to accept new animals, environmental cultures can be especially useful. They might also be useful before spending money on pricey treatments like duct cleaning; cultures might show that these are not required. Clean Swiffers can be used to sample the environment, detect contamination, and confirm decontamination to obtain cultures.

Cut a Swiffer sheet into four pieces and place each section in a resealable plastic bag in a clean, well-known area free of contamination. Every potentially contaminated room or area should have multiple locations sampled, making sure to include areas where hair or dust is likely to accumulate or where infected animals are known to frequent. Wipe an area with Swiffer sheets until it becomes clearly soiled. Samples should also be taken of any items that might be particularly challenging to clean. After that, by firmly pressing the Swiffer sections onto the fungal culture plate, they can be used to inoculate fungal culture media. Using a toothbrush, fabric-covered furniture is frequently easier to clean. To assess the relative levels of contamination in each area, utilize the P-score system previously discussed. Disease transmission should be minimized and prevented when designing a room. Their furniture should be made of plastic or vinyl that is easily cleaned with a hose when necessary. They should also use washable and soft bedding, as well as disposable items like cardboard or a simple log that can be used as a scratching post. Smooth, well-maintained, and easily sanitized walls, floors, cages, and kennels are ideal. Spores can accumulate and elude disinfection efforts due to numerous opportunities provided by cracked walls and baseboards. Performing visual/physical examinations, Wood's lamp/direct examination screening, and/or fungal cultures on higher-risk animals prior to their admission to unclean spaces like foster homes or community rooms are important aspects of preventive management. It should also involve routinely cleaning and removing hair from the surroundings as well as keeping an eye on any animals living in those areas in case lesions develop later than expected. Low-stress housing that promotes normal behavioral expression is also beneficial because stress and improper grooming are significant risk factors for dermatophyte infection. Controlled studies on the effects of external exposure are lacking.

2.10. OUTBREAK MANAGEMENT AND RESPONSE

To effectively manage dermatophytosis, a continuous preventative process utilizing the environmental prevention practices and systematic screening protocol previously discussed in this chapter must be implemented. When outbreaks do happen, decisions about diagnosis, treatment, and environmental cleanup are guided by the same systematic screening protocol. It can be challenging to determine the precise number of cases or set of circumstances that may qualify as an outbreak, so the first step is frequently addressing the question, "Is this an outbreak?" To determine the number of animals impacted and stop additional transmission, some kind of intervention and coordinated response are required whenever in-shelter transmission occurs. While outbreaks can affect either or both species as well as in-contact humans, they are far more common in shelter cats than in shelter dogs, as well as in shelters or other facilities that house cats.

It is advised to follow a methodical procedure and use the fundamental screening instruments previously discussed. All potentially exposed animals should be screened and cultured in the event that an outbreak is suspected. Verifying the diagnosis through fungal culture should always be the first step in responding to a true dermatophytosis outbreak, as it will require significant resources and time. In case there is any uncertainty regarding the precision of the fungal culture conducted in-house or the proficiency in conducting and interpreting cultures, samples ought to be forwarded to a commercial laboratory for confirmation. To effectively control outbreaks of dermatophytosis, every attempt should be made to quickly assign available resources. Without a thorough screening of all at-risk animals, evaluating or responding to an outbreak or possible outbreak is unlikely to be resolved and could lead to needless treatment or euthanasia. Certain situations may result in needless euthanasia due to inadequate training, delayed, incomplete, or erroneous diagnostics. For every organization, responding to an outbreak is expensive, time-consuming, and emotionally taxing. It will be virtually impossible to attempt to respond without utilizing all of the resources at hand. It is equally important to understand that proactive preventative measures need to be implemented in tandem with outbreak responses. If not, the cycle might start up again as soon as the previous outbreak is contained and an unidentified infected animal enters the facility (Polak et al., 2014).

REFERENCES

1. Baldo, A., Monod, M., Mathy, A., Cambier, L., Bagut, E. T., Defaweux, V., Symoens, F., et al. (2012). Mechanisms of skin adherence and invasion by dermatophytes. *Mycoses, 55*, 218–223.
2. Ben-Ziony, Y., & Arzi, B. (2000). Use of lufenuron for treating fungal infections of dogs and cats: 297 cases (1997–1999). *Journal of the American Veterinary Medical Association, 217*, 1510–1513.
3. Cafarchia, C., Romito, D., Sasanelli, M., et al. (2004). Epidemiology of canine and feline dermatophytosis in southern Italy. *Mycoses, 47*, 508–513.
4. Carlotti, D. N., Guinot, P., Meissonnier, E., et al. (2010). Eradication of feline dermatophytosis in a shelter: A field study. *Veterinary Dermatology, 21*(3), 259–266.
5. Cerundolo, C. (2004). Generalized *Microsporum canis* dermatophytosis in six Yorkshire terrier dogs. *Veterinary Dermatology, 15*, 181–187.
6. DeBoer, D. J., & Moriello, K. A. (1993). Humoral and cellular immune responses to *Microsporum canis* in naturally occurring feline dermatophytosis. *Journal of Medical and Veterinary Mycology, 31*, 121–132.
7. DeBoer, D. J., & Moriello, K. A. (1994). Development of an experimental model of *Microsporum canis* infection in cats. *Veterinary Microbiology, 42*, 289–295.
8. DeBoer, D. J., & Moriello, K. A. (1995). Inability of two topical treatments to influence the course of experimentally induced dermatophytosis in cats. *Journal of the American Veterinary Medical Association, 207*, 52–57.
9. DeBoer, D. J., Moriello, K. A., Blum, J. L., et al. (2003). Effects of lufenuron treatment in cats on the establishment and course of *Microsporum canis* infection following exposure to infected cats. *Journal of the American Veterinary Medical Association, 222*, 1216–1220.
10. Diesel, A., Verbrugge, M., & Moriello, K. A. (2011). Efficacy of eight commercial formulations of lime sulphur on *in vitro* growth inhibition of *Microsporum canis*. *Veterinary Dermatology, 22*(2), 197–201.
11. Foust, A. L., Marsella, R., Akucewich, L. H., et al. (2007). Evaluation of persistence of terbinafine in the hair of normal cats after 14 days of therapy. *Veterinary Dermatology, 18*, 246–251.
12. Heinrich, K., Newbury, S., & Verbrugge, M. (2005). Detection of environmental contamination with *Microsporum canis* arthrospores in exposed homes and efficacy of the triple cleaning decontamination technique [Abstract]. *Veterinary Dermatology, 16*, 192.

13. Jacobson, L. S., McIntyre, L., & Mykusz, J. (2018a). Comparison of real-time PCR with fungal culture for the diagnosis of *Microsporum canis* dermatophytosis in shelter cats: A field study. *Journal of Feline Medicine and Surgery, 20*(2), 103–107.
14. Jacobson, L. S., McIntyre, L., & Mykusz, J. (2018b). Assessment of real-time PCR cycle threshold values in *Microsporum canis* culture-positive and culture-negative cats in an animal shelter: A field study. *Journal of Feline Medicine and Surgery, 20*(2), 108–113.
15. Kaufmann, R., Blum, S. E., Elad, D., & Zur, G. (2016). Comparison between point-of-care dermatophyte test medium and mycology laboratory culture for diagnosis of dermatophytosis in dogs and cats. *Veterinary Dermatology, 27*, 284–e68.
16. Mancianti, F., Nardoni, S., Corazza, M., et al. (2003). Environmental detection of *Microsporum canis* arthrospores in households of infected cats and dogs. *Journal of Feline Medicine and Surgery, 5*, 323–328.
17. Moriello, K. A. (2015). Kennel disinfectants for *Microsporum canis* and *Trichophyton* sp. *Veterinary Medicine International*, 853937. https://doi.org/10.1155/2015/853937.
18. Moriello, K. A., & DeBoer, D. J. (1991a). Fungal flora of pet cats. *American Journal of Veterinary Research, 52*, 602–606.
19. Moriello, K. A., & DeBoer, D. J. (1991b). Fungal flora of the hair coat of cats with and without dermatophytosis. *Journal of Medical and Veterinary Mycology, 29*, 285–292.
20. Moriello, K. A., & DeBoer, D. J. (1995). Efficacy of griseofulvin and itraconazole in the treatment of experimentally induced dermatophytosis in cats. *Journal of the American Veterinary Medical Association, 207*, 439–444.
21. Moriello, K. A., & Hondzo, H. (2014). Efficacy of disinfectants containing accelerated hydrogen peroxide against conidial arthrospores and isolated infective spores of *Microsporum canis* and *Trichophyton* sp. *Veterinary Dermatology, 25*(3), 191–e148.
22. Moriello, K. A., & Newbury, S. (2009). Dermatophytosis. In L. Miller & K. F. Hurley (Eds.), *Infectious Disease Management in Animal Shelters* (pp. 243–274). Ames: Wiley-Blackwell.
23. Moriello, K. A., & Verbrugge, M. (2007). Use of isolated infected spores to determine the sporicidal efficacy of two commercial antifungal rinses against *Microsporum canis*. *Veterinary Dermatology, 18*, 55–58.
24. Moriello, K. A., Coyner, K., Trimmer, A., et al. (2013). Treatment of shelter cats with oral terbinafine and concurrent lime sulphur rinses. *Veterinary*

Dermatology, 24(6), 618–620, e149–150.

25. Moriello, K. A., DeBoer, D. J., Schenker, R., et al. (2004b). Efficacy of pre-treatment with lufenuron for the prevention of *Microsporum canis* infection in a feline direct topical challenge model. *Veterinary Dermatology, 15*, 357–362.
26. Moriello, K. A., DeBoer, D. J., Volk, L. M., et al. (2004a). Development of an *in vitro* isolated infected spore testing model for disinfectant testing of *Microsporum canis* isolates. *Veterinary Dermatology, 15*, 175–180.
27. Newbury, S., & Moriello, K. A. (2014). Feline dermatophytosis: Steps for investigation of a suspected shelter outbreak. *Journal of Feline Medicine and Surgery, 20*(5), 407–418.
28. Newbury, S., Moriello, K. A., Coyner, K., et al. (2015). Management of endemic *Microsporum canis* dermatophytosis in an open admission shelter: A field study. *Journal of Feline Medicine and Surgery, 17*(4), 342–347.
29. Newbury, S., Moriello, K. A., Kwochka, K. W., et al. (2011). Use of itraconazole and either lime sulphur or Malaseb Concentrate Rinse® to treat shelter cats naturally infected with *Microsporum canis*: An open field trial. *Veterinary Dermatology, 22*(1), 75–79.
30. Newbury, S., Moriello, K., Verbrugge, M., et al. (2007). Use of lime sulphur and itraconazole to treat shelter cats naturally infected with *Microsporum canis* in an annex facility: An open field trial. *Veterinary Dermatology, 18*(5), 324–331.
31. Newbury, S., Verbrugge, M., & Moriello, K. A. (2005). Management of naturally occurring dermatophytosis in an open shelter: Part 1: Development of a cost-effective screening and monitoring program [Abstract]. *Veterinary Dermatology, 16*, 192.
32. Newbury, S., Verbrugge, M., & Steffen, T. (2005). Management of naturally occurring dermatophytosis in an open shelter: Part 2: Treatment of cats in an offsite facility [Abstract]. *Veterinary Dermatology, 16*, 193.
33. Polak, K. C., Levy, J. K., Crawford, P. C., et al. (2014). Infectious diseases in large-scale cat hoarding investigations. *The Veterinary Journal, 201*(2), 189–195.
34. Sparkes, A. H., Gruffydd-Jones, T. J., & Stokes, C. R. (1996). Acquired immunity in experimental feline *Microsporum canis* infection. *Research in Veterinary Science, 61*, 165–168.
35. Stuntebeck, R., Moriello, K. A., & Verbrugge, M. (2017). Evaluation of incubation time for *Microsporum canis* dermatophyte cultures. *Journal of Feline Medicine and Surgery, 20*(10), 997–1000.

36. Verbrugge, M., Kesting, R., & Moriello, K. A. (2007). Effects of light exposure and temperature on time to growth of dermatophytes using commercial fungal culture media [Abstract]. *Veterinary Dermatology, 19*, 110.

37. Verbrugge, M., Moriello, K. A., & Newbury, S. (2006). Correlation of skin lesions and dermatophyte culture status in cats at the time of admission to a shelter [Abstract]. *Veterinary Dermatology, 17*, 213.

38. White-Weithers, N., & Medleau, L. (1995). Evaluation of topical therapies for the treatment of dermatophyte-infected hairs from dogs and cats. *Journal of the American Animal Hospital Association, 31*, 250–253.

CHAPTER 3

SMALL RODENTS: COMMON DISEASES

Contents

Introduction.. 102

3.1. Overview of Rodents 103

3.2. Husbandry.. 106

3.3. The Diagnostic Challenge.. 108

3.4. Diseases of Small Rodents... 109

3.5. Disorders and Diseases of Rats... 111

3.6. Clinical Techniques... 121

References... 129

INTRODUCTION

Veterinarians receive many complaints regarding small pet rodents, including rats, hamsters, and gerbils, for a range of issues. The care of pet rodents and common diseases are very different from those of laboratory-bred rodents. Since most small pet rodents live long lives, neoplasia, and degenerative diseases are more common than in lab rodents. However, the diets of pet rodents are frequently inadequate, and obesity and other diet-related illnesses are widespread. Veterinarians must be knowledgeable about the dietary, husbandry, and natural history of animals to give pet owners the best advice. Veterinarians will be able to diagnose and treat small pet rodents with greater accuracy if they are able to identify common diseases in them.

With the exception of Syrian hamsters, the majority of small pet rodents are gregarious creatures and should, if at all feasible, be housed in groups to promote normal behavior and overall wellbeing. Nonetheless, male neutering or housing same-gender groups together is the best way to stop unchecked reproduction. Compared to rats, neutering hamsters is more difficult because they lack a protruding scrotum and their testicles are free to pass through the inguinal rings. Regardless of whether they are housed with male rats or not, spaying female rats is advised because they are less likely to develop uterine diseases like endometritis and mammary neoplasia. It significantly increases the life expectancy of rats if they are avoided. Long-acting hormone implants, such as GnRH agonists (deslorelin), are commonly used to induce infertility as an alternative to surgical sterilization. Since Syrian (golden) hamsters are solitary creatures, groups of them should never be kept. Dwarf hamsters, on the other hand, get along with others and can live in groups.

Small rodents require similar housing, so it's important to make sure they have enough room and ventilation. Because wood shavings are foraging material and absorbent, they make excellent bedding for small rodents. To create an environment that is stimulating, a range of products are available, including cardboard tubes, wooden sticks, and chewing blocks. Offering enrichment is a crucial aspect. Since small rodents are prey animals, it's critical to provide them with shelter in the form of huts, tubes, or boxes. Drafts should be avoided and the cage should be kept in a place with little traffic. Different species of small rodents have different diets. Because they are omnivores, rats, gerbils, and hamsters should be fed commercial pelleted diets that can occasionally be supplemented with different fruits and vegetables as well as animal protein like cooked chicken breast, egg white, or low-fat cheese. For small rodents, vitamin or mineral supplements are not required if a balanced diet is provided. There should always be fresh water available for drinking.

3.1. OVERVIEW OF RODENTS

Mammals of the order Rodentia, or rodents (from the Latin rodere, "to gnaw") are distinguished by having a single pair of continuously growing incisors in both the upper and lower jaws. Of all the mammal species, rodents make about 40%. They have been brought to the majority of these land masses by human activity, but they are native to all major land masses with the exception of Antarctica and a few oceanic islands.

Rodents can be found in nearly every terrestrial habitat, including places that have been created by humans. They have incredibly diverse ecosystems and lifestyles. A species may be semiaquatic, arboreal, saltatorial/ricochetal (leaping on their hind legs), or fossorial (burrowing). All rodents, however, have a few morphological traits in common, such as having a single pair of ever-growing incisors on both the upper and lower jaws. Among the well-known rodents are guinea pigs, hamsters, mice, rats, squirrels, prairie dogs, porcupines, and beavers. Although they formerly shared the same incisors as rodents, rabbits, hares, and pikas—which have two pairs of upper incisors rather than one—are now classified as belonging to a different order, the Lagomorpha. Despite this, Rodentia, and Lagomorpha are sister groups that belong to the clade of Glires and have a single common ancestor.

The majority of rodents are small, robust creatures with long tails and short limbs. Their razor-sharp incisors are used for defense, digging burrows, and chewing food. While the majority consume seeds and other plant matter, some have more diverse diets. Animals have a tendency to be gregarious, and many species have sophisticated social structures for communication. Rodents exhibit a range of mating behaviors, including promiscuity, polygyny, and monogamy. Some are precocial (relatively well developed) at birth, while many have litters of underdeveloped, altricial young.

The history of rodent fossils on the Laurasia supercontinent extends back to the Paleocene. During the Eocene, rodent diversity increased dramatically as they dispersed over continents and occasionally even over oceans. Before the emergence of Homo sapiens, rodents were the only terrestrial placental mammals to travel from Africa to South America and Madagascar and settle in Australia. Rats have been kept as pets, food, clothing, and lab animals for scientific studies. Certain species, particularly the house mouse, black rat, and brown rat, are major pests that spread illness and eat and spoil food that has been stored by people. Rat species that were unintentionally brought into the area are frequently regarded as invasive and have been responsible for the extinction of many species, including island birds–the dodo is one such example–that were previously protected from predators on land (Harkness et al., 2010).

3.1.1. General Characteristics of Rodents

Based on variations in the masseter muscle's anatomy and function, the hundreds of species that make up the order Rodentia are divided into three suborders. These three groups are called the Sciuromorpha (like a squirrel), the Myomorpha (like a mouse or rat), and the Caviomorpha (like guinea pigs). Despite wide variations in body size, rodents generally have a similar body structure with a variety of adaptations. Although many small rodent species kept as pets share a similar medical approach, uncommon species are occasionally encountered in practice and more detailed information about these species can be found elsewhere.

Many species of small rodents found in pet stores, such as rats, mice, hamsters, and gerbils, belong to the suborder Myomorpha. This group of species has anelodont cheek teeth and elodont incisors. The inguinal canal is open, allowing the testicles to freely travel from the abdomen to the scrotum. The testicles and scrotum are typically large in comparison to the size of the entire body. Squirrels and chipmunks are examples of simian species that are less frequently kept.

3.1.1.1. Rats

Because of their larger size and peaceful disposition, rats (*Rattus norvegicus*) are popular pets and regarded as some of the best rodent pets available. The hooded color varieties of rats, where the coat color is limited to the head and shoulders, are the most popular in the pet industry. Rats are hardy creatures in general when they are young, but as they get older, they can develop mammary tumors, obesity, and chronic respiratory conditions. Rats rarely bite, and they are big enough for kids to easily grasp. Rats have been known to "escape" from their cages and then return to them, though some may become agitated and bolt. Male rats may be present while females are caring for their litters. Rats are gregarious animals that can live in mixed-sex groups. It is possible to introduce strangers successfully in a neutral setting. Rats can be trained to come when a treat is called for; they are fairly intelligent and show interest in people. Rare in the pet trade are giant Gambian pouched rats (*Cricetomys gambianus*), which can weigh upto 1.5 kg.

The U.S. Department of Agriculture (USDA) issued a temporary ban on their importation after monkey poxvirus was identified in a group of Gambian rats shipped to the United States in 2003. In 2008, the prohibition was lifted. It was found that the Florida Keys had become a breeding colony for Gambian pouched rats in 2004. There are attempts in progress to manage their populations.

3.1.1.2. Mice

Standard lab mice (*Mus musculus*), available in a variety of coat and color variations, are frequently kept as pets. A mature mouse weighs about 30 grams. Although they can move quickly and may be too big for younger children to handle, mice make excellent pets for older kids (10 years old and up). Despite being primarily nocturnal, mice are easily agitated. Because intact male mice have a strong odor, it is recommended to use female mice or castrated male mice. Compared to other popular rodent pets, mice tend to be more solitary individuals. While males that are intact cannot be kept together due to the possibility of aggression, harm, and death, female mice typically get along well. Pet mice are resilient and rarely contract infectious diseases, but they frequently get mite infestations, which can be challenging to treat. Mammary tumors and pneumonia are also common.

Some uncommon species are kept as pets, such as a variety of African spiny mice (*Heteromys* species), though children find them more challenging to handle due to their smaller stature. Spiny mice, distinguished by their dorsal, inflexible hairs resembling spines, are probably closer to gerbils. Spicy mice are omnivores in the wild.

3.1.1.3. Hamsters

There are various groups of hamsters that are commonly kept as pets. When they are first purchased, hamsters are the least resilient of the small rodents, and proliferative ileitis and other stress-related illnesses are frequent. Although hamsters are nocturnal animals, they will typically wake up without biting if they are gently picked up with two hands. A hamster is likely to react with startle or threat if you touch its back with your finger in an attempt to rouse it. When handling excited hamsters, exercise the necessary caution because they frequently jump from hands or tables. The capacity for escape is a well-known trait of all hamsters. If a child is taking care of a hamster, an adult should frequently check the pet's cage for indications of wear. Cleaned or damaged cage parts should be replaced right away. During the summer, try to keep your hamsters in a cool part of the house, as they can become stressed in hot and muggy conditions.

The Russian dwarf or Djungarian hamster (*Phodopus sungorus*), the closely related Campbell's Russian dwarf (*Phodopus campbelli*), and the Roborovskii or desert hamster (*Phodopus roborovskii*) are examples of dwarf hamsters. Their average weight is between 25 and 50 grams, and they have white underparts, short, furred tails, and, depending on the species, a grayish to tan dorsal surface with or without a dorsal stripe. Compared to golden hamsters, they are more agitated, harder to control, and more likely to bite when restrained. Compared to golden hamsters, dwarf hamsters are more gregarious and inclined to live in families. Though they are smaller than dwarf hamsters, Chinese hamsters (*Cricetulus griseus*) are less gregarious and should be housed separately.

3.1.1.4. Gerbils

Gerbils (*Meriones unguiculatus*) are desert-dwelling animals with effective kidneys for water conservation. There are white, black, buff, gray, and spotted varieties of pet gerbils available. Due to their extreme activity, they may be challenging for younger children to handle; if their tail is caught or gripped too tightly, they may slough the skin off of it. Although they are social animals that live in families, gerbils are also territorial and frequently will not accept the introduction of strangers (trying to keep incompatible pairs or males together can lead to cannibalism). Despite the fact that older gerbils may experience a range of degenerative and neoplastic conditions, gerbils are more disease-resistant than hamsters. Although it's not common in many pet strains, epilepsy has been documented in gerbils.

3.2. HUSBANDRY

3.2.1. Housing and Equipment

Rat enclosures that work well should be easy to clean and have no escape routes. Multilevel cages with tubes, wheels, and hide boxes may be eye-catching and colorful, but they may be so difficult to dismantle that basic maintenance is overlooked. It is rarely done, but the manufacturers advise completely disassembling and cleaning the cage. More recent multilevel wire cages can be divided to make cleaning easier and rest on a plastic base. The extra advantage of this kind of cage is that living space is kept apart from feces and urine. Large doors for convenient pet access, secure locking mechanisms for each cage opening, adequate ventilation (aquariums are not ideal due to poor ventilation), easy-to-remove or slide-out bottoms for ease of cleaning, and bottoms with high sides to contain bedding are all characteristics of optimal enclosures.

Maintaining a clean cage is essential for caring for rodent pets. Ammonia builds up in the cage due to improper cleaning, which also increases stress and disease. *Mycoplasma pulmonis* organisms multiply more quickly in mice and rats when there is 50–100 ppm of ammonia present. The size of the cage and the quantity of animals housed determine how often the cage needs to be cleaned. Encourage owners to pay attention to the bedding's aroma; if it doesn't smell like fresh litter, the cage needs to be cleaned. To prevent the food containers from toppling over, serve food in substantial crocks or food dispensers.

Shredded chapter, corncob products, recycled chapter products, and shavings of pine and other hardwoods are among the options for bedding. The usage of hardwood and aromatic shavings, like cedar, is hotly debated; there are anecdotal reports linking it to respiratory and skin conditions. Although these products cost more than wood shavings, chapter bedding is typically preferred. The study examined the levels of endotoxins, dust, and coliform in 20 different types of rodent bedding. The results showed that chapter bedding had the lowest levels of endotoxins and coliform, and these products were suggested as a way to lower the risk of respiratory diseases and immune suppression in laboratory rats. When rats were housed on aspen shavings as opposed to chapter bedding, the incidence of lung pathology and the rate of sneezing were higher. On the other hand, a study conducted on male CD-1 mice maintained on Care FRESH (Absorption Corp.) revealed no variation in the mice's growth, food intake, oxygen consumption, IgE antibody concentrations, or overall appearance and behavior over a four month period using the original bedding, cedar shavings, or pine shavings from Ferndale, Washington, U.S.A. In a study assessing the toxic effects of juniper and cedar on skin.

The development of facial lesions is indirectly related to the housing of gerbils on shavings. Dust bathing is believed to be a regular part of gerbils' grooming routine because cages with shavings lack sand or other dust that would typically be present in their natural habitat. Much like with chinchillas, gerbils can also benefit from having sandboxes. Giving mental stimulation is referred to as enrichment, and it is suitable for all captive animals. Rodents are given enrichment in the form of exercise wheels, hide boxes, shredder materials, toys with treats concealed inside or wrapped in chapter, and time spent with the owner outside the enclosure. Hide boxes can be made from easily accessible materials like cardboard boxes or polyvinyl chloride (PVC) pipe, or they can be bought commercially. The added benefit of cardboard is that it is a material that can be disposed of, destroyed, and shredded when it gets dirty. Larger rodents have been successfully housed with smaller cat treat balls. A lot of small rodents like using exercise wheels, particularly hamsters. There are virtually silent plastic exercise wheels available, but some hamsters might chew on them if they are made of softer materials, like soft wool.

3.2.2. Diet and Feeding

Food is naturally hoarded by rodents, so it's hard to tell how much food is really being eaten by how quickly it disappears from the feeder. While bowls of water are a possibility, high-quality water bottles are usually recommended as they will keep bedding out of the water. Over time, blockages or leaks may occur from malfunctioning water bottles. Teach owners to regularly test and replace their water bottles.

In the wild, the majority of rodents are omnivores, frequently consuming grasses, seeds, grains, and occasionally invertebrates. The food needs of various species in lab environments are well-established. The best way to meet needs in a pet environment is to use a prepared diet that is enriched with tiny amounts of fresh foods and seeds to add interest and variety. For rodents, seed mixes are a popular choice, but these diets frequently result in selective feeding. Dietary imbalance occurs when animals ignore formulated pellets and instead eat high-calorie seeds like sunflower. Although dietary deficiencies are rarely noticed in rodents because many of them live such short lives, obesity is the most frequent negative outcome in rats. Several studies showing longer lifespans and fewer instances of specific diseases in rats fed a diet low in calories are also noteworthy. Gerbils are sensitive to high-fat, high-cholesterol diets, which cause changes in blood cholesterol levels. For this reason, they have been used extensively in nutritional studies of dietary fat. The percentage of protein needed by rodents varies: hamsters need between 14 and 17%; rats and mice need between 14 and 16%; and gerbils need upto 22%. These needs should be reflected in the diets designed for gerbils, rats, mice, and hamsters. Rats should eat diets higher in protein when they are reproducing (Harkness et al., 2010).

3.3. THE DIAGNOSTIC CHALLENGE

3.3.1. Pet Rodent Etiquette

A good and productive physical examination can be achieved by teaching and acquainting the veterinary staff with a few basic principles of "pet rodent etiquette." Because of their preparation, the clinician is less likely to experience the unnecessary anxiety that can occur when they are faced with an overly protective and worried owner. The vet should be prepared to treat the patient by using their knowledge and therapeutic abilities. Treatment and recovery for the rodent are considerably aided by the owner and clinician having open lines of communication.

3.3.2. Scheduling an Appointment

Essentially nocturnal creatures, rats, and hamsters are rarely active during the day. Mice and gerbils, on the other hand, are active day and night. In general, healthy rodents are active during the awake phase of their regular circadian cycle. Subtle symptoms of illnesses might go unnoticed when a sleep-deprived, sleepy, and agitated animal is brought to the veterinarian. This is particularly valid for hamsters. Receptionists ought to aim to book rat and hamster appointments for early evenings whenever feasible. For gerbils and mice, appointment timings are less important. If a client is reluctant to schedule an evening appointment, taking the time to explain the rationale behind it frequently changes their mind and establishes a positive rapport between the veterinarian and the client.

Tell the patient to bring the rodent in its own cage to the hospital. Taking a good clinical history requires good hygiene and husbandry. An understanding of the rodent's environment can only be gained by the clinician by looking at the cage, water source, feed containers, bedding, and food. It is advisable to gently advise clients against cleaning the rodent's housing prior to the appointment, as this could unintentionally destroy crucial information for diagnosis and treatment.

3.3.3. Reception Area

Owners of pet rodents should have a quiet waiting area. When a pet rodent is due for a checkup, receptionists should refrain from scheduling appointments for outdoor cats or hunting dogs. The world of rodents is full of olfactory stimuli and pheromonal cues, and their sense of smell is highly developed. The effects of heat affect rodents more than those of cold. Despite living in desert environments, wild golden hamsters and gerbils primarily use burrowing or searching for cool spots to escape the heat. Particularly sensitive to the effects of heat are mice. Rodents should have relatively cool waiting areas. The National Institutes of Health recommends a temperature range of 26□–64□ (4°F–18°F) for dry bulbs, which is perfect for housing rodents. Also, a quiet and secure waiting area is the only place where a conditioned rodent can find a place to nest next to its owner's familiar scent. It is well worth the time investment to teach patients and receptionists how to make a trip to the clinic less stressful. A rodent that is willing to be examined by the veterinarian rather than one that is ready to fight, run, or recoil will be seen if adequate attention is paid to education.

3.4. DISEASES OF SMALL RODENTS

Small rodent diseases in practice differ greatly from those in research settings in terms of both prevalence and kind. Despite the fact that this may seem fairly obvious, a large portion of the literature discussing the illnesses that affect pets has been drawn indiscriminately from conditions observed in lab rats. Rather than managing the health of rodents from a research colony, the diagnosis and treatment of pet rodents entails the assessment and care of a single animal in a home. Infectious diseases, age, and nutrition related issues, trauma-related injuries, and genetic disorders are the most common derangements observed in practice. Natural infections that would be rare in a laboratory animal colony are frequently passed from children and other pets to rodent pets. For instance, humans are the primary natural host of Streptococcus pneumoniae and Streptococcus pyogenes, and pet animals other than rodents are a major reservoir of dermatophytes. The environments in which research rats are kept are strictly regulated to lessen the influence of unfavorable factors in animal experiments. However, a wide variety of foods, numerous microorganisms carried by humans and animals, and different handling techniques are all commonplace for pet rodents. They are also exposed to changes in temperature, humidity, and light cycle. Consequently, compared to rodents used in research, pet rodents display a greater variety of physiologic and pathologic responses. As a result, many pet rodents' disease presentations do not match the traditional description of the illness.

3.4.1. Significant Diseases and Life Spans

Pets such as degus, rats, gerbils, hamsters, and mice can have a few naturally occurring medical issues. The most frequent spontaneous disease outbreaks are brought on by, or at the very least exacerbated by, poor management and nutrition. A selection of these issues, specific to each species, are enumerated and categorized according to the main organ system they impact. Additionally provided is the average lifespan for each species. In these species, pneumonia, enteropathies, and dermatopathies account for the majority of issues. There are some issues that are not discussed, like trauma, hypothermia, and malnourishment, which are also frequently observed in small exotic animal practice but are not always specific to a species (Figure 3.1).

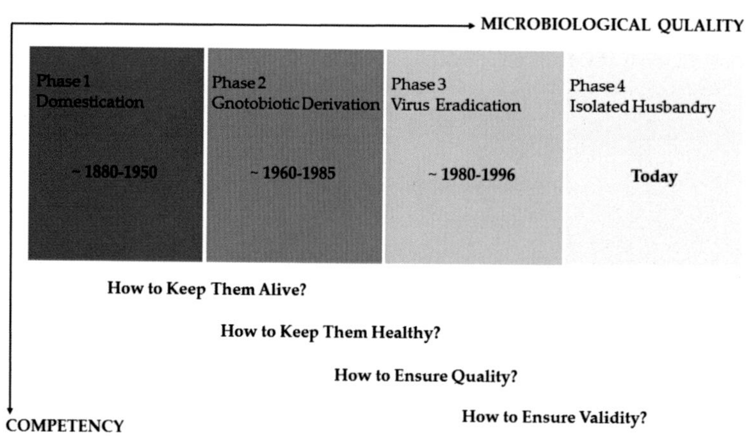

Figure 3.1. *Rodent health.*

Source: Buchheister S, Bleich A. Health Monitoring of Laboratory Rodent Colonies—Talking about (R)evolution. Animals. *2021; 11(5):1410. https://doi.org/10.3390/ani11051410.*

3.4.2. Prophylaxis for Small Rodents

In rodents, disease prevention is far more effective than treatment. Healthy, genetically sound animals should be purchased; balanced, fresh food with the right amount of protein and calories should be provided; clean, fresh water should be available; adequate shelter should be provided, including shade from direct sunlight; drafts and drastic temperature or humidity changes should be avoided; excess feces and urine should not be allowed to accumulate in cages; sick animals should be isolated from the group while receiving treatment; and vulnerable animals should be shielded from more aggressive members of the group (e.g., male hamsters from female hamsters, young animals from older animals, or from natural predators residing in the same home (e.g., mice that come from cats). To stop the spread of interspecies diseases, housing different species apart is another wise husbandry practice (e.g., rats' nasopharyngeal cavities are home to *Streptobacillus moniliformis*, which causes septicemia in mice), and lowering obesity through dietary restriction and the provision of play and exploration-promoting cage accessories (e.g., ramps, tunnels, and exercise wheels). Unlike larger companion animals, pet rodents are not vaccinated. The introduction of ivermectin, although not approved for use in any rodent species, has allowed routine systemic treatment of pet rodents with pinworms and mites.

3.5. DISORDERS AND DISEASES OF RATS

Rat illnesses and ailments of the most common kinds include cancer, parasite infections, and other diseases. Your rat can be kept in the best possible condition with regular veterinary care, a balanced diet, and appropriate housing.

3.5.1. Digestive Disorders of Rats

Intestinal parasites and bacterial infections are the main causes of the digestive problems that affect rats most frequently.

1. **Intestinal Parasites:** Rats frequently contract pinworms as intestinal parasites. They spread by contaminated excrement and only need a single host, in this case, rats. The majority of infected rats show no symptoms, but intestinal inflammation brought on by a severe pinworm infection can result in diarrhea. The disease is identified by looking for worms or their eggs in contaminated excrement or on the rat's anus. Pinworm infections can be managed with the right medications that a veterinarian prescribes. It's critical to regularly clean and disinfect the cage because pinworm eggs are light and have the potential to float in the air.

Rats rarely get infected with tapeworms, and when they do, the infection usually shows no symptoms. On the other hand, severe infestation may result in diarrhea and weight loss. If consumed, the dwarf tapeworm has the potential to infect people. Indirect transmission of tapeworms occurs via fleas, beetles, or cockroaches. Finding tapeworm eggs in feces that have been infected is how the infection is identified. The appropriate antiparasitic medications are used to treat the worm infection. It is necessary to clean and sanitize the cage. *Taenia taeniaeformis*, the cat tapeworm, uses rats as an intermediary host. Rats can contract cat tapeworms. Rats become infected when they consume feed or come into contact with bedding that has been tainted by cat excrement. When tapeworm cysts lodge in the rat's liver, the organ swells. For people who own cats and rodents, getting rid of possible infection sources is crucial.

2. **Protozoa:** In the gastrointestinal tract of rats, a variety of protozoa (microorganisms) are typically present but do not typically cause illness. On the other hand, these protozoa can result in intestinal infections in younger or stressed rats. Rats that are infected experience diarrhea, lethargy, a rough coat, weight loss, and, in extreme situations, heavy bleeding that may result in death. The infection is spread by contaminated excrement. With the right medications, the protozoa can be managed but not always completely eradicated.

3.5.2. Lung and Airway Disorders of Rats

Bacterial infection is the cause of murine respiratory mycoplasmosis, a chronic respiratory disease. In rats, it results in various issues as well as short- and long-term respiratory symptoms. Sexual contact, airborne bacteria, and direct contact are the three ways that the infection is spread. It can also be transferred during childbirth from a mother to her children. Sneezing, sniffling, rough hair coat, fatigue, labored breathing, weight loss, head tilt, and reddish-brown staining around the eyes and nose are some of the symptoms of infection. The lungs are affected as the illness worsens.

When other bacterial and viral infections are present, the infection may worsen. Females with long-term respiratory disorders may get infections in their ovaries and uterus. This condition has no known cure. Antibiotics may help with infection symptoms. Rarely do rats with persistent respiratory conditions survive for more than two years. The best defenses against this illness are keeping your rat's living quarters clean, especially by lowering the ammonia levels in the cage and treating the infection as soon as it arises. Rats can develop pneumonia and respiratory infections from many different bacteria and viruses. Sneezing, sniffling, labored breathing, rough hair coat, inactivity, weight loss, lack of appetite, and discharge from the eyes or nose are all symptoms that are similar to each other. You should take your rat to the vet if you observe any of these symptoms.

Depending on the organism, there are a few ways that the diseases can spread amongst rats, such as direct contact with infected animals, contaminated feces, or sneezing or coughing. Antibiotics may be prescribed by your veterinarian to help lessen the severity of illness, even though the majority of these infections have no known cure. It will also help to provide supportive care and maintain a clean environment for the rat. To stop the disease from spreading, rats exhibiting respiratory infection symptoms should be housed apart from other rats. The bacteria *Streptobacillus moniliformis*, which causes rat bite fever in humans, naturally inhabits rats. These bacteria are present in the upper respiratory tracts (snake and throat) of rats. Although the bacteria do not infect rats, they can seriously harm humans.

3.5.2.1. Sialodacryoadenitis and Rat Coronavirus Infection

Rats' salivary glands, lungs, Harderian glands (beyond the eyes), and nasal cavities are all infected by these closely related viruses. They can spread through direct contact, airborne virus particles (like those from an infected rat's sneeze), or contact with contaminated bedding, feces, or other items in the cage. They are extremely contagious. It's crucial to wash your hands after handling animals at a pet store or a friend's house to stop the spread of these illnesses before handling your rat. A typical infection lasts two to three weeks. Rats that are affected will sneeze and may try to stay out of direct or bright light. Sometimes one can feel enlarged lymph nodes and salivary glands. A rat could appear to have the mumps. Although not all infected animals exhibit these symptoms, reddish-brown pigments and discharge may be seen around the eyes, as well as eye involvement such as inflammation of the cornea or conjunctiva. Although there is no cure, affected animals typically recover and become less susceptible to these viruses in the future.

3.5.3. Skin Disorders of Rats

Skin disorders are common in rats. They can be caused by parasites, bacteria, or injury.

1. **Barbering:** Groups of male or female rats occasionally exhibit this abnormal grooming behavior. Rats with less dominance have their whiskers and hair chewed by dominant members of the group. The word "barbering" comes from the fact that the rat chews its hair so close to the skin that it appears clean-shaven. This behavior can be caused by stress, boredom, or even genetics. Rats can also self-shave. The stomach and front legs, if the barbering is due to

self-grooming, or the muzzle, head, or shoulders of a cage mate, are the most common places on the body. In most cases, the skin is unaffected and appears normal, showing no signs of cuts, inflammation, or irritation. Treatment is not necessary for this condition unless irritation appears. For the sake of the other rats in the cage, the dominant rat should be removed if barbering is the result of its presence.

2. **Fight Wounds:** Male rats frequently fight, injuring each other's faces, backs, and genitalia. There will be scabs and areas of hair loss on the skin. Bacteria can infect such wounds, resulting in the development of abscesses. Gangrene can result from tail biting. Affected rats occasionally pass away and lose weight. Treating fight wounds involves cleaning them with an antiseptic solution, draining the abscesses, and using the right antibiotic ointments that your veterinarian has prescribed or advised. Frequent fighters among rats ought to be divided.
3. **Fleas:** Although they are rare in pet rats, rodent fleas can occasionally be found in pets that come into contact with wild rodents. By identifying them on the infected rats, fleas can be diagnosed. Inquire about treatment from your veterinarian. Thoroughly clean and disinfect the cage to avoid reinfestation. It is advised that you wash and change into new clothes before handling your own rats if you are holding or playing with rats that are not your own.
4. **Lice:** Bloodsucking lice infestations are uncommon in pet rats but common in wild rats. Rat lice do not spread from one species to another; hence they have no effect on humans. Rats with severe infestations exhibit extreme itching, agitation, anemia (low red blood cell count), and weakness. Diagnosing an infestation involves locating adult lice or eggs on the fur of the rat. Treating lice is similar to treating mites.
5. **Mites:** Rats' skin and fur can become infested with a variety of mite species. External parasites called mites frequently cause no symptoms at all in rats. Rats with severe infestations may exhibit skin inflammation, and the mites can be observed as white dust particles on their hair follicles. Furthermore, severe itching brought on by mites may result in scabbing on the face, neck, and shoulders. Neither humans nor other animals are infested by rat fur mites. By identifying the mites or their eggs from the rat's skin and hair, an infestation can be determined. Medication applied topically or taken orally is used to treat mites. The best course of action will be recommended by your veterinarian. In normal circumstances, mites

are scarce and do not cause any discomfort to their host. But when the rat is under stress, has weakened immunity from other diseases, or can't control its population growth with regular grooming, its numbers rise. It is crucial that you give your pet the care they need, which includes keeping an eye out for any illnesses. Because unhatched eggs can result in reinfection, the rat's cage and all of its components need to be properly cleaned and disinfected after treatment.

6. **Ringtail Syndrome:** Ringtail is the term for the ring-shaped constriction of the tail that develops in young rats as a result of low humidity, high temperatures, and drafts. The toes or feet may also be affected by this condition. Rats kept as pets are less likely to have ring tail, which is more common in lab rats. Ringtail can cause swelling and, in the end, the tail portion below the constriction to disappear. It is frequently necessary to remove the tail entirely or in part surgically, and tail stumps heal without much difficulty. Preventing ringtail involves keeping the cage temperature between 64°F and 79°F (18□ and 26□), minimizing drafts, and maintaining an environmental humidity of 30% to 70%.

7. **Ringworm:** Fungi known as dermatophytes that parasitize the skin are the cause of ringworm. Both humans and other animals can contract the infection through direct contact or contaminated bedding, litter, or cage supplies. In rats, ringworm rarely occurs. Rats that are infected might not exhibit any symptoms. But some of the afflicted rats have reddened, irritated, or flaky skin, along with patches of hair loss. Your veterinarian should prescribe the course of treatment, which may involve taking oral or topical antifungal medications to eradicate the fungus. A fungal infection may go away on its own in a few weeks, but treatment can lessen the likelihood that the afflicted animal will infect humans or other animals.

8. **Staphylococcal Infection:** Staphylococcus bacteria, which are typically found on the skin of most animals, including rats, are the cause of this infection. When bite wounds or scratches cause damage to the skin, infection results. Immune compromised rats are more vulnerable to infection. On the head and neck, there may be sores and inflammatory skin. The abscesses that result from these lesions may grow and spread beneath the skin, forming lumps (tumors) all over the face and head. Antibiotics or steroid/antibiotic ointments applied as prescribed are used to treat the infection. The hind toenails should be clipped to stop more damage from scratches.

3.5.4. Kidney and Urinary Disorders of Rats

An prevalent condition affecting older rats is glomerulonephrosis, or chronic progressive nephrosis. It involves the kidney's blood vessels becoming inflamed. The sex, genetic heritage, and dietary components of the rat, such as the amount of protein and total calories consumed, all affect the disease and how severe it is. In comparison to female rats, male rats experience the disease earlier and more severely. Rats that are affected exhibit lethargy, weight loss, and possibly increased thirst and urination. Regretfully, there is no known cure for the illness, and it is always fatal. Treatment with support could be able to lessen the symptoms.

The most common cases of leptospirosis, a bacterial urinary tract infection, are in wild rats and mice. It might be spread to people and house rats alike. Blood tests or urine bacterial isolation are used to make the diagnosis. Treatment is not advised due to the possibility of infection in humans. The bladder of the rat can occasionally become infected with specific roundworms (nematodes), but these parasites are rarely documented. The course of treatment is comparable to that for intestinal pinworm cases. Uroliths, or stones, develop in the bladder and kidneys of elderly rats. Rats that are affected may exhibit no symptoms at all or may exhibit blood in their urine along with other indications of bladder inflammation or infection. To get rid of the stones, surgery might be required. Death could result from kidney damage or urine passage obstruction caused by stones.

3.5.5. Disorders Affecting Multiple Body Systems of Rats

A rat's illness may impact more than one body system. Other names for these illnesses include systemic or generalized disorders. Below is a list of the disorders that affect various body systems in rats.

1. **Lymphocytic Choriomeningitis Virus Infection:** Rats can become infected with this virus on occasion. Rats can contract the disease from wild rodents like house mice or from interacting with other infected rodents at pet stores, such as mice, guinea pigs, or hamsters. Coughing, sneezing, or coming into close contact with the urine or saliva of an infected animal can all spread the infection. Most infected rats don't exhibit any symptoms. But some rats are virus carriers, and they excrete large amounts of the virus in their urine. There's no cure that works. The affected animals must be put to sleep, and the cage needs to be thoroughly cleaned and sanitized.

This virus may spread from infected rats to humans, where it may result in life-threatening disease. It may result in viral meningitis, an inflammation of the brain and spinal cord, or influenza-like symptoms. Nonetheless, it is uncommon for rodent pets to infect humans.

2. **Parvovirus Infection:** Rats kept as pets rarely have this viral infection. Infected rats experience small litters, stillbirth, infertility, and runting when the infection is at its most active. Mortal bleeding and cell death in the brain and gonads are possible symptoms in infected rats. Every parvovirus has a high rate of spread. Transmission can happen by direct contact with contaminated urine or feces or by contaminating environmental objects like bedding. The cage must be thoroughly cleaned to eradicate the virus. At this time, there is no cure.
3. **Polyarteritis Nodosa:** Inflammation of the artery walls is a feature of polyarteritis nodosa, which can impact multiple organs. It is unknown what causes the illness. Aneurysms and heart attacks are possible in rats suffering from polyarteritis nodosa.
4. **Salmonellosis:** In pet rats, this illness, which is brought on by Salmonella bacteria, is rare. On the other hand, young rats and pregnant women are more likely to become infected. Eating food tainted with excrement spreads the infection, which is frequently linked to bedding, water, or food tainted by wild rodents. Rats that are afflicted may exhibit weight loss, a bloated belly, diarrhea, dehydration, rough fur, depression, and unexpected death. Pregnant rats may experience miscarriage. People can contract the disease from infected rats as well. There isn't a cure. To remove any possible cause of infection, the afflicted rats should be kept apart and their cage cleaned and sanitized.

3.5.6. Cancers and Tumors of Rats

Rats are particularly prone to the growth of tumors. Because tumors have the potential to grow and spread to other parts of the body, your veterinarian will probably advise surgical removal of the tumor. The best result is possible with the lowest likelihood of complications and recurrence when the removal is done early.

Skin tumors that are benign are called keratoacanthomas. They grow on the back, the tail, or the chest.

The most prevalent tumors in rats are called mammary fibroadenomas. Rats' extensively dispersed mammary (breast) tissue means that tumors can be discovered beneath the skin anywhere on the abdomen side of the body, from the chin to the tail. Tumors can develop in male or female rats equally. These growths are usually round, soft, or somewhat flat tumors that are pliable and movable under strong pressure. The tumor will be surgically removed by your veterinarian, but it is typical for it to return in other body parts. The majority of these tumors do not progress to malignancy. Rats are prone to pituitary tumors, particularly in female rats. The pituitary gland is a gland that regulates hormone secretion and is connected to the brain. Consuming a high-calorie diet accelerates the development of these tumors. Rats that are affected exhibit head tilt and depression may pass away quickly. In rats, benign testicular tumors predominate. When necessary, surgical testicle removal is the advised course of action. Zymbal gland tumors are uncommon in rats. In older rats, they form at the base of the ear. These growths might be cancerous.

3.5.7. Special Considerations for Rats

Domesticated rats make excellent pets because they are gregarious, lively, and intelligent. Due to their nocturnal nature, they typically sleep for the majority of the day and are very active at night. Pet rats generally don't bite. Grasping a rat gently over the shoulders and gently lifting it up while supporting its body with your hand is the optimal method for handling rats. A rat should not be grabbed by the tail. A rat is more likely to bite if it rears up to confront your hand as it approaches, so proceed cautiously to prevent startling it. Because they are gregarious and social creatures, rats should ideally be housed in groups. Because they are prolific breeders, it is advised to house only rats of the same sex in one cage. Domesticated rats that grow up as littermates rarely become aggressive. However, because fighting amongst newly housed groups of animals can result in injuries, it is important to keep a close eye on them.

Rats are highly susceptible to both cold and heat. They do not perspire or pant, and their ability to control body temperature is restricted. They use the dilation and constriction of their tail veins to control their body temperature. Rats should be housed between 64°F and 79°F (18□ and 26□) with between 30 and 70% humidity. Heat exhaustion can occur at temperatures higher than 86°F (30□), particularly in overcrowded cages. Rats should not be exposed to direct sunlight, despite the importance of adequate lighting.

3.5.8. Selecting a Rat

Make sure all of the rats in a pet shop or animal shelter appear healthy and alert, and that their living quarters are tidy and comfortable, before choosing one. The rats may have been exposed to an infectious disease if any of them appear ill, if there is evidence of diarrhea, or if the cage smells bad. Each rat should have shiny, clean fur, and there should be no discharge coming from its eyes or nose. After being handled, rats should not be overly fearful of people and are frequently inherently curious about them. Rats are nocturnal animals, so it's best to observe them in the early evening when they're more likely to be awake and focused.

3.5.9. Providing a Home for a Rat

To stay healthy, your pet rat needs a suitable cage, a well-balanced diet, and the chance to interact with you and get exercise.

3.5.9.1. Housing

Pet rats can be kept in a variety of cages. A cage should, in general, be escape-proof, have adequate ventilation, and be simple to clean. It should have areas for feeding and nesting, as well as enough room for exercise. If multiple rats are housed together, then additional space should be provided. Rats' feet and legs can be harmed by wire mesh or grid flooring, so the cage floor should be solid. Rats should not be kept in wooden cages because their urine will seep into the wood. This results in an accumulation of ammonia, which may cause respiratory disorders. It's possible for rats to chew through the wood and get away. It could also splinter, which would be dangerous.

If the animals are to climb on wire shelves in the cage, the wire should have openings no larger than ½ by ½ inch. Any larger gap could cause the rat's hind foot to become trapped, breaking its leg, among other injuries. Rats seem to like climbing on the different levels of the new powder-coated wire rat cages, which have solid metal shelves and plastic cat pan bottoms. These cages are also the easiest to clean. Both the carpet and the bedding need to be absorbent. Soft bedding for rats can be made from wood shavings or prepared litter. Hay might be included. Bedding made from recycled chapter is secure and simple to clean. The use of cedar shavings is still debatable because they contain oils that could be harmful to small animals. It is possible to use a water bottle with a hanger so that rats can easily drink water. It is best to stay away from a water dish because rats might push shavings or spill water into it. There are water bottle holders with chew guards available. These enable the bottle to be hung inside the cage. In the event that you use a food dish, it should be easily cleaned and sufficiently sturdy to avoid tipping.

Just like any other pet, those kept in unfavorable conditions will eventually experience health issues. Rats are more susceptible to viral and bacterial infections when their cage is dirty and their bedding is of low quality. It is recommended to clean the cages using hot water once a week and to replace the bedding at least twice a week. Prevent the accumulation of ammonia (caused by urine), which arises from inadequate cage litter changes.

3.5.9.2. Diet

Rats in the wild consume a variety of foods. A diet that is balanced in terms of carbs, protein, and fat, as well as specific vitamins and minerals, is necessary for domesticated rats. Pet stores sell commercially prepared food that is full of all the vitamins and minerals required by rats, along with pellets that are tough enough to wear down the rat's ever-growing incisor teeth. Rats will often pick out the seeds and leave the pellets in mixtures that contain pellets mixed with seeds, so these should be avoided. If you are switching from a seed mixture to pellets, it is best to introduce the new diet gradually because rats can be picky eaters. Fruit, sunflower seeds, and fresh veggies can all be offered as treats in moderation. Given that older rats are more likely to be obese, fatty treats should be avoided.

Food should always be available, but before fresh food is given (at least once a day), any uneaten food should be taken out of the cage. Use sturdy food containers that are difficult for rats to topple over. Water from a bottle that is hung or fixed on the cage should always be available. Every day, all food and water containers need to be thoroughly cleaned.

3.5.9.3. Exercise

All animals require regular exercise to maintain good health. Rats love to play and exercise because they are such active animals. Having access to climbing blocks, ladders, tubes, and exercise wheels contributes to a balanced lifestyle and offers stimulation. Also, it might aid in preventing obesity. If you purchase an exercise wheel for your rat, make sure its diameter is at least 12 to 14 inches. A few of these cage toys should fit in the cage, and there should be enough room for you to move around them comfortably.

3.5.9.4. Temperament

Rats are highly gregarious animals. They enjoy socializing with others and will frequently dedicate a significant amount of time to grooming either themselves or their cagemates. Unless they grew up together, male rats are prone to fight. There are no issues when females are housed in the same cage. Usually, rats are great with children but they should be kept away from other pets in the house such as dogs, cats, birds, and other rodents. Remember that rats are the natural predators or prey for several of these animals. Rats can be very active at night because they are nocturnal animals. If possible, locate the cage so that its activities won't wake up any family members who are sleeping.

3.5.10. Routine Health Care of Rats

Maintaining proper hygiene, providing a healthy diet, appropriate housing, and thoughtful care can reduce your rat's risk of illness. By keeping your pet in a cage free of hazards and by handling it with care and gentleness at all times, you can prevent injuries. Rats and other small rodents are not currently covered by any vaccinations. You should check your rat on a regular basis for any general indicators of disease, trauma, or illness. These include signs of trauma (bite, wound, limping), hunched posture, discharge from the eyes or nose, hair loss, matted or fluffed fur, and general dullness.

1. **Signs of Illness:** Early symptoms of illness include dull coat, puffiness of the coat, labored breathing, loud or raspy breathing, and loss of energy or appetite. A change in your rat's urine or feces' color, consistency, smell, or volume could also be a sign of illness. If your rat exhibits any of these symptoms, it is imperative that you take it to the vet right away. A reddish-brown material that lubricates the eye and eyelids is secreted by the Harderian gland, which is located behind the rat's eyeball. Though it does not actually contain blood, this secretion–also called red tears occasionally–gives the impression of blood. Occasionally observed around the eyes and nose, it could indicate that your rat is sick or under stress. Rats' head tilt, also referred to as "wry neck," is another indicator of disease. Although pituitary tumors, respiratory infections, or even strokes can cause it, inner ear infections are the usual cause. A veterinarian should examine the rat to determine the best course of action.
2. **Dental Care:** Their permanent incisor teeth grow continuously during their lifetime. They must be given the proper materials to gnaw on to wear down their developing teeth. Wooden chewing blocks are an excellent way to sharpen teeth; however, avoid using wood that has been treated with potentially harmful chemicals. Hard dog biscuits or rawhide chew sticks can also be utilized. Rats especially love cooked soup bones. When required, your veterinarian will trim the growing teeth on a regular basis. In pet rats, overgrown incisors are a common issue that can cause mouth trauma, weight loss, difficulty eating, and dehydration. The teeth may grow into the nasal cavity if they are not kept trimmed (Harkness et al., 2010).

3.6. CLINICAL TECHNIQUES

3.6.1. Handling and Restraint

Small rodents can be difficult to handle and confine. Maximizing diagnostic information while preserving patient and examiner safety is crucial. Release the animal right away and devise a different strategy if it ever seems distressed. To observe husbandry conditions and assess fecal output, many owners bring their pet rodents to the clinic in their typical enclosures. But it's best to avoid and take a lot of time trying to get a rodent out of colorful caging tubes. To facilitate capture and restraint, we kindly ask that owners confine their pets to a separate carrier or a small travel cage inside the cage. Depending on the patient's size, temperament, species, and general state, different restraint techniques are used. Most small rodent pets don't mind being held loosely in the palm of your hand. The only exceptions are animals that are in pain or discomfort and rodents that have not been properly accustomed to handling. For instance, a hamster that has intestinal gas and enteritis will seem to be in a lot of pain; it will roll onto its back, bite, and be resistant to being handled.

After the initial discomfort has passed, the same animal is frequently calm and receptive to handling. Three general techniques can be used to achieve restraint: (1) assessing the animal while it is free on the table or in your hand; (2) using tiny towels or cloths to restrain the animal; or (3) using some sort of scruffing technique. Combinations of the aforementioned techniques are employed occasionally. Rats can be slowed down while they are walking freely on the table by the examiner by gripping firmly at the base of the tail. Use of this method on rodents with furry tails is not advised due to the possibility of a skin injury known as "tail slip." Except for rats, who cannot stand this method, many species–including mice, hamsters, and gerbils–benefit from scratching. Using your thumb and forefinger, grab a large amount of loose skin from the lateral aspects of your neck and over your shoulders to cause scruffing. Insufficient skin allows the animal to turn around and bite the fingers of the person holding it in place. Huge, calm rodents can be confined by securing your hands around their shoulders and thorax.

For examination or diagnostic testing, mild sedation is beneficial for extremely agitated or distressed rodents. But be cautious when dealing with ill or disabled patients. Generally speaking, an animal will tolerate sedation if it is conscious enough to defend itself violently. When used in conjunction with an opioid, such as butorphanol (0.1–0.4 mg/kg) or hydromorphone (0.1 mg/kg), midazolam (0.25–0.5 mg/kg) is generally safe and effective for mild sedation. After gently securing the rodent with a cloth, inject it rapidly with a small syringe and a 27 g needle while its hind limb is still exposed. Then, release the animal and put it back in its enclosure. If more sedation is required, consider short-term general anesthesia or give ketamine. The small rodent should be directly placed into a small to medium-sized face mask to induce anesthesia. The mask should then be replaced for maintenance with a rodent-sized anesthetic cone or modified syringe case. Although large induction chambers should be avoided due to the increased volume of escaped anesthetic gas into the clinic environment and consequent staff exposure, very small induction chambers can also be used. Large rodent intubations have been documented, and endoscopy or SPE can be used to facilitate the procedure.

3.6.2. Sample Collection

The advantages of diagnostic testing in other species can also apply to rodents. Sample collection in rats is made feasible and practicable by improved methods and the capacity to obtain significant findings from small sample sizes. The age, sex, strain, and husbandry of the animals, as well as the technique of sample collection and laboratory methodology employed to derive the values, all affect reference values, which are frequently obtained from laboratory populations. Due to the fact that many published reference ranges are based on extremely small sample sizes, they should be used and interpreted cautiously (Carolyn Cray, personal communication).

3.6.2.1. Blood Sample Collection

Blood collection is frequently postponed until the animal's condition is stabilized because venipuncture should only be attempted in normovolemic, normothermic patients. The physician must be aware of the maximum volume of blood that can be safely drawn from the patient as well as the sample volume needed for the particular test before taking a sample. The majority of rodents have blood volumes that range from 6% to 7% of their total body weight; it is usually safe to remove no more than 10% of this blood. Assuming, for instance, that a 100 g gerbil has a total blood volume of 6 to 7 mL, no more than 0.7 ml should be collected. Please adapt these guidelines based on the overall health of the patient; in all cases, draw only the amount of blood needed for testing.

If submitting to a reference laboratory, find out from the lab what the minimal sample requirements are and whether they prefer serum or plasma samples. While some laboratories routinely analyze plasma samples to maximize sample volume, others require serum for biochemical analysis. In small rodent patients, it might not be feasible to obtain the volume of blood needed for both an automated complete blood count (CBC) and a biochemical analysis. If the laboratory is prepared to estimate the CBC from a blood film or carry out a manual count, the necessary volume can be decreased.

Blood sample analysis done in-house is advantageous for two main reasons: it can be done quickly, which is particularly helpful for sick rodent patients, and it can use very small sample volumes. Testing is now possible for even the tiniest rodent patients thanks to the Abaxis VetScan, which can perform a preanesthetic or diagnostic biochemical panel on 0.13 ml of whole blood. An estimated white blood cell (WBC) count, differential, hematocrit, and estimation of total serum solids can all be obtained from a single drop of blood collected into a hematocrit tube. These methods can be used to obtain a modified CBC and biochemical panel in a 25 g mouse (6% of body weight = 1 point 5 ml total blood volume; $10\% \times 1.5$ ml = 0.15 ml).

The size, temperament, and familiarity of the clinician with the patient determine the collection technique. The method that consistently yields samples of optimal patient safety and diagnostic quality is considered "correct." Hemolysis and poor sample quality are common outcomes of repeated venipuncture attempts and the use of excessive negative pressure, which can skew the results. Samples can be taken using small-gauge needles (25–27 g) and tiny syringes (1 ml or less), depending on the size of the patient. Consider using a clear-hubbed needle exclusively to puncture very small patients or vessels where negative pressure leads to the vessel collapsing, and collect blood using a heparinized hematocrit tube as it enters the needle hub.

The majority of rodents are unable to withstand the degree of restraint necessary to securely gather enough blood for analysis. Pet rodents may not benefit from the same techniques used to restrain laboratory rats for venipuncture. In many animals, simple sedation as previously mentioned, along with the addition of brief general anesthesia if necessary, greatly facilitates blood collection. A single sedation procedure can help with full examination, diagnostic testing, and some therapeutic procedures if it is planned carefully.

In rodent species, the largest readily accessible vessel is the cranial vena cava. Generally speaking, it is situated dorsal to the cranial most part of the manubrium, right inside the thoracic cavity; species-specific differences exist in the vessel's depth and lateral excursion. For most rodent species, this vessel is the first option and is typically successfully accessed. Nevertheless, the animal must be sedated or given general anesthesia prior to cranial vena cava venipuncture.

The animal must be in a dorsal recumbent position while the area above the manubrium is ready for venipuncture to draw blood from the vena cava. Approach the vessel from the right or left of the manubrium, angling slightly medially and dorsally, using a 25–27 g needle and a 1 ml or smaller syringe. Certain species (gerbils, hamsters) have a well-developed clavicle, which can make positioning the needle more difficult. Use light negative pressure as you move the needle forward. A burst of blood will indicate the cranial vena cava's entry. The needle may have passed through or been oblique to the vessel, or the bevel may be in contact with the vessel wall, if blood does not flow freely. Turn or reroute a little. Since the vessel is just below the manubrium, deep needle penetration is neither required nor advised. Vessel rupture and exsanguination are possible side effects, although they seem uncommon and are probably related to excessive patient movement.

The ventral tail artery of the rat, the tarsal veins of larger rodents, and the lateral saphenous or lateral tail vein in the rat, gerbil, or mouse are additional sites for blood sample collection. The superficial lateral tail veins are visible on both sides of the tail and are most noticeable in young animals and albino rats. Ideally, apply a tourniquet around the base of the tail and gently warm it to increase blood flow and occlude the veins; a rubber band and mosquito hemostat work well for this. Enter the skin at a shallow angle, about one-third of the way down the tail, using a needle that is the right gauge for the species. Try again at a location nearer the base of the tail if the first attempt at collection proves to be unsuccessful. Use a small volume syringe to remove the sample or gather the blood as it flows freely from the needle hub into a microhematocrit tube to prevent the vessel from collapsing. It's also possible to use a modified butterfly catheter, which has all of the tubing except the proximal 5 mm removed.

An additional option in the rat is the ventral tail artery. Although it is not as superficial as the lateral tail veins, the artery runs along the ventromedial aspect of the tail. Assume dorsal recumbency for the anesthetized or sedated animal. Utilize a 3 ml syringe that has had the plunger removed along with a 22 gauge needle. As an alternative, some practitioners like to use a 3 ml syringe coupled to a short piece of tubing and a 23 gauge butterfly needle. At a point one-third of the way down the tail's length, attempt the first puncture. With the needle's bevel facing up, puncture the skin at a 20 to 30□ angle to the tail. The artery has typically been entered when there is a noticeable "pop," which is soon followed by the syringe (or butterfly tubing) being rapidly filled with blood. The negative pressure created by the plunger's withdrawal is unnecessary because of the vessel's high blood pressure. In fact, it is more difficult to recognize when the syringe has been properly penetrated by mistake when the plunger is inside the case. Once the necessary amount of blood has been drawn, remove the needle and press against the puncture site. Keep in mind that more time is needed to halt the tails bleeding.

Larger, lighter-colored rodents frequently have visible lateral saphenous and tarsal veins running across the surface of their lateral legs or tarsi. Since these vessels are so tiny, it is best to puncture them with a needle alone. Samples should then be taken and placed straight into heparinized hematocrit tubes. A method has been described for the minimally invasive, anesthetic-free saphenous venipuncture of laboratory rodents. With the right restraint, one person can perform this technique, and it can be repeated several times at the same location. The process is used when relatively small amounts of blood are needed, and it does require skill. To use in a mouse, insert the mouse head first into a 50 ml syringe case that has had one or more tiny breathing holes pierced through the tip.

Depending on which leg is being bled and which hand is being used, firmly grasp the skin just in front of or just behind the knee to extend an exposed hind limb. To better expose the superficial vein, shave the hair off the area over the lateral tarsal and moisten the skin with alcohol. Use a needle of the proper size to puncture the vessel at a 90□ angle to the skin (a 25 gauge needle is usually sufficient in the mouse). Immediately after, a drop of blood appears, which can be collected into a microtube container or a standard microcapillary tube. Release the skin once the desired volume has been achieved. Usually, this is enough to halt the bleeding; if not, gently press the puncture site. This method might be too taxing for sick or disabled animals. Because of the high rate of complications and the availability of better options, it is not recommended to take blood samples from the orbital plexus or by cardiac puncture in pet rodents.

3.6.2.2. Bone Marrow Collection

Small rodents can also be used to obtain a bone marrow sample. For intraosseous catheterization of the femur or tibia, the same method as detailed below may be applied. Symptoms include nonregenerative anemia of unknown etiology and are the same as for any other species. As soon as samples are collected, prepare them for best quality.

3.6.2.3. Urine Collection

When handled, the majority of rodents easily urinate, especially when they're excited. These samples are helpful for testing with a urinalysis test strip and determining specific gravity in other species, just like free catch samples are. However, urine culture and sensitivity testing require samples that are sterilely collected through cystocentesis. Deep sedation or complete anesthesia is advised for this procedure. All rodents, with the exception of the most obese ones, have palpably felt bladders. Using a very small gauge needle, insert it caudally at an oblique angle through the abdominal wall and into the bladder while the animal is in a dorsal recumbent position. As an alternative, locate the bladder for cystocentesis using ultrasound. Occasionally, it is possible to obtain very small urine samples (suitable for culture) by gently expressing the bladder and collecting the sample into a hematocrit tube from the cleaned urinary orifice. Since many rodents will urinate when handled, sedated, or given anesthesia, it can be challenging to gather a sufficient sample size.

3.6.2.4. Fecal Sample Collection

Since most rodents produce a large number of fecal pellets, collecting a fecal sample is easy. Because the mucosa of the rodent colon is so delicate, it is best to avoid inserting fecal collection loops or devices into the rectum.

3.6.2.5. Miscellaneous Sample Collection

Rodents can be used to safely collect a wide range of samples, such as skin scrapings, hair samples, thoracic fluid samples, fine-needle aspirates, or biopsy samples of external masses, abdominal masses, or cysts. Give careful thought to the kind of restraint (manual vs chemical), differences in anatomy, and the dimensions of the collection tools (syringes and needles). When gathering a diagnostic sample, make use of the least intrusive and compact equipment available.

3.6.3. Other Diagnostic Testing Procedures

Small pet rodents are frequently used for radiology and ultrasound procedures. In summary, ideal film and setting combinations, along with a properly positioned immobile patient, are necessary to produce diagnostic-quality images. Most of the time, this calls for anesthesia or sedation (Bober, 1988).

3.6.4. Hospitalization

Hospitalizations for advanced care or diagnostic testing of rodents are common. There should be no audible or visual cues of possible predators in the hospital room, and it should be quiet. Even if a patient doesn't seem to be able to try to escape, hospital enclosures need to be impenetrable. The majority of rodents can be kept in pet carriers made of clear plastic squares that snap on top. For temperature control, these carriers can be inserted into the majority of commercial pet incubators. Light-colored chapter or towels make the best bedding because they make it simple to identify excrement, urine, and unusual discharges like blood. A hide box is beneficial for hospitalized rodents and can be easily made from chapter or a small cardboard box. When feeding hospitalized animals, keep a variety of rodent foods on hand or ask the pet's owner to provide some before the animal is admitted.

3.6.5. Therapeutics

The use of therapeutic agents in rodents is not authorized and is regarded as off-label, similar to the majority of exotic species. It might be necessary to dilute or compound certain medications if they are too concentrated for small rodents. Depending on the patient's overall condition, there are four different ways to administer fluids: subcutaneous, intravenous, intraosseous, and intraperitoneal. Patients who are unstable and severely dehydrated are best served by intravenous or intraosseous administration. Intraosseous catheterization is a preferable option because most rodent patients have difficulty placing an intravenous catheter. Studies conducted on a variety of species have demonstrated that administering drugs and fluids intraosseously produces a similar peak blood level and time of action as administering them intravenously.

Although needles made especially for intraosseous catheterization are available, basic injection needles (22–27 gauge, depending on patient size) are suitable for the majority of small rodents. The tibia is the site most frequently used, though the femur has also been reported to be used. Since the animal finds the insertion of an intraosseous catheter uncomfortable, either complete anesthesia or sedation with a local block is necessary. To insert a catheter into the tibia, inject 1 to 2 mg/kg of lidocaine into the skin, subcutaneous tissues, and periosteum, then give them time to heal–at least 10 minutes. To prevent the catheter from entering the joint, insert it at the patellar ligament's insertion site, which is located on the tibial crest. The precise angle of placement is determined by the size and shape of the proximal tibia, and radiography in two views provides the best confirmation of correct placement.

Subcutaneous or oral fluid administration can be used to meet the fluid needs of rodent patients who are stable but only mildly dehydrated. In the area of loose skin between the shoulder blades, administer subcutaneous fluids. The majority of parenteral medications are injected intramuscularly into the back leg or epaxial musculature, or subcutaneously into the loose skin above the shoulder blades. The sciatic nerve may be harmed by injections into the lateral leg muscles, which could lead to self-mutilation or lameness. The smallest needle size feasible should be used to enable medication to pass through and penetrate the skin. Because there are alternative sites for administration and there is a chance of harm to abdominal structures or accidental injection into the gastrointestinal tract (particularly the cecum), intraperitoneal administration of therapeutics is not recommended. Medication intravenous administration is rarely indicated, but if done carefully, it is possible in a larger patient who is sedated or anesthetized. Using an intraosseous catheter for injection is an alternative.

For the most part, oral medication administration in rodents is a straightforward process that involves gently restraining the animal and inserting a syringe into its mouth. Since the amount of medication is usually tiny, insulin syringes without a needle enable precise measurement and convenient administration. Compounding drugs with sweet sugar bases frequently results in higher compliance rates. Because rodents have a propensity to cache food and mistrust new tastes, administering medications in food or water can be problematic due to their general inability to control dosage and frequency. Medication added to water may cause one to drink less water. Direct medication administration is recommended for these reasons. Rodents have a high metabolic rate, so hand feeding is crucial for anorectic animals. Using a food supplement that can be administered via a tiny syringe makes hand feeding generally simple. A convenient and nutrient-balanced hand-feeding solution for herbivores. If necessary, strained vegetable baby foods like sweet potatoes or mixed greens can be added to increase palatability (Rivard et al., 2006).

REFERENCES

1. Altmann, D., Schwendenwein, I., & Wagner, K. (1994). Zu Biologie, Haltung, Ernährung und Erkrankungen des Degus (*Octodon degus*). *Erkrankungen der Zootiere, 36*, 277–292.
2. Baker, H. J., Lindsey, J. R., & Weisbroth, S. H. (1979). *The Laboratory Rat. Volume 1: Biology and Diseases.* New York: Academic Press.
3. Barthold, S. W., Bhatt, P. N., & Johnson, E. A. (1987). Further evidence for papovavirus as the probable etiology of transmissible lymphoma of Syrian hamsters. *Laboratory Animal Science, 37*, 283–288.
4. Bauck, L., Orr, J. P., & Lawrence, K. H. (1984). Hyperadrenocorticism in three teddy bear hamsters. *Canadian Veterinary Journal, 25*, 247–250.
5. Becker, C. E., Mathur, C. F., & Rehnberg, B. G. (2010). The effects of chronic exposure to common bedding materials on the metabolic rate and overall health of male CD-1 mice. *Journal of Applied Animal Welfare Science, 13*(1), 46–55.
6. Bober, R. (1988). Technical review: Drawing blood from the tail artery of a rat. *Laboratory Animals, 17*, 33–34.
7. Boorman, G. A., Eutis, S. L., & Elwell, M. R. (1990). *Pathology of the Fischer Rat: Reference and Atlas.* San Diego: Academic Press.
8. Burn, C. C., Peters, A., Day, M. J., et al. (2006). Long-term effects of cage-cleaning frequency and bedding type on laboratory rat health, welfare, and handleability: A cross-laboratory study. *Laboratory Animals, 40*(4), 353–370.
9. Cantrell, C. A., & Padovan, D. (1987). Biology, care, and use in research. In G. L. Van Hoosier & C. W. McPherson (Eds.), *Laboratory Hamsters* (pp. 369–387). Orlando: Academic Press.
10. Capello, V., & Lennox, A. M. (2008). *Radiology of Exotic Companion Mammals.* Ames: Wiley-Blackwell.
11. Capello, V., Gracis, M., & Lennox, A. M. (2005). *Rabbit and Rodent Dentistry Handbook.* Ames: Wiley-Blackwell.
12. Carter, G. R., Whitenack, D. L., & Julius, L. A. (1969). Natural Tyzzer's disease in Mongolian gerbils (*Meriones unguiculatus*). *Laboratory Animal Care, 19*, 648–651.
13. Chevret, P., Denys, C., Jaeger, J., et al. (1993). Molecular evidence that the spiny mouse (*Acomys*) is more closely related to gerbils (Gerbillinae) than to true mice (Murinae). *Proceedings of the National Academy of Sciences of the USA, 90*(8), 3433–3436.
14. Craig, A. M., Karchesy, J. J., Blythe, L. L., et al. (2004). Toxicity studies

on western juniper oil (*Juniperus occidentalis*) and Port-Orford-cedar oil (*Chamaecyparis lawsoniana*) extracts utilizing local lymph node and acute dermal irritation assays. *Toxicology Letters, 154*(3), 217–224.

15. de Matos, R. (2009). Rodents: Therapeutics. In E. Keeble & A. Meredith (Eds.), *BSAVA Manual of Rodents and Ferrets* (pp. 52–62). Gloucester, UK: British Small Animal Veterinary Association.
16. Donnelly, T. M., & Stark, D. M. (1985). Susceptibility of laboratory rats, hamsters, and mice to wound infection with *Staphylococcus aureus*. *American Journal of Veterinary Research, 46*, 2634–2638.
17. Field, K. J., & Sibold, A. L. (1999). *The Laboratory Hamster and Gerbil.* New York: CRC Press.
18. Flynn, B. M., Brown, P. A., & Eckstein, J. M. (1989). Treatment of *Syphacia obvelata* in mice using ivermectin. *Laboratory Animal Science, 39*, 461–463.
19. Foster, H. L., Small, J. D., & Fox, J. G. (1983). *The Mouse in Biomedical Research. Volume III: Normative Biology, Immunology, and Husbandry.* New York: Academic Press.
20. Fox, J. G., Cohen, B. J., & Loew, F. M. (1984). *Laboratory Animal Medicine.* Orlando: Academic Press.
21. Harkness, J. E., & Wagner, J. E. (1995). *The Biology and Medicine of Rabbits and Rodents* (4th ed.). Baltimore: Williams & Wilkins.
22. Harkness, J. E., Turner, P. V., VandeWoude, S., et al. (2010). *Harkness and Wagner's Biology and Medicine of Rabbits and Rodents* (5th ed.). Ames: Wiley-Blackwell.
23. Harvey, R. G., Whitbread, T. J., & Ferrer, L. (1992). Epidermotropic cutaneous T-cell lymphoma (*Mycosis fungoides*) in Syrian hamsters (*Mesocricetus auratus*): A report of six cases and the demonstration of T-cell specificity. *Veterinary Dermatology, 3*, 13–19.
24. Hem, A., Smith, A. J., & Solberg, P. (1998). Saphenous vein puncture for blood sampling of the mouse, rat, hamster, gerbil, guinea pig, ferret, and mink. *Laboratory Animals, 32*, 364–368.
25. Hoffman, R. A., Robinson, P. F., & Magalhaes, H. (1968). *The Golden Hamster: Its Biology and use in Medical Research.* Ames: Iowa State University Press.
26. Johnson-Delaney, C. (2009). Rodents: Biology, husbandry and clinical techniques in more unusual pet species. In E. Keeble & A. Meredith (Eds.), *BSAVA Manual of Rodents and Ferrets* (pp. 96–106). Gloucester, UK: British Small Animal Veterinary Association.
27. Keeble, E. (2009). Rodents: Biology and husbandry. In E. Keeble &

A. Meredith (Eds.), *BSAVA Manual of Rodents and Ferrets* (pp. 1–17). Gloucester, UK: British Small Animal Veterinary Association.

28. Kohn, D. F., Wixson, S. K., White, W. J., et al. (1997). *Anesthesia and Analgesia in Laboratory Animals.* New York: Academic Press.
29. Laber-Laird, K., Swindle, M. M., & Flecknell, P. (1996). *Handbook of Rodent and Rabbit Medicine.* Oxford, UK: Elsevier Science.
30. Lennox, A. M. (2003). Firsthand encounter with monkeypox. *Exotic DVM, 5*(4), 15–17.
31. Lennox, A. M. (2007). Venipuncture in small exotic companion mammals. *NAVC Clinician's Brief, October,* 23–25. Retrieved from: http://www.cliniciansbrief.com/topic/146 (accessed on 6 December 2024).
32. Lennox, A. M. (2008). Intraosseous catheterization of exotic animals. *Journal of Exotic Pet Medicine, 17*(4), 300–306.
33. Masoro, E. J. (2009). Caloric restriction-induced life extension of rats and mice: A critique of proposed mechanisms. *Biochimica et Biophysica Acta, 1790*(10), 1040–1048.
34. Montedonico, S., Godoy, J., Mate, A., et al. (1999). Muscular architecture and manometric image of the gastroesophageal barrier in the rat. *Digestive Diseases and Sciences, 44*(12), 2449–2455.
35. Perry, N. D., Hanson, B., Hobgood, W., et al. (2006). New invasive species in southern Florida: Gambian rat (*Cricetomys gambianus*). *Journal of Mammalogy, 87*(2), 262–264.
36. Pickering, M., & Jones, J. F. X. (2002). The diaphragm: Two physiological muscles in one. *Journal of Anatomy, 201*(4), 305–312.
37. Rivard, A., Simura, K., Mohamme, S., et al. (2006). Rat intubation and ventilation for surgical research. *Journal of Investigative Surgery, 19*(4), 267–274.
38. Saito, M., Nakayama, K., Muto, T., et al. (1982). Effect of gaseous ammonia on *Mycoplasma pulmonis* infection in mice and rats. *Jikken Dobutsu, 31*(3), 203–206.
39. Spenser, E. L. (2004). Compounding, extra-label drug use, and other pharmaceutical quagmires in avian and exotics practice. *Journal of Exotic Pet Medicine, 13*(1), 16–24.
40. Waren, D. W., Kissoon, N., Mattar, A., et al. (1994). Pharmacokinetics from multiple intraosseous and peripheral intravenous site injections in normovolemic and hypovolemic pigs. *Critical Care Medicine, 22*(5), 838–843.
41. Wesche, P. (2009). Rodents: Clinical pathology. In E. Keeble & A.

Meredith (Eds.), *BSAVA Manual of Rodents and Ferrets* (pp. 42–51). Gloucester, UK: British Small Animal Veterinary Association.

42. Whiteside, T. E., Thigpen, J. E., Kissling, G. E., et al. (2010). Endotoxin, coliform, and dust levels in various types of rodent bedding. *Journal of the American Association for Laboratory Animal Science, 49*(2), 184–189.

CHAPTER 4

CANINE AND FELINE VACCINATIONS AND IMMUNOLOGY

Contents

Introduction 134

4.1. Vaccinations and Immunity 134

4.2. Vaccine Characteristics: Efficacy, Onset, and Duration of Immunity For Modified Live, Killed, and Other Vaccine Types 136

4.3. Vaccination Protocols for Shelters 138

4.4. Vaccine Handling, Administration, and Adverse Reactions 141

4.5. Core Vaccines for Shelter Dogs and Cats 143

4.6. Optional Vaccines for Cats or Dogs in a Shelter 153

4.7. Juvenile Animals 155

4.8. Use of Hyperimmune Serum for Disease Prevention 160

4.9. Risk Assessment and Evaluation Of Serologic Immunity for Canine Distemper, Canine Parvovirus, and Feline Panleukopenia 161

4.10. Vaccine Effects on Diagnostic Testing 165

References 169

INTRODUCTION

Vaccination is a crucial part of programs for animal shelters' preventative health management. Strong immunization campaigns, along with stress-reduction and pathogen-exposure minimization techniques in husbandry, help maintain an animal's health or lessen the severity of a clinical illness. This chapter will discuss immunity in relation to both individuals and populations of animals housed in the shelter. It will also address specific issues with immunity pertaining to young animals. We will talk about certain dog and cat vaccinations within the framework of immunization plans created for shelter settings.

4.1. VACCINATIONS AND IMMUNITY

4.1.1. Importance of Immunity and Vaccination

Many major, potentially fatal viral diseases, like feline panleukopenia (parvovirus) (FPV), canine parvovirus (CPV-2), and canine distemper (CDV), are "vaccine preventable" if animals receive effective vaccinations before being exposed. Every animal entering an animal shelter needs to be vaccinated. Vaccines are one component of a foundation for prevention in a shelter environment, where susceptibility to these diseases is common and opportunities for exposure frequently arise at the time of admission or very soon thereafter. It is not possible for vaccinations to shield all animals from disease on their own. Even in cases where every animal receives vaccinations, diseases like the feline respiratory disease complex (upper respiratory infection) and canine respiratory disease complex (kennel cough) will frequently arise and have the potential to be extremely severe (Schultz and Conklin, 1998; Schultz, 1998, 1999). These diseases are not preventable by vaccination. However, immunization lessens the intensity and frequency of clinical symptoms, which is a useful addition to treatment even for these illnesses.

4.1.2. Community Immunity

The researchers estimate that just 50% of all puppies in the United States receive the distemper (CDV), parvovirus (CPV-2), adenovirus (CAV-2), and rabies (core) vaccines, and that less than 25% of kittens receive the panleukopenia (FPV), calicivirus (FCV), herpes virus (FHV-1), and rabies (core) vaccines. The United States is likely the country with the highest vaccination rate in the world for both dogs and cats. The majority of communities must create initiatives to vaccinate a far higher proportion of dogs and cats with the core vaccines.

Most dogs and cats would be permanently immune to all core diseases, with the exception of rabies, if they were to receive only one dose of the modified live virus (MLV) core vaccines after 16 weeks of age. They would, however, have a far higher chance of being protected for at least a few years even in the event of rabies (e.g., three years) than if they received no vaccinations at all. All animal species that are vulnerable to these core diseases would benefit greatly from increased vaccination rates as they would have a significant impact on "herd immunity." Additionally, it would improve public health by preventing rabies.

One crucial method of immunizing more animals in the population than those who are currently receiving vaccinations is through shelter vaccination clinics. Animals who are not vaccinated are typically not clients of veterinary clinics. When pet owners realize how crucial and durable a one-time-only vaccination with the core vaccines is for safeguarding their dogs and cats, even those who do not frequently visit veterinary clinics are frequently willing to have their animals vaccinated. Participating in community immunization clinics can frequently strengthen the bond between people and animals or increase the owner's perception of the animal's value, which may increase the likelihood that the pet will need veterinary care down the road.

4.1.3. Immune Status of Animals Entering Shelters

When animals come into shelters, they are either (a) immune naïve and can get infected or develop a disease if they are exposed to pathogens; (b) already immune due to natural immunization (e.g., recuperation from illness or infection) or prior immunization; or (c) already afflicted, manifesting symptoms or potentially spreading the infection to other vulnerable animals by shedding infectious organisms. When animals enter shelters, a significant proportion of them lack immunity to vaccine-preventable diseases. The researchers discovered that roughly half (± 20%) of dogs in serologic surveys entering shelters across the country are immunologically naïve (antibody negative) for CDV, and roughly 30% (± 15%) of dogs are immunologically naïve for CPV-2. Based on more restricted studies of cats entering shelters, the researchers discovered that roughly 50% of cats are FPV naïve, while more than 75% of cats have antibodies to FCV and FHV-1, many from prior vaccinations and some from natural infection (Schultz et al., 2007). Since many of the cats lacked the FPV antibody, it is evident that many of them acquired the FCV/FHV-1 antibody as a result of a spontaneous infection. Antibody responses against all three viruses should have developed if they had received the vaccination.

4.2. VACCINE CHARACTERISTICS: EFFICACY, ONSET, AND DURATION OF IMMUNITY FOR MODIFIED LIVE, KILLED, AND OTHER VACCINE TYPES

4.2.1. Vaccine Efficacy

It is never possible to anticipate that a vaccine will result in a more robust or long-lasting protective immunity than what happens when an individual is naturally exposed to an infection (virulent disease developing or not). Generally speaking, corresponding vaccine products will affect the immune system similarly for viral diseases like CDV, CPV-2, and FPV for which exposure to the virus and recovery from it confer lifetime immunity. Regarding illnesses that don't produce permanent or full immunity (e.g., feline herpesvirus, *Bordetella bronchiseptica*), or cross-protection for different strains (e.g., feline calicivirus), the matching vaccinations will probably, at most, provide a similar degree of restricted immunity (Schultz, 1999). The vaccinations that prevent CDV, CPV-2, and CAV-1 (infectious canine hepatitis) in dogs as well as the FPV vaccine (which prevents panleukopenia in cats) are the most effective ones. When given in the best possible conditions, these vaccinations totally prevent infection and clinical disease, for example, both prior to before exposure and in compliance with the manufacturer's guidelines.

On the other hand, the feline respiratory disease complex (FRDC) cannot be completely prevented by vaccination. Environmental variables, stress, and several infectious agents–including some bacterial pathogens and mycoplasma that are not covered by vaccines–combine to cause this syndrome. Furthermore, as previously mentioned, there is a limited natural immune response against these pathogens. Finally, pathogens linked to FRDC, such as calicivirus and herpesvirus, are frequently present in cats that are brought to shelters already infected. Similarly, this multifactorial disease cannot be prevented by vaccinations intended to help prevent canine respiratory disease complex (CRDC). Immunizations should assist in lessening the intensity or length of illness in certain animals, but not all of them, when combined with appropriate management techniques, such as maintaining an eye on the environment and animal husbandry.

4.2.2. Onset of Immunity for Modified Live Versus Killed Vaccines

Rapid onset of immunity is crucial to maintaining each animal's health and establishing a sufficient level of immunity within the group to prevent outbreaks, as many animals entering shelters are susceptible to the "core" viral diseases and the risk of exposure is very high. In the absence of maternally derived antibodies (MDAs), parenterally administered MLV "core" vaccines offer quick immunity against CDV, CPV-2, and FPV with a single dose. Immunity to CDV vaccines, including MLV and recombinant viral vector vaccines, develops hours after vaccination. After receiving one dose of the MLV vaccine, 98–99% of dogs without MDA who were challenged experimentally within three days showed immunity against CPV-2. In a previous study, cats that received MLV parenteral vaccination almost immediately afterward showed immunity to FPV. As opposed to an injection-based experimental challenge, the exposure dose in this instance was lower because the cats were reintroduced into a contaminated environment after receiving their vaccination. This type of exposure more likely mimics a shelter setting (Brun et al., 1979; Carmichael et al., 1983).

Generally speaking, killed viral vaccines for CDV, CPV-2, and FPV shouldn't be used in shelter settings because protective immunity to these vaccines develops much more slowly than it does for MLV vaccines. The majority of killed vaccines, such as those for FPV and canine parvovirus, need to be administered at least twice to induce a protective immune response. Two to four weeks after the first dose, which primes the immune system, comes the second dose, which confers immunity. Protection prompted by the killed vaccine typically takes a week or more to manifest itself after the second dosage. Furthermore, it is frequently necessary to restart the series for most killed vaccines if the second dose is not administered within a maximum of six weeks of the first dose, as the priming effect may be lost and the second dose will not immunize. Every dose of killed vaccines after that should strengthen the immune system. Most animals are protected with a single injection for rabies virus vaccines, which are lethal and exempt from the need for two doses spaced six weeks apart. Nonetheless, it is highly advised and typically mandated by law that all animals receive a second dose of the rabies vaccine one year following the first.

4.2.3. Intranasal Modified Live Vaccines

All intranasal vaccines are modified live products that may provide local immunity more quickly than parenteral vaccines. They are commonly used to help prevent "upper respiratory infection" (FRDC) in cats and "kennel cough" (CRDC) in dogs. Since maternal antibody interference is avoided when these vaccines are given at the mucosal surface, puppies, and kittens may benefit from them at a young age. Panleukopenia immunity is not included in the early age at which intranasal vaccination is expected to be effective. It is advised to receive parenteral vaccines for panleukopenia.

The "kennel cough" vaccines with modified live *Bordetella bronchiseptica* and canine parainfluenza (CPI), with or without canine adenovirus-2, will stimulate nonspecific immunity within hours after administration, due primarily to the adjuvant effect of *B. bronchiseptica*. A "shower of cytokines" is frequently released from respiratory tract cells by bacteria such as Bordetella. The Type I interferons (IFN), such as alpha and beta IFN, would be one of the cytokines released. Certain viruses can have their replication suppressed by interferons, offering broad-spectrum, immediate protection against viruses that cause CRDC. Within three to seven days, the respiratory tract will develop specific humoral and cellular immunity, resulting in upper respiratory tract local immunity. Therefore, it is advised to use intranasal products to protect against respiratory pathogens, particularly in animals with naïve immune systems. Anamnestic (secondary) immunity should develop quickly in animals that have already been naturally immunized due to infection with the organisms causing CRDC and FRDC, which is likely to occur frequently in adult animals entering a shelter. This immunity is acquired through intranasal or parenteral vaccination. This is most likely the reason the researchers found it challenging, if not impossible, to show a difference in the protection given to newly adopted shelter animals receiving a parenteral vaccine (either whole cell killed or antigen extract) versus an intranasal CRDC vaccine or an MLV parenteral vaccine versus an intranasal FRDC vaccine. In those investigations, we were unable to demonstrate a difference in the course or intensity of the illness based on the kind of vaccination administered (Newbury et al., 2007).

4.2.4. Duration of Immunity

Since immunity far outlasts the typical length of shelter stay, most shelters do not find it necessary to consider immunity duration once it has been established. Years of protection are offered by the canine core (CDV, CPV-2, and CAV-2) and feline core (FPV, FCV, and FHV-1) MLV vaccines following one or two vaccination doses in an animal older than 16 weeks or following the completion of the initial kitten or puppy series at 16 weeks or older (Abdelmagid et al., 2004; Mouzin et al., 2004a, 2004b; Scott and Geisinger, 1999; Schultz 2006).

4.3. VACCINATION PROTOCOLS FOR SHELTERS

4.3.1. Core Vaccines for Shelter Cats and Dogs

Both dogs and cats can receive a variety of immunizations. But out of all the vaccines available, only a select handful are deemed significant enough to warrant being given to every dog and cat as soon as they enter a shelter. The term "core vaccines" refers to the set of vaccinations that are recommended for all cats and dogs. "Vaccinations against feline parvovirus (panleukopenia), feline calicivirus (FCV), and feline herpesvirus type 1 (FHV-1) should be administered to all cats entering the shelter. CDV, CPV-2, CAV-2, CPI, and *Bordetella bronchiseptica* should all be vaccinated against in dogs.

Vaccines needed to stop or lessen the severity of viral-caused disease are frequently found in combination products. You can find the canine *Bordetella bronchiseptica* vaccine either as a modified live intranasal product or as a killed parenteral product, or you can find it in combination with MLV CPI with or without MLV CAV-2.

4.3.2. Addition of Noncore Components to Core Vaccines

The use of products containing any other vaccine components that might lead to the immune system developing a less effective response to the core pathogens should be avoided due to the type of immune response induced by MLV vaccines for the core viruses and the requirement for a rapid onset of immunity in the face of almost certain exposure. To provide the fastest onset and most effective long-term immunity, specific immune system cell types and components should be stimulated at the time of vaccination. Only MLV core vaccines should be included in combination vaccine products intended to prevent or lessen the severity of disease in shelters. Since rabies vaccines are typically monovalent, they are not combined with other essential immunizations. Agents like canine coronavirus, feline *Bordetella bronchiseptica*, *Leptospira serovars*, Lyme vaccine, and *Chlamydophila felis* shouldn't be regularly included in "core combination vaccines" for shelter animals. In the event that a disease is diagnosed and confirmed, the short-term use of a noncore vaccine may be appropriate for known outbreaks of feline Bordetella or chlamydophila. Noncore vaccines should be administered separately (ideally at a different time or location in the animal) whenever it is determined that doing so will be advantageous or necessary. The final user should never mix vaccines in a way that the vaccine manufacturer did not intend.

4.3.3. Timing of Vaccination: Vaccination on Entry Is Essential

The evidently high degree of viral pathogen susceptibility that dogs and cats have, which is likely to exist in animal shelters, emphasizes, and validates the urgent need for vaccination-provided protection. The feline vaccination guidelines of the American Association of Feline Practitioners (AAFP) and the canine vaccination guidelines of the American Animal Hospital Association (AAHA) both recommend vaccination on intake for animals entering animal shelters, and this should be regarded as the standard of care for shelters. The timing of a vaccine is crucial because it can affect whether or not susceptible (naïve) animals develop immunity. Hours can have this effect. The aim is vaccination before exposure.

The most crucial action that can be taken to save lives is to vaccinate susceptible animals; immunizing immune animals offers no advantages. A vaccination won't make the illness worse, but it also won't help an animal that is already vaccinated. Given that shelter animals most likely lack a verified vaccination history, immunizing every animal contributes to both population protection (herd immunity) and the development of a safety net for each individual animal. Particulars about revaccination for adults and juveniles are provided below and in the sections on individual core pathogens.

4.3.4. Special Considerations for Vaccination

4.3.4.1. Vaccination of Puppies and Kittens

Animals that are younger than four months of age are not as well-suited to receive vaccinations as adults are. From a very young age (e.g., less than two to four weeks of age), modified live vaccines cannot be administered to animals without risk because of their immature immune systems. Maternal antibody interference may impede effective immunization from four to about 16 weeks of age. Puppies and kittens in shelters should generally receive their core parenteral MLV vaccinations between four and six weeks of age, and intranasal respiratory vaccines labeled for this use as early as two weeks of age. Until they are at least 16 weeks old, juvenile animals in shelters should receive core parenteral vaccinations every two weeks. Furthermore, throughout their stay in a shelter, young animals should be physically separated to prevent them from coming into contact with pathogens. More information about immunization and protection of puppies and kittens, along with the reasoning behind these recommendations, can be found in the section on juvenile animals. Each main section on vaccines includes a description of specific tactics or issues, where relevant.

4.3.4.2. Vaccination of Sick or Injured Animals

Regardless of their health at the time of arrival, all animals older than four to five weeks should receive a vaccination before or at the time of arrival at a shelter. Immunizations are not expected to be harmful, but living in a shelter increases the risk of contracting a virulent virus. Animals shouldn't be kept in shelter settings if their conditions are so bad that vaccinations are deemed risky.

It is not likely, but possible, that an arriving sick animal won't be able to mount an immune response. The vaccination is likely to offer much-needed protection, and it is unlikely to have a negative effect on the animal. After recovering, these animals might receive another vaccination as an added precaution (and no less than two weeks after the previous vaccine).

4.3.4.3. Vaccination of Pregnant and Nursing Animals

The advantages and disadvantages of vaccination must be carefully considered before accepting a pregnant animal into a shelter. Fetuses may be at risk from vaccine viruses, but as animals enter the shelter, the risk from virulent viruses typically increases. The risk to the unvaccinated dam when she enters a shelter setting, followed by an almost certain exposure, could be very high, even though vaccination of the dam (if she has not been immunized previously) may present a risk to the unborn puppies or kittens. The mother and the puppies or kittens may all die from illness in an attempt to shield the fetuses from the harmful effects of vaccination.

Choosing to vaccinate a pregnant animal may depend on the animal's legal situation. It is necessary to balance the benefits and risks. To determine the mother's risk, every effort should be made to secure vaccination records or permission before impounded pregnant animals are used in court. One could argue that immunizing the mother with an unknown vaccination history is a reasonable course of action if shelter admission is unavoidable, even in cases where animals have an unclear legal status. One method of assessing risk before vaccination would be to measure the levels of antibodies in pregnant animals. Dogs who are pregnant and carry the antibody will not contract the vaccine virus, protecting the embryos and fetuses from harm. But when a product is a combination, one of the antigens (e.g., CAV-2) that make it difficult to assess titers may infect embryos and fetuses and result in absorption or abortion.

When using a combination product, one must always weigh the advantages and possible drawbacks of the individual vaccines as well as the overall immediate risk from the particular virulent virus in the shelter. This is particularly valid when you're pregnant. In the event that pregnant animals are not vaccinated, every precaution must be taken to keep them physically safe from exposure, preferably by housing them off-site or providing careful isolation. Every pregnant animal that is scheduled for a spay or abortion should receive a vaccination as soon as they are brought in. Core vaccinations should be given to nursing animals as normal. While the offspring will not be protected in any way by this, the mother will be protected and the offspring will not suffer any harm.

4.4. VACCINE HANDLING, ADMINISTRATION, AND ADVERSE REACTIONS

4.4.1. Care and Handling of Vaccines

MLV vaccines should make up the majority of the core vaccinations administered in shelters. MLV vaccinations are known as infectious vaccines because the animal must become infected with the live vaccine virus or bacteria in order for the vaccine to immunize. Conversely, the killed (K) vaccines are referred to as noninfectious vaccines since they are not contagious. MLV vaccinations need to remain infectious, so managing the shots to keep them that way is essential to the immunization regimen. Prior to use, MLV vaccines should always be kept refrigerated (e.g., between 1□ and 3□, or 33 to 35°F). They must stay lyophilized (a dried cake) for long-term storage; do not reconstitute them with diluents. An MLV vaccine should never be frozen, either before or after it has been reconstituted using sterile diluent. The vaccine should only be kept in the refrigerator for a maximum of four days after it has been reconstitution. The vaccines should not be used after two to four hours at ambient (room) temperatures between 70°F (21□) and above 80°F (26□). Single-dose vials are advised because multidose vials carry a risk of contamination from bacteria that can grow at refrigerator temperatures. Ideally, vaccines should be reconstituted and used within the first few hours. Reconstituted live modified vaccines should only be made using the sterile diluent that comes with the vaccine, if at all possible. If the diluent contains an unneeded or undesirable component (e.g., leptospira), an alternate sterile diluent (e.g., water or saline) can be substituted for the diluent supplied with the vaccine.

Almost all killed vaccines are sold as liquids rather than lyophilized, as are MLV vaccines. Killed vaccines are more stable than MLV vaccines and can be refrigerated to extend the product's shelf life. However, once taken out of the refrigerator, killed vaccines should also be used within four to six hours, particularly in warm weather (above 80°F/26□). Again, single-dose vaccine vials are preferable because multiple-dose vials have the potential to become contaminated. Since some killed vaccines contain preservatives that can inactivate the MLV vaccine, killed vaccines should never be combined with MLV vaccines unless they are a part of a commercial combination product. It is best to avoid combining any kind of vaccine in one syringe with another. It is not recommended to use MLV or KV vaccines after the product's expiration date.

4.4.2. Vaccination Site Guidelines for Cats

It is advisable to give vaccinations to cats following the AAFP site guidelines, if at all possible. These site guidelines assist in tracking possible causes of vaccine-associated sarcomas and provide individual animals that may develop these invasive tumors with better treatment options, including the possibility of tumor removal. The right forelimb should receive FVRCP, the right hindlimb should receive rabies, and the left hindlimb should receive FeLV. It is no longer advised to inject in the scapular region because it is rarely possible to remove an invasive tumor from this location. Even in cases where adherence to these guidelines isn't always possible, the advantages of administering a parenteral vaccination subcutaneously at any suitable site far exceed the danger of developing sarcoma.

4.4.3. Adverse Reactions

Adverse reactions to all vaccines can range in severity from minor (e.g., stiffness, sluggishness) to extreme (e.g., cancer, allergic reaction, autoimmunity, and demise). While killed adjuvanted vaccines are among the most reactogenic, the MLV core vaccines that are advised for shelter animals are among the least likely to result in adverse reactions. Furthermore, the risk of an adverse vaccine reaction would never outweigh or negate the benefits of vaccination with the core vaccines, given the large number of animals in shelters that are susceptible to serious disease and death caused by CDV, CPV-2, and FPV. As a result, each animal that enters a shelter needs to receive the necessary vaccinations either before or right away. It would be best to keep an animal out of a shelter where it could be exposed to more virulent wild-type agents if it is determined that its immunosuppression is so severe that it should not receive the MLV vaccine. If housing and shelter entry are unavoidable, vaccination probably carries less risk than admission without immunization. Veterinarians should give written protocols and specialized training to the shelter staff who will administer vaccine products, explaining how to handle adverse reactions if they arise, even though they are uncommon when using only the core vaccine products. Adverse reactions must be thoroughly recorded in the animal's long-term medical file, and prospective adopters must be informed when a reaction takes place.

4.5. CORE VACCINES FOR SHELTER DOGS AND CATS

The following sections address each component of the core vaccines individually to describe the expected onset, duration, and efficacy of immunity.

4.5.1. The Canine Shelter Core Vaccines

4.5.1.1. Canine Distemper Virus Vaccine

If MDAs are not present, immunity to CDV has been shown to develop almost immediately following a single vaccination with an MLV or recombinant product. This immunity has been shown to shield the animal against serious clinical symptoms of CDV, such as neurologic illness and death. However, it is possible for animals exposed to CDV in the first few days after vaccination to contract the virus, exhibiting minimal or no clinical symptoms, and shedding the virus into the surrounding environment. Dogs with these less severe infections may also have immunosuppressive symptoms, such as pneumonia, and may exhibit other clinical indications of canine reproductive disease. To fully prevent infection and virus transmission, animals should ideally receive their vaccinations two to three days before exposure, although substantial benefits and protection against a challenge have been shown as early as minutes to hours after vaccination (Appel 1987, 1999; Larson and Schultz, 2006).

Puppies that are not able to receive an effective vaccination upon intake due to MDAs are likely to exhibit clinical symptoms that eventually lead to neurologic disease or even death when CDV transmission is a serious issue in shelters that already offer vaccinations at admission. In these cases, adults may develop mild to severe respiratory disease initially, but they may also develop severe neurologic disease. Throughout their time in shelters, puppies younger than five months old need to be kept safe from exposure through physical separation and cautious handling, particularly during outbreaks. Dogs do not need to be vaccinated against CDV more frequently than every three years after finishing the puppy series at 16 weeks of age or older, as immunity to the virus can last for the duration of the dog's life.

Safer than modified live viral vaccines, the canarypox vectored recombinant vaccine offers immunity comparable to that of the former. It can be administered to very young puppies, ferrets, wildlife, and exotic animals that are prone to contracting virulent CDV (e.g., wolves, pandas, foxes, coyotes, big cats, etc.) Puppies with MDAs benefit from the recombinant CDV (rCDV) vaccine, which can immunize them up to four weeks ahead of time compared to MLV CDV vaccines. PureVax Ferret CDV is the name of the monovalent rCDV vaccine designed specifically for ferrets. As early as two weeks of age, very young puppies who are at high risk of contracting CDV can receive the monovalent product rather than a combination product. You can obtain up to three doses per 1 ml vial by using just 0.3–0.5 ml of the 1 ml PureVax Ferret CDV. This product ought to be administered every two weeks until a combination product containing MLV CPV-2, rCDV, and additional viruses (e.g., CAV-2, and CPI) can begin as early as five or six weeks. Both the MLV and rCDV vaccines are very effective when given to dogs at the time of intake or before infection, and they will work similarly in older dogs. Among all vaccines, the canine distemper vaccine is one of the most effective. Even in a shelter where there is a severe CDV outbreak, vaccinated dogs with detectable antibodies will be able to fend off reinfections and illness.

4.5.2. Canine Parvovirus Type 2 Vaccine

Three or more days following a single MLV vaccination, dogs can exhibit immunity to CPV that lasts for the duration of the dog's life in the absence of MDAs. However, young dogs are susceptible to infection and fulminant clinical disease if they are exposed to the virus before or within three to five days following vaccination. When compared to dogs under a year old, older susceptible dogs have a lower chance of developing a serious illness. It is crucial to shield all dogs from fomite transmission or exposure to parvo-contaminated environments during this early stage of immunity development because the environmental persistence of CPV-2 increases the likelihood of exposure.

Regardless of the number of vaccinations received, all puppies under 16 weeks of age should be regarded as susceptible, even though some are probably adequately protected from MDAs or vaccination. The parvoviral vaccine does not cause any early protection before immunity develops. Unlike the early partial immunity brought on by CDV vaccinations, parvoviral immunity is either total or nonexistent. When adult dogs are immunologically naïve to CPV-2, they can contract the virus and shed it without displaying any symptoms. When a young, vulnerable dog contracts the virulent virus, it can lead to serious illness or even death.

Similar to CDV, CPV-2 vaccinations offer lifetime immunity, so dogs do not require booster shots more frequently than every three years following the completion of the puppy series at 16 weeks of age or older, and booster shots one year later.

Current CPV-2 vaccines from the major manufacturers have been demonstrated by researchers to induce protective immunity for all current field types of CPV found clinically. CPV-2c has been introduced (2005–2006) to the U.S., probably from Europe, as it was detected there in 2001. No vaccine-resistant variants have been found (Quesada et al., 2007).

4.5.2.1. Vaccination of Puppies for Canine Parvovirus

Regarding the ability of the current CPV-2 vaccines to immunize puppies with MDAs at or less than 16 weeks of age, there are still a few small but noteworthy differences. Field observations and experimental studies indicate that there may be some CPV-2 vaccines that are more effective for puppies living in a shelter or commercial colony where CPV-2 is a major disease problem, even though the differences are difficult to prove.

Neopar is a CPV-2 product that has done exceptionally well in high-risk kennels or colonies and is not produced by one of the major veterinary biological manufacturers. Neotech is a Tennessee-based company that manufactures this product. Dog breeders have been using this well-known vaccine for many years. According to comparative studies of CPV-2 vaccines, it is still one of the best products for immunizing puppies at a younger age than many other vaccines (Larson and Schultz, 1997).

To guarantee protection against the other core diseases, an additional vaccination is required when using a monovalent product like Neopar. It is imperative to provide staff education, training, and close monitoring of compliance when implementing this or any other vaccination protocol. Shelters ought to carefully assess their staff members' capacity to adhere to any vaccination regimen that complicates the fundamental core vaccinations recommended at intake. This is covered in Section 68 of Principles of Disease Management. Adult dogs and cats do not require the use of special programs designed for puppies and kittens.

The monovalent canarypox (from Merial), the monovalent Progard CPV-2 (from Intervet), or the monovalent Neopar (from Neotech) can be administered to all puppies beginning at four weeks of age and again every two weeks until the puppies are at least 10 to 12 weeks old when both CDV and CPV-2 outbreaks are occurring concurrently. Subsequently, up to and including 16 weeks of age, a combination core vaccine product containing CDV, CPV-2, and CAV-2 (with or without CPI) should be administered at 2-week intervals. The earliest and most effective vaccination protection against CDV and CPV-2 will be offered by this vaccination schedule. Despite following this vaccination schedule, certain puppies may still contract CDV and/or CPV-2, which can even be fatal because these pathogens particularly parvovirus are frequently found in most shelter settings.

4.5.2.2. Canine Adenovirus Type 2 Vaccine

Infectious canine hepatitis, or CAV-1 virus, is a necessary vaccination for shelter dogs. Immunity can be shown when the dog is challenged with the CAV-1 virus (infectious canine hepatitis) as early as seven days following vaccination. It most likely requires a comparable length of time (i.e., seven days) for immunity to grow and to shield the dog from CAV-2-related respiratory illness. According to Greene and Schultz (2006), CAV-2 alone can cause pneumonia in dogs and contributes to CRDC. The duration of immunity (DOI) for protection from CAV-2, one of the numerous factors that cause CRDC, is roughly three years, but the DOI against CAV-1 is lifetime (Schultz, 2006). A combination intranasal CRDC vaccine that contains CAV-2 in addition to the core parenteral combination viral vaccine may be a useful adjunct to control the problem if significant CRDC is an issue in a shelter, especially if CAV-2 is discovered on diagnostic samples.

4.5.2.3. Canine Parainfluenza Virus Vaccine

In shelter dogs, the canine parainfluenza (CPI) virus is a necessary vaccination. For effective protection against this virus that causes CRDC, intranasal administration of the CPI vaccine is recommended. In the case of CPI, immunity lasts for three years. CPI is frequently included in intranasal CRDC vaccinations, either with or without CAV-2, that contain *Bordetella bronchiseptica*. Given that the *Bordetella bronchiseptica* component only confers immunity for a year or less, the combination intranasal vaccinations need to be administered at least once a year.

4.5.3. Bordetella Bronchiseptica Vaccine

For dogs going into shelters, *Bordetella bronchiseptica* is regarded as a necessary vaccination; however, it is not a necessary vaccination for cats. For dogs, there are parenteral and intranasal vaccinations available, but only an intranasal vaccine is available for cats. MLV intranasal vaccinations, such as *Bordetella bronchiseptica* alone or in combination with CPI, with or without CAV-2, are advised for use in shelters. Every intranasal product that is presently offered is changed in real time.

Just as the canine intranasal Bordetella vaccine has only been tested for safety and efficacy in the appropriate species, the feline monovalent intranasal vaccine should not be administered to cats. For dogs, there is a monovalent injectable vaccine called *Bordetella bronchiseptica* cellular antigen extract. Since the safety and effectiveness of this injectable canine product have not been studied in cats, it should not be used on cats.

When it comes to *Bordetella bronchiseptica* vaccinations, care must be taken. The parenteral product should never be administered intravenously as it will not offer specific immunity against Bordetella. More importantly, parenteral administration of intranasal products containing MLV Bordetella is contraindicated due to the possibility of a severe local reaction or, in rare cases, death from severe acute hepatic failure. In the event that an injection of the intranasal Bordetella vaccine is accidentally administered, it ought to be treated as an emergency. The injection site may experience swelling and pain as a result of a localized inflammatory response. The ASPCA Poison Control Center advises the following course of action: injectable gentamicin sulfate should be diluted in 10 to 30 ml of saline (depending on the size of the dog) and injected into the affected area at a standard dose of 2 to four mg/kg q six to eight hours. Additionally, an oral antibiotic should be started right away, such as tetracycline, trimethoprim sulfa, or doxycycline. The majority of animals only experience pain and swelling at the injection site, but some have also been known to experience hepatic necrosis or injection site abscesses. Animals that vomit, have diarrhea, or become uninterested should be closely observed. Hospitalization, intravenous (IV) fluids, and other supportive care may be necessary if liver disease develops. Liver enzyme abnormalities may persist for several months if liver damage does occur. There have been some cases of dog deaths.

Canine respiratory vaccines administered intranasally are MLV products that are administered at the mucous membrane, where MDA interference is not present. As a result, these vaccines can be given to dogs at a young age (i.e., four to eight weeks) without the need for a second vaccination. Because it might be challenging to inject the vaccine into the puppy's nose, package inserts frequently advise revaccining puppies two to four weeks after the initial intranasal vaccination. It is not necessary immunologically, but a second dose might help protect puppies who were not vaccinated due to problems with vaccine delivery. Revaccination would not be helpful because most puppies kept in shelters would have been exposed to the pathogens that the intranasal products protect against naturally after two to four weeks. Adult dogs can receive intranasal products once during intake; additional administration is not necessary.

As immunity to *Bordetella bronchiseptica* wanes, revaccination may be required in privately owned animals as frequently as every nine months to a year. Since natural exposure to Bordetella occurs frequently in shelter settings, immunity is likely to last for most animals at least the average or standard length of stay. Revaccination with intranasal vaccinations every year or every nine months may be beneficial for animals that have not had much exposure to recently arrived dogs, such as dogs in quarantine or legal holding. When using the killed parenteral *Bordetella bronchiseptica* product, two doses spaced two to four weeks apart must be administered. The time between those doses shouldn't be longer than six weeks, just like with other killed products (Appel and Bemis, 1978).

According to recently released studies, intranasal products significantly outperform parenteral products in a research setting when it comes to reducing clinical signs after a challenge. That study actually demonstrated that the dogs in the injectable product vaccination group did not develop immunity; instead, they responded to a challenge in a manner similar to those of the dogs given a placebo (saline).

4.5.4. The Feline Shelter Core Vaccines

4.5.4.1. Feline Panleukopenia Virus Vaccine

The feline panleukopenia (parvovirus) vaccine offers quick and durable immunity, much like the CPV vaccines. One study showed that immunization against FPV confers immediate protective benefits on susceptible kittens. When susceptible kittens were given a parenteral MLV FPV vaccination and then placed in a contaminated environment right away, they showed no clinical symptoms of illness, but unvaccinated kittens suffered serious side effects.

Because the researchers have shown that kittens challenged experimentally by the intranasal/oral route must receive their vaccination at least three days before challenge in order for protection to occur, these results most likely represent delayed infection (Schultz, unpublished). The results of FPV challenge infection with the closely related canine CPV-2 virus are comparable to those of vaccination. Similar to canine parvovirus, kittens in shelters are susceptible to FPV infection. As such, they should receive their first vaccination no earlier than four weeks of age, and then receive booster shots every two weeks (or more frequently) until they are 16 weeks old. At 16 weeks of age or older, the final vaccination dose needs to be administered. Cats who receive their final dose of vaccination after 16 weeks of age and who are vaccinated again a year later do not require booster shots more frequently than every three years.

The researchers advise against using killed or intraperitoneally delivered MLV FPV vaccinations in shelter settings because they are not as effective at immunizing cats as parenteral (e.g., MLV vaccinations (intramuscular or subcutaneous). All of these vaccinations have been shown to eventually produce strong protection against FPV, but the rate at which protection begins to take effect after using the killed vaccine may not be quick enough to stop outbreaks in a shelter environment. It was discovered that fewer cats, particularly those with MDAs, are immunized by the intranasal vaccine. When the vaccine is administered intranasally, there might not be enough virus to replicate in susceptible cells. For MLV vaccines to induce immunity, the animal must become infected and replicate.

CPV-2 can cause a panleukopenia like disease in cats, but MLV parenteral FPV vaccination of cats will prevent infection with both FPV and CPV-2 (Truyen, Evermann et al., 1996; Olsen, Larson, Schultz, 1998).

4.5.4.2. Feline Calicivirus Vaccines

After receiving an FCV vaccination, immunity can begin to develop as soon as seven days later and last for at least three years. However, immunity alone cannot completely protect against infection or disease, especially in high-risk settings like shelters. Due to the wide variety of calicivirus strains and/or variants, immunity is never fully guaranteed. The FCV vaccine does not prevent infection, in contrast to protection from FPV, which does; however, it is hoped that it will prevent or lessen the severity of disease. Regretfully, vaccination offers only patchy protection against a growing number of FCV strains (vaccine-resistant strains). The diverse clinical manifestations of fluorococcal virus (FCV) are caused by the same mutational tendency that also leads to the constant emergence of vaccine-resistant strains. In fact, the majority of current FCV vaccines contain strains that have not been updated in years.

It is wise to assume that even vaccinated cats will be susceptible to infection and potentially severe respiratory disease from some strains of calicivirus, especially in the shelter environment, even though new vaccine strategies, such as polyvalent vaccines, are being investigated. Vaccination alone will not prevent FRDC, so shelters must implement specific management practices to help lessen its severity. When adopting or caring for foster cats from shelters who exhibit symptoms of upper respiratory infections (URIs), owners of pets should also exercise caution when isolating them. This is especially true for any outbreak that affects otherwise healthy adult cats who have received their vaccinations.

A product to immunize cats against the virulent systemic feline calicivirus (VSFCV) disease has recently become available. Since this product is a killed vaccine, its effectiveness in a shelter setting is unlikely due to the requirement for multiple doses and the protracted period of time it takes for immunity to develop. Furthermore, VSFCV is a disease that has emerged in unrelated regions as a result of isolated mutations in various respiratory calicivirus strains. Rather than being the product of regional spread, the pathogen in every outbreak in which VSFCV has been identified to date has either evolved into a new strain or undergone a mutation. There is only one strain of VSFCV in the VSFCV vaccine. There is no reason to think that the single VSFCV strain included in the vaccine would be protective for other novel mutations that might emerge in shelters where cats are infected with multiple strains of FCV, given the variation among strains and the low chance of cross-protection. Additionally, the combination product with the VSFCV vaccine also contains killed vaccines for the other components (FPV, FVR, and FCV). As a result, this product would not meet the requirements for a quick immune response, particularly against the FPV virus. Administrators of shelters should administer an MLV FPV in addition to the VSFCV vaccine if they choose to use FPV in conjunction with it to prevent panleukopenia, a serious and widespread illness that will kill a significant number of susceptible cats. The illness known as VSFCV has not been frequently confirmed to date (Camero, Cavalli et al., 2004).

4.5.4.3. Feline Herpesvirus Type 1 Vaccines

Herpesvirus infection in cats was also formerly known by the term feline viral rhinotracheitis (FVR). After vaccination, immunity to FHV-1 can occur as soon as seven days later, and it lasts for at least three years. However, immunity is never total since field virus infection can still affect any animal that has received vaccinations. When this happens, almost all cats develop latent infections, while some may exhibit clinical symptoms. This is not surprising since cats that have already contracted FHV-1 can re-infect themselves when they come into contact with the challenge virus, even after a natural infection. However, vaccinated cats have a higher chance of avoiding serious illnesses than unvaccinated ones (Lappin et al., 2006; Scott and Geisinger, 1999).

Stress can reactivate latently infected cat FHV-1 virus, shedding the virulent virus to both vaccinated and unshielded animals. Regardless of vaccination status, recrudescence of the disease in stressed animals can result from FHV-1 shedding and reactivation in latently infected cats. When unvaccinated kittens and cats share an environment, there is a significant risk of FHV-1 shedding.

4.5.4.4. Intranasal Combination Products for Cats

There are intranasal combination vaccines (with or without FPV) that contain both FCV and FHV-1. Both shelter cats and pet cats frequently use these products. When administered to naïve cats and kittens, these vaccines have the advantage of producing local immunity more quickly than the MLV parenteral products. Additionally, they promote immunity against FCV and FHV-1 at a younger age in kittens with MDAs. In terms of FPV, these intranasal vaccinations offer no advantages. Even when the three-way intranasal product is used, young, and adult cats in shelters should also receive a parenteral MLV FPV, as advised in the FPV section. Use of the MLV FCV/FVH-1/FPV parenteral vaccine alone has been shown to be as effective as a combination of both MLV parenteral and intranasal vaccines for protection against respiratory disease in a shelter setting, even though the intranasal route of administration is theoretically optimal for FCV and FHV-1 vaccine viruses because it provides better local cellular and humoral immunity (Newbury et al., 2007).

4.5.5. Rabies Virus Vaccines for Dogs and Cats

The rabies virus vaccine is not regarded as a necessary vaccination for animals in shelters. For cats and dogs in long-term care facilities as well as those in short-term care facilities, rabies vaccination is advised at the time of intake or soon after adoption. The adjuvanted, killed rabies vaccines can induce immunity, which can take up to three weeks to manifest. According to challenge studies, immunity lasts for at least three years after the majority of vaccinations and up to four years after receiving one feline vaccine. It is permissible to administer the rabies vaccine at the time of adoption (or when a veterinarian is available, if necessary), even if the animal was vaccinated on entry with the core vaccines less than two weeks prior, even though core vaccinations shouldn't be given at intervals of less than two weeks.

Regional laws differ, but where allowed, authorized personnel should administer vaccinations under the direct or indirect supervision of a veterinarian. In the U.S., the only rabies vaccine for dogs is the killed vaccine. This product is highly immunogenic due to a combination of adjuvant and high immunogenicity of rabies glycoprotein G. Cats can also receive a one-year recombinant viral vectored feline rabies vaccine in addition to killed rabies vaccines. This recombinant rabies product is ineffective and shouldn't be administered to dogs.

The best vaccines to use in shelters are the one or three year versions of the rabies vaccine. Both vaccinations should cost about the same. In most communities, it is legal to revaccinate animals who have not received a previous rabies vaccination within a year, regardless of whether a one or three year product was used initially. Then, animals need to be vaccinated again every three years, or more frequently if the state or municipality in which they are found requires it. However, if a three year product is used instead of some one year products, the dog or cat's immunity may last longer if it is not re-vaccinated within a year of the initial dose of the rabies vaccine. The three year product is recommended since a longer duration of immunity would be advantageous to both the individual animal and the community as a whole, and many dogs do not receive a second vaccination after being adopted. Staff members must receive education and training to effectively explain to adopters the necessity of a one-year vaccination rather than a three year one. Comparing the three year product to the one year product, it is equally safe or safer.

Cats can safely use both the four year product and the one year non-adjuvanted recombinant feline rabies vaccine. Similar to dogs, cats who have received an initial vaccination without a record of previous vaccinations will typically need to be vaccinated again within a year, regardless of the product used. A product that combines the main vaccines with a rabies vaccine component is available for cats. Although the rabies component of this product shouldn't be administered at regular intervals of two weeks, it can be given as early as eight weeks of age.

Local or state laws specify the age at which a person must receive their first vaccination as well as the necessary intervals between vaccinations. Regardless of state law, a rabies product that is labeled as a one-year product should be administered within that time frame.

Local laws frequently apply to documentation of rabies vaccinations. The lot and serial numbers of the rabies vaccines are typically required to be included in the vaccination record. The majority of regions demand a veterinarian's signature in addition to their presence or actual administration of the vaccine. All animals should ideally receive at least two doses of the rabies vaccine, with the interval between doses being no less than three weeks and no more than a year, although local laws and vaccine manufacturers' recommendations may differ. This complies with the majority of local ordinances, which mandate that an animal be vaccinated again within a year of its first vaccination or at the age of one year. This procedure will enable excellent protection through re-vaccination after three or more years. Any animal that has never received a rabies vaccination that has been properly documented should be vaccinated as an initial step. There are surprisingly few examples of unprotected animals contracting rabies after receiving two vaccinations. Nevertheless, rabies has been acquired by a greater number of animals who only received one dose of the vaccine. Additionally, a sizable portion of dogs do not produce sufficient antibody titers within the first year following their first vaccination, and antibodies are not found until after the second dose. To protect against the rabies virus, antibodies are essential.

4.6. OPTIONAL VACCINES FOR CATS OR DOGS IN A SHELTER

Supplemental vaccinations may undermine the immune system's ability to fight off the core pathogens. As such, the rationale behind administering extra vaccinations needs to be properly thought through. The most recent AAFP shelter guidelines for cats and AAHA shelter guidelines for dogs should be useful in long-stay shelters when determining when to use optional, noncore products. The Canine and Feline Vaccination Guidelines, which were first published in 2006, contain vaccination recommendations for cats and dogs living in shelter environments. These recommendations are reviewed annually or more frequently as needed.

4.6.1. Feline Leukemia Virus (Fez) Vaccines

The FeLV vaccine is one optional vaccination that might be used in shelters. This is especially crucial if the kittens will be housed in a group shelter for an extended length of time, as there is a chance that the cats may be exposed to FeLV even if they are tested at intake (since some early cases may go unnoticed). In these cases, kittens benefit from vaccination, but only after undergoing FeLV diagnostic testing. FeLV-negative kittens and kittens born to FeLV-negative queens should receive two vaccinations: one at eight to nine weeks of age and another at three to four weeks of age using a vaccine from a reputable vaccine producer. For FeLV vaccines to be effective, revaccination must be administered at the prescribed interval of three to four weeks. When trying to overcome MDAs, age at the time of the last vaccination is not as important to take into account as it is for other vaccines. Rather, with the FeLV vaccine, the second dose, or revaccination, needs to happen no later than four weeks after the first shot. Persistent viremia may result from exposure to a FeLV viremic cat prior to, during, or up to two weeks following the second dose of vaccination; however, protection should be more than 90% effective after that point. If the cat is still living in a high-risk environment, a second vaccination should be given either a year after the first one. A vaccination shouldn't be administered more frequently than every three years. FeLV vaccinations should not be used unless two doses can be administered three to four weeks apart since a single dose offers no protection.

4.6.2. Feline Immunodeficiency Virus (FIV) Vaccines

It is not advised for shelter cats to receive the FIV vaccine. With current diagnostic tests, there is no discernible difference between the immune response to this vaccine and the antibody response to a natural infection. Kindly refer to the section regarding diagnostic testing. For best results, this vaccine needs to be administered in three doses at least, separated by two to four weeks. Cats may be up to 70% protected against certain strains (clades) of FIV when this program is followed, but they may be little to not protected against other strains. Shelter cats are not advised to receive this vaccine due to the low risk of FIV infection that most cats, including those in shelters, have, the requirement for multiple doses, a delayed onset of partial immunity, and interference with diagnostic testing.

4.6.3. Bordetella bronchiseptica and C. felis Vaccines

Bordetella bronchiseptica and *C. felis* vaccines are not recommended for most shelter cats. These vaccinations, however, might be beneficial in certain situations in shelters dealing with extreme FRDC. Use of these vaccines is indicated when laboratory testing has clearly identified these pathogens as suspect or contributing factors in disease outbreaks, or when comparisons between vaccinated and nonvaccinated cat groups demonstrate that the vaccines shorten the duration or severity of disease. Before these vaccines can be widely advised, more research in this field is required.

4.6.4. Other Noncore Vaccines

Giardia and feline infectious peritonitis (FIP) vaccines are generally not recommended for any cats by the AAFP guidelines. However, as with the Bordetella and *C. felis* products, if a benefit can be shown using these vaccines in a particular shelter, through controlled studies, that information should be published. Until more information is available, these vaccines are not recommended. The use of canine coronavirus, Giardia, *Porphyromonas* spp. (peridontal), and rattlesnake vaccines is not recommended for most shelter dogs. Four-way leptospira and Lyme vaccines are listed in the AAHA Guidelines as optional vaccines for pet dogs at high risk, but not generally recommended for shelter dogs.

4.7. JUVENILE ANIMALS

4.7.1. Maternal Antibody Interference with Vaccination

When compared to adult animals, puppies, and kittens in a shelter setting present unique challenges in terms of disease prevention through vaccination. This is particularly true for FPV, CPV-2, and CDV.

Newborn juvenile animals obtain varying amounts of maternal antibodies from their mothers' colostrum during nursing within the initial three days. The quantity of these antibodies transferred is subject to various factors such as the size of the litter, the frequency of early feedings, and the mother's antibody concentration. These antibodies typically have a half-life, implying that about half of them degrade, every two weeks, ranging from 8 to 13 days.

A "window of vulnerability," also called a "window of susceptibility," appears in young animals that received MDAs and is characterized by a decline in the levels of these antibodies. A virulent virus has a higher chance of defeating MDA defense than a vaccine virus. MDA levels can drop to a point where a virulent virus can infect a person, but they can also stay too high to effectively immunize against it. The window of susceptibility is created by the difference between the time at which a person becomes susceptible to a virulent virus and the time at which the vaccine virus can overcome MDAs to immunize. The term "window" refers to the time or age at which the MDA in the serum of puppies or kittens is no longer able to protect the animal against infection by the wild-type (virulent) virus; however, MDAs continue to prevent the development of active protective immunity after the animal is vaccinated.

The window of susceptibility for some of the original (pre-1997) CPV-2 vaccines lasted up to 12 weeks in individual animals. While the MDA only offered protection from infection against the virulent virus up to 10 to 12 weeks of age, it was possible to demonstrate with some of the "old generation" CPV-two vaccines that puppies born to a dam with a very high level of the CPV-2 antibody could not be actively immunized until 20 to 22 weeks of age (Larson and Schultz, 1997). This meant that because the last dose of vaccines needed to be administered at 20 to 22 weeks of age, puppies with very high MDAs never developed an immune response from the initial vaccination series, even when the puppy vaccination was started at four to six weeks of age and continued every two to three weeks until 16 to 18 weeks of age. After 1997, the less effective vaccines were either taken off the market or replaced by more effective vaccines (Larson and Schultz, 1997).

Since the introduction of the MLV CPV-2 and FPV vaccines, each species' window of susceptibility has rarely been extended beyond two to four weeks. This smaller window decreases infection overall but does not eliminate the need to isolate young animals from potentially harmful virus exposure for the full 16 week (or, in rare cases, longer) period during which the 2–4 week window may have opened. This is because it is rarely known when that period will occur. Although viral disease may still affect puppies and kittens during their first juvenile vaccination series, very few, if any, will have MDAs remaining at levels that can prevent the development of an immune response to current parvovirus vaccines when the final dose of CPV-2 or FPV vaccine is given to shelter puppies or kittens at 16 weeks, respectively.

If administered, killed vaccines (apart from rabies vaccines) can also be blocked by MDAs, though this is not advised. The second dose won't immunize and won't offer protective immunity if the first dose is inhibited. It merely serves to stimulate the immune system; a third dosage is needed to achieve immunization. As the third dose serves as the priming dose and the fourth as the immunizing dose, it will actually take a fourth dose to provide protection if the second dose is also blocked by MDAs. This means that immunizing an animal can take weeks or even more than a month. With a commercially killed CPV-2 vaccine that is no longer available, the researchers discovered this. The same probably holds true for vaccines against killed FPV.

4.7.2. Variation in Maternal Antibody Levels

A significant portion of animals that come into shelters have never received a vaccination or have had any maternal exposure to CDV, CPV, or FPV. As a result, these animals do not have any maternal antibodies to pass on to their progeny. It is therefore likely that passive immunity for CPV, FPV, and CDV has diminished or was absent at the time of initial presentation for a large number of puppies and kittens that come into shelters. Then, regardless of age, the first dose of the MLV vaccine should successfully immunize them. To induce complete immunity against these three pathogens, one dose of a modified live vaccine is sufficient when MDAs are absent or low enough not to prevent vaccination.

Even for puppies and kittens who did receive MDAs, each individual will differ in the amount and timing of MDA receipt. Individual antibody levels (titers) can vary enough even within a litter of puppies or kittens that some will become susceptible to infection sooner than others because they have lower levels of MDAs. A litter's infection may be prolonged by MDAs. For instance, because of their higher level of MDAs, some puppies in the litter may be protected by MDAs for up to two weeks after the other puppies succumb to an infection. If these puppies are not properly immunized prior to infection, they may become infected later and may even die because MDAs high enough to prevent illness will also have prevented earlier immunization.

Additionally, the rate at which maternal antibodies degrade in response to various antigens varies. Immune reactions to various vaccinations (e.g., CDV, and CPV-2) in the combination product will, therefore, typically happen at different ages because it is uncommon for a puppy or kitten to have antibody levels to drop to the point where simultaneous immunization is possible. By 16 weeks of age, the vast majority of juveniles should have MDA levels to all of the core vaccine components low enough for immunization to be effective. It's never clear which puppies or kittens have MDAs high enough to interfere with immunization, so it's best to give all young animals the entire juvenile vaccination series.

4.7.3. Risk of Vaccine Use in Very Young Animals

Modified live canine and feline vaccines should not be given during the first two weeks after birth, even though protection against CDV can be safely achieved by actively immunizing puppies as early as two weeks of age with a monovalent canarypox vectored recombinant vaccine. The immune system is not likely to develop a protective response to many vaccines before the age of two weeks, and the vaccine virus can still cause illness and even death at this young age.

Research conducted years ago by author Schultz showed that altered live canine viruses can quickly infiltrate the brains of newborns (six to seven days old), puppies devoid of colostrum, or puppies receiving colostrum but lacking an antibody against the viruses. Studies of a similar nature in orphaned, colostrum-deprived kittens have not been conducted, but if vaccinated at a young age (less than two weeks), one would assume that modified live FPV, FCV, and/or FHV-1 vaccine viruses could cause a similar infection of the brain.

Limited thermoregulatory control is present in both puppies and kittens for the first seven or more days of life. For certain puppies and kittens, thermoregulation may not fully develop until after two weeks of age. These newborns' bodies hardly ever reach temperatures suitable for the immune system to operate at its best. As a result, puppies, and kittens are left without a healthy, active immune system for at least one week, and frequently for up to two weeks or longer. Consequently, it would not be safe to administer MLV vaccines, and it may not be immunogenic to administer killed vaccines, so neither should be given. Immunizations do not offer protection against the wide range of possible pathogens that passively acquired MDA may cover.

4.7.4. Other Issues with Vaccinating Juvenile Animals

It makes no difference how long the puppy or kitten is allowed to nurse after that point because the only antibody that prevents active immunity is IgG absorbed from the intestine into the puppy or kitten's blood during the first three days after birth. Moreover, vaccinations given to nursing mothers will not provide any protection to juveniles (though nursing mothers should still be vaccinated for their own protection). Thus, as long as they don't have MDAs that would prevent vaccination, young animals can be vaccinated without difficulty before being weaned. Puppies that are vaccinated while MDAs are present will not be at greater risk because vaccination will not cause MDAs to decline faster while the vaccine virus is being reacted to. Delaying vaccinations in an attempt to protect MDAs is not a good idea.

4.7.5. Vaccine Schedule for S Helter Puppies and Kittens

To guarantee that at least one dose is administered at the age when the animal's levels of MDAs no longer inhibit the immune response to the vaccine, multiple doses of core vaccines should be given to kittens and puppies every two weeks beginning at four to six weeks of age and ending at 16 weeks. In an attempt to immunize as soon as possible after MDA interference can be overcome, vaccinations are administered every two weeks. Maintaining the narrowest possible window of susceptibility can be achieved by administering vaccinations every two weeks. The vaccinations should never be administered more frequently than every two weeks as this will reduce their effectiveness. An animal's risk of negative reactions, such as hypersensitivity or allergy to vaccines, is also increased by over vaccination.

4.7.6. Physical Protection of Juvenile Animals

It is impossible to predict which puppies and kittens will be exposed to MDAs, when the window of susceptibility will open, or which puppies and kittens will receive vaccinations before the age of four months. For this reason, it is still critical to implement management practices that shield juveniles from infectious agents, particularly potentially fatal viruses like CDV, CPV-2, and FPV. Moving puppies and kittens from the shelter to other low-risk settings, like hygienic foster homes with little traffic, or well-segregated facilities within the shelter, with separate supplies, equipment, and protective gear for volunteers and staff, is frequently the best way to keep them safe from illness. For animals too young to be placed for adoption, foster care is frequently a better option than isolation inside the shelter for disease prevention and optimal socialization. After that, until they are ready for adoption, puppies, and kittens can be raised and vaccinated in a disease-free environment. This shortens the time that the animal is in the shelter, which lowers the risk of exposure. Nevertheless, since this is frequently not feasible, puppies, and kittens raised in shelters should receive vaccinations every two weeks until they are 16 weeks or older.

4.7.7. Colostrum Replacement for Orphans

It's possible that abandoned newborn puppies and kittens who come to shelters were unable to nurse and develop passive immunity through MDAs. MDAs do offer vital protection against a variety of pathogens, even though passive immunity can be problematic since it often interferes with attempts to immunize through vaccination. When passive transfer fails, kittens, and puppies are highly susceptible to contracting a wide range of infectious diseases, particularly in the first four weeks of life. Frequently, even minor pathogens can result in illness and even death (Levy et al., 2001).

Passive immunity, which is akin to the immunity found in the colostrum that newborns normally receive from their mothers, is the most effective type of immunity for orphaned puppies or kittens. In addition to the viral diseases covered in the sections on core vaccines, passive immunity is necessary to shield young animals from bacterial infections that could cause sepsis. A significant portion of the neonatal mortality observed in orphaned puppies and kittens is probably caused by a breakdown in the passive transfer of antibodies. Failure happens when the puppy or kitten cannot absorb the colostral antibody through its intestinal tract, or when it receives little or no colostrum during the first few days of life.

For up to three days following birth, the newborn kitten or puppy's digestive tract can actively transfer IgG antibodies from colostrum to their blood; however, this absorption stops once colostral feeding has occurred and sufficient amounts of IgG have been absorbed. When kittens or puppies do not get colostrum during the first few days (e.g., pups or kittens are separated, a mother will not allow animals to suckle, or the mother has no milk or dies at parturition), "artificial colostrum" can be made and administered to correct this failure of passive antibody transfer. While it may be beneficial to foster orphaned kittens to nursing bitches or pups to lactating queens, bitches, and queens that are mid-lactation or even earlier will not have enough antibodies in their milk to make up for this passive transfer failure. Furthermore, for replacement colostrum to be absorbed, it needs to be given within 72 hours of delivery.

Serum from the mother or other adult animals can be obtained, ideally from the same area where the pups or kittens will be kept, and used as a source of replacement antibodies to give colostrum-deprived juveniles. Only healthy, fully immunized adult animals should be used to obtain serum. FeLV and FIV negative adult cats are required. It might be possible to build a donor base of well-known, healthy animals that are free of contagious diseases, but doing so requires following strict guidelines and receiving veterinary care and supervision. Home environments can provide a pool of potential donors who are willing to donate, as long as their owners screen and select them beforehand. Offering a complimentary health examination could serve as a motivator for involvement. Every donor animal should have a veterinary health examination done at least once a year. Allow the blood to clot after collection, then separate the serum and pool it if not enough serum can be extracted from a single donor.

You can freeze serum (at –20□) for future use. Commercial versions of immune canine serum are also available.

To administer orally, incorporate 1 volume of serum (e.g., 10 ml) to a corresponding amount (10 ml) of milk substitute (e.g., Esbilac) and give the kittens or puppies the artificial colostrum for the first three days of their lives. Passive immunity from the artificial colostrum will offer some protection.

Parenterally injected serum (preferably subcutaneous injection, or IP) into colostrum-deprived kittens has been shown to have significant benefits (Levy et al., 2001). For most shelter workers, the SQ route would be the best option because IP injection is more complicated and could be hazardous for individuals who are not experienced with the process. When compared to oral administration, injecting antibodies allows for a longer passive transfer period because the transfer is not reliant on gastrointestinal absorption. Three 5 ml doses of the serum were given to the kittens: one at birth, one at 12 hours, and one 24 hours later. Nonetheless, immune serum can be administered subcutaneously at any age because, in contrast to oral administration, it is not reliant on gut absorption. Serum antibody concentrations following this parenteral administration were comparable to those of kittens who nursed nonstop while the queen was present. Orphaned puppies could benefit from similar protocols, which involve giving 5 to 10 ml, based on the puppy's size, every 12 to 18 hours for a total of three to five doses to produce antibody levels that will offer transient protection for upto two to four weeks. Vaccinations ought to continue as normal once the puppy or kitten reaches the proper age (e.g., four to six weeks in shelter settings).

4.8. USE OF HYPERIMMUNE SERUM FOR DISEASE PREVENTION

It has been discovered that hyperimmune serum can successfully fend off infection in immunologically naïve dogs that have been exposed to canine herpesvirus (CHV-1), CDV, CPV-2, and CAV-1. Hyperimmune serum has prevented FPV infection in naïve cats. Hyperimmune serum is injected intravenously (plasma) or subcutaneously (serum) and is given either before or at the time of exposure. It has also been suggested that animals who have recently been exposed use hyperimmune serum before they exhibit symptoms. Although using serum from immunized animals in the environment instead of commercially produced hyperimmune serum may be advantageous, this has not been proven to be successful. Passive immunization is not as common or advised as it once was because immune serum or plasma is expensive and hard to come by commercially. If exposure has potentially already occurred, as is frequently the case with animals in shelters, animals should receive an MLV vaccination right away if it hasn't already. This will provide immunity in three days or less for CDV, CPV-2, and FPV. It won't stop infection from exposure before vaccination, but it won't hurt animals that have already been exposed, and it might stop sickness from exposure in the future. The most crucial stage of immunization for a newborn puppy or kitten is passive immunization, which can occur naturally through the mother's natural antibodies found in colostrum or through artificial injection or feeding.

4.9. RISK ASSESSMENT AND EVALUATION OF SEROLOGIC IMMUNITY FOR CANINE DISTEMPER, CANINE PARVOVIRUS, AND FELINE PANLEUKOPENIA

Veterinarians and pet owners have grown more wary in recent years of the potential for "overvaccinating" their animals. As a result, measuring antibody titer levels has become more and more common as a means of assessing immunity to specific diseases for which antibody levels strongly correlate with immunity. Revaccination is usually deemed unnecessary in pets for which protective levels have been documented. Titer level measurement is useful in this population even though it was not intended or tested for risk assessment of animals that have recently been exposed.

After a possible exposure to canine distemper, canine parvovirus, or feline panleukopenia, an individual animal's risk can be assessed using the following description of a systematic risk assessment process, which is based on evaluation for serologic immunity. It is reasonable to assume that healthy dogs and cats with protective titer levels and no clinical signs of illness are unlikely to contract an infection or become ill even if exposed, as antibody levels are closely correlated with protection against these three specific diseases. Certain animals may be subclinically infected even though they show no outward symptoms of illness and have normal titers. Although these animals might release viruses, they haven't caused any clinical issues when outbreaks have been contained using this risk assessment method.

Although there is never a 100% guarantee that a given animal will contract an infection, defining the risk level for individual animals and subgroups can greatly aid in decision-making. By using serologic risk assessment, an outbreak can be effectively controlled with fewer instances of euthanasia and other extreme or expensive measures. Limiting the quantity of animals undergoing quarantine, isolation, or special rescue is another benefit of creating risk categories for animals that have been exposed. Certain animals can be placed in a low-risk group and any extra precautions should only be directed at higher-risk animals because they may have strong, demonstrable immunity. Avoiding diagnostic or titer testing in an effort to save money could eventually result in even higher resource costs and euthanasia.

4.9.1. Options for Antibody Titer Testing

Antibody testing can be done using an in-house test kit or by sending serum samples to a diagnostic laboratory that has been verified. While in-house tests only yield qualitative (positive or negative) results, diagnostic laboratories typically return quantitative data. Challenge testing must be used to validate all tests by a laboratory to establish cutoff points for risk categories based on numerical values.

TiterCheck™ is the only test kit currently available for in-house titer evaluation. It has been confirmed that these kits can withstand testing. The TiterCheck™ kit is a well-style kit intended for testing canine serum for CDV or CPV antibodies. The authors' research indicates that feline antibodies to FPV may also be tested for using the CPV test wells, though the results may not be comparable in terms of sensitivity and specificity. A positive result is indicative of a titer level that was protective against challenge because the TiterCheck ™ kits have been validated (for canine viruses). There might be more tests like this available soon.

As of this writing, TiterCheck™ costs about $11,00 per animal (not including staff time) when all 14 sample wells are run at once, which makes it a cost-effective tool for many shelters' outbreak response plans. When administered by technicians with prior experience administering comparable well-type tests, the test yields the most accurate results. The writers believe that the washing step is one of the most important. Inadequate washing increases the likelihood of false positive results. Clinically, the test and related risk category designations have proven to be helpful tools for shelters dealing with outbreaks, despite the possibility of false positives. Serologic evaluation is best performed in a validated commercial laboratory if staff lacks the necessary time or training to perform the test. However, using a commercial lab may often be unfeasible due to the longer turnaround times for test results, particularly during an outbreak. The same interpretational guidelines apply whether an in-house test or a commercial laboratory is used.

4.9.2. Initiale Valuation for Clinical Signs

Only animals who are completely free of clinical signs of the illness in question at the time of testing can have their serology used for risk assessment interpreted. Positive titers may not indicate a preexisting protective titer level; rather, they may indicate an active immune response to infection in animals exhibiting any symptoms of disease. Results in animals that have just recovered from illness should also be interpreted cautiously. Positive titers in these animals might indicate an immune reaction to a recent infection. Even though there is little chance that these animals will ever experience a serious illness related to the condition, they might still be shedding and thus present a risk of infection to other animals. This is especially concerning for illnesses like canine distemper that may cause prolonged post-recovery shedding.

It is vitally important to conduct an early evaluation for illness. All dogs exhibiting clinical signs even suggestive of distemper, such as respiratory illness, anorexia, nasal or ocular discharge, or unexplained gastrointestinal disease, or, in the case of parvoviruses, dogs or cats exhibiting vomiting, diarrhea, lethargy, or inappetance, must be deemed high risk in the event of an outbreak. Close observation via daily monitoring sheets or direct observation of fecal quality and output, appetite, attitude, and other clinical signs are all necessary components of the evaluation process. When attempting risk assessment, an infected animal may be mistakenly identified as immune if clinical signs are ignored.

4.9.3. Interpretation of Positive Titer Results in Adult Dogs and Cats

A dog or cat may be placed in the low-risk category if they do not exhibit any clinical symptoms, test positive for antibodies on the TiterCheck™ kit, or exceed the protective cutoff for samples sent to a diagnostic laboratory. It is more likely that the animals' antibodies are a reflection of immunity rather than infection because they have the antibody and no clinical symptoms. This low-risk group of dogs and cats is not likely to contract an infection or show symptoms of the disease. Dogs classified as low-risk (positive titers) for CDV are marginally more at risk than those classified as low-risk for CPV due to the longer incubation period of CDV compared to parvoviruses. This is because titers for CDV may rise before clinical signs in response to infection. However, this theoretically increased risk has not resulted in a clinical issue when this system has been used to manage multiple outbreaks.

Positive titer results allow for the classification of animals as low risk, which allows them to be adopted and stay in the general population relatively safely. Potential adopters should be informed about the outbreak and the possibility of animal exposure in the shelter, along with the details of the risk assessment and testing that have been completed. A new pet from a multianimal background, like a shelter or pet store, always carries some risk; low risk does not equate to no risk. To the best of the authors' knowledge, this tactic has been effective in controlling a number of shelter outbreaks.

4.9.4. Interpretation of Negative Titer Results in Adult Dogs and Cats

Dogs and cats that are below the protective cutoff for samples sent to a validated diagnostic laboratory or who show no clinical signs but have negative test results on the TiterCheck™ kits must be considered at high risk of infection in the event of an outbreak, as they were likely susceptible at the time of exposure. Not every titer-negative animal will get sick, especially if they received their vaccinations prior to or at the time of intake. But during the incubation period, these animals have a comparatively high chance of contracting a disease at any point. These animals need to be taken out of the general shelter population right now.

4.9.5. Interpretation of Titer Results in Juvenile Animals

It is impossible to distinguish between passive, maternally derived immunity and actively produced antibody immunity in puppies and kittens younger than five months of age. While titers resulting from passively acquired antibodies would only represent those antibodies in circulation, titers resulting from active immunity reflect the presence of associated memory cells in addition to the presence of circulating antibodies. In the face of a challenge, memory cells, which are a component of active immunity, boost immune response. A higher detectable antibody titer may be required for defense against challenge because passively acquired antibody titers do not indicate any such reinforcements. Furthermore, antibodies that are passively acquired (MDAs) are always diminishing, meaning that an animal that was protected during testing may soon lose that protection.

For juveniles, a higher threshold for titer level would ideally be set to indicate low risk. The TiterCheck™ in-house tests are qualitative assessments that, when performed on adult dogs with active immunity, are limited to providing a positive or negative result based on a titer level that has been demonstrated to be protective. Because a higher cutoff point can be used, quantitative titer testing may be a more accurate indicator of risk for puppies or kittens. Less than 16 weeks old puppies and kittens with positive antibody titers on the qualitative TiterCheck™ kit, however, will probably have sufficient antibody levels to shield them from challenge. Assigning puppies and kittens without clinical signs and receiving a qualitatively positive TiterCheck™ antibody test result to the low-risk group is most likely the safest and easiest course of action.

However, it should be noted that the categorization of juvenile animals into low-risk groups may not be based on active immunity; rather, it may be based on MDA titers, which does not reflect the same level of certainty as for an adult. Furthermore, as maternal antibodies decline, documented "low-risk" status could quickly change. The shelter system should process puppies and kittens with positive titers as soon as feasible.

Outside laboratory testing is frequently impractical due to lengthy wait times for results, differences in testing types and cutoff points used by each lab, and other factors. Even though quantitative antibody titer levels are the best way to assess risk in puppies and kittens, using the TiterCheck™ kit is still generally a better option than waiting for test results, dividing up the risk into categories without seeing them, or treating every puppy equally high risk after exposure.

4.10. VACCINE EFFECTS ON DIAGNOSTIC TESTING

4.10.1. FeLV and FIV Vaccine Effects on Testing

The most often run diagnostic tests in shelters are probably FeLV and FIV, either alone or in combination. These assays are highly specific and sensitive. FeLV tests identify viral antigens; if the test is positive, the virus can be seen in the blood or other secretions, depending on the sample that was examined. Test results are unaffected by vaccination with FeLV vaccines (Richards, 2003).

The FIV test, on the other hand, finds antibodies. As antibodies are generated both during vaccination and from infection with a virulent virus, there is currently no test that can distinguish between animals that have received vaccinations and those that have not. It appears that an enzyme-linked immunosorbent assay (ELISA) test can effectively differentiate between cats who have received vaccinations and those who have not. There are currently no commercial versions of this ELISA, but more testing is being done.

4.10.2. Effect of Vaccination on Testing for CPV-2 and FPV Using Parvovirus Fecal Antigen ELISA Test

Diagnosing CPV-2 disease in dogs is aided greatly by a test that looks for the CPV-2 antigen in their feces. When checking for the FPV in cats, this test can also be used on cats. The two viruses have similar antigens, and FPV antigens will elicit a response from the antibody used to identify the CPV-2 virus (Quesada et al., 2007).

The probability of positive FPV and CPV-2 tests after vaccination has been studied in some recent studies conducted by the authors' laboratory and by others. The results show that true positives from vaccination are incredibly rare. All of the commercially available tests varied to some extent, but the Idexx™ tests produced the fewest positive results following vaccination.

In a study conducted by the authors' lab, the Idexx Snap test was used to check for fecal parvovirus antigen in over 600 recently vaccinated dogs. Not one of those dogs tested positive. Less than 1% of the samples had CPV-2 detected by the Witness TM test. With cats, similar outcomes were discovered. Upon testing more than 600 cats after vaccination, the Idexx™ test yielded only one positive result. That cat, which was thought to be a non-responder to FPV, passed away following its FPV vaccination.

One of the 64 vaccinated, 10-week-old, specific pathogen-free (SPF) kittens tested positive (weakly) for the Idexx Snap test after vaccination, according to a similar study conducted by other researchers. Other test brands yielded more frequently occurring positive results in this investigation.

Since many shelters presently adhere to the recommendation that all animals be vaccinated upon intake, this information is crucial for them. A chance to break the cycle of transmission may be lost if a positive or weak positive test in an infected animal is mistakenly identified as a false positive test due to recent vaccinations. Additionally, necessary care and intervention may be postponed because it may be assumed the animal does not have virulent parvovirus. Although it is known that vaccinated animals will probably shed the vaccine virus afterward, the amount of virus shed is less than what would typically result in a positive result on fecal antigen tests. After vaccination, only a small proportion of animals will have detectable levels of virus in their feces, even with more sensitive methods like hemagglutination (HA) or viral isolation (VI) being used. No matter how recently a person has received a CPV-2 or FPV vaccination, a positive result with antigen detection tests like hemagglutination or ELISA is highly suggestive of virulent viral infection.

On the other hand, because of the assay's high sensitivity, many fecal samples from vaccinated animals test positive for CPV-2 or FPV when CPV-2 or FPV is detected in feces using the polymerase chain reaction (PCR). Regretfully, there's no easy way to tell if a virus is virulent or vaccine-derived. As a result, negative parvovirus PCR test results are significant, but positive results from fecal samples of animals that received vaccinations within the previous two weeks could be the consequence of either vaccination or actual infection.

4.10.3. Canine Distemper Vaccine Effects on Testing

Any common antigen detection techniques, such as PCR, virus isolation, fluorescent antibody (FA), and immunohistochemistry (IHC), can identify the CDV vaccine virus. Due to the MLV vaccine's ability to replicate in a range of cell types and its presence in leukocytes throughout the body, vaccine virus may cause problems with diagnostic testing up to two weeks after vaccination. Although it depends on the individual animal, vaccine viruses should not be found in different tissues for longer than two weeks. When the canarypox vectored rCDV product is used, interference with testing and false positives would not occur with some tests (virus isolation, FA, IHC), and they are unlikely to occur with PCR testing. Infectious CDV is not present in the rCDV vaccine (Noon et al., 1980).

Additionally, vaccination may readily impede the use of paired antibody titer testing as a diagnostic method for CPV-2 and/or CDV infection. A dog's antibody titer should increase noticeably during the three weeks that follow vaccination with the CDV or CPV-2 vaccines.

4.10.4. Feline Herpesvirus/Feline Calicivirus Vaccine Effects on Testing

A two-week period may see the shedding of feline herpesvirus and feline calicivirus from cats that have recently received an MLV intranasal vaccination. After vaccination, the first three to seven days are when viral shedding occurs most frequently. Shedding would obstruct viral isolation and PCR, producing favorable outcomes. Although reverse transcriptase-polymerase chain reaction (RT-PCR) can distinguish vaccine from field strain FCV, this test is not commercially available. In most cases, field strain viruses are difficult to distinguish from vaccine viruses.

4.10.5. Bordetella/CAV-2/CPI (Intranasal Vaccine) Effects On Testing

Since all of the intranasal Bordetella vaccinations currently on the market are MLV, detectable levels of the virus can be shed, usually for no longer than two weeks after vaccination. In the course of this shedding period, vaccine pathogens may produce false positive results in PCR or culture tests.

4.10.6. Canine Lyme Vaccine Effects on Testing

Dogs who are likely to be infected or who may have previously been infected with Borrelia burgdorferi can be identified at the shelter using a rapid test (4Dx, Idexx Laboratory). Positive results for Borrelia can sometimes cause confusion when it comes to interpretation, which makes them a barrier to adoption, transfer, or transportation. Many dogs that test positive don't require any kind of care. It is occasionally required to carry out a laboratory-based quantitative test to separate the infected dogs from the vaccinated dogs because dogs who have received prior vaccinations may also test positive on the screening test. If the test is positive, the animal may be infected and may benefit from antibiotic treatment, particularly if it is exhibiting symptoms like arthritis. Along with antibodies to *Ehrlichia canis* and *Anaplasma phagocytophilum*, this test also finds the heartworm antigen.

4.10.7. Antibody Testing for Diagnosis of Canine Distemper, Canine Parvovirus, and Feline Panleukopenia

Antibody testing for FPV, CPV, and CDV has been discussed in many chapters on the vaccination of pregnant animals as well as risk assessment. Antibody titers are employed in these situations to evaluate a patient's immunity to an infection or past vaccination history. In general, it is not advised to use CDV, CPV, or FPV antibody titers as diagnostic tools in shelters because, ideally, MLV vaccinations will be administered upon intake. Following an MLV vaccination, an increase in antibody titers would be anticipated. Examining paired titers over time would reveal little difference between vaccine response and infection.

REFERENCES

1. Abdelmagid, O. Y., Larson, L., et al. (2004). Evaluation of the efficacy and duration of immunity of a canine combination vaccine against virulent parvovirus, infectious canine hepatitis virus, and distemper virus experimental challenges. *Veterinary Therapeutics, 5*(3), 173–186.
2. Appel, M. (1987). Canine distemper virus. In M. Appel (Ed.), *Virus Infections of Carnivores* (pp. 133–159). New York: Elsevier.
3. Appel, M., & Bemis, D. A. (1978). The canine contagious respiratory disease complex (kennel cough). *Cornell Veterinarian, 68*(7), 70–75.
4. Brun, A. G., Chappuis, G., et al. (1979). Immunization against panleukopenia: Early development of immunity. *Comparative Immunology, Microbiology and Infectious Diseases, 1*(4), 335–339.
5. Buonavoglia, C., Martella, V., et al. (2001). Evidence for evolution of canine parvovirus type 2 in Italy. *Journal of General Virology, 82*, 3021–3025.
6. Camero, M., Cavalli, A., et al. (2004). A severe dual infection by feline panleukopenia virus and feline calicivirus in an adult cat. *New Microbiologica, 27*(1), 79–82.
7. Carmichael, L. E., Joubert, J. C., et al. (1983). A modified live canine parvovirus vaccine. II. Immune response. *Cornell Veterinarian, 73*(1), 13–29.
8. Cocker, F. M., Newby, T. J., Gaskell, R. M., et al. (1986). Responses of cats to nasal vaccination with a live, modified feline herpesvirus type 1. *Research in Veterinary Science, 41*, 323–330.
9. Davis, R., Jayappa, H., et al. (2007). Comparison of the mucosal immune response in dogs vaccinated with either an intranasal avirulent live culture or a subcutaneous antigen extract vaccine of *Bordetella bronchiseptica*. *Veterinary Therapeutics, 8*(1), 32–40.
10. Dawson, S., McArdle, F., Bennett, D., et al. (1993). Investigation of vaccine reactions and breakdowns after feline calicivirus vaccination. *Veterinary Record, 132*, 346–350.
11. Duration of immunity for canine and feline vaccines: A review. *Veterinary Microbiology, 117*(1), 75–79.
12. Edinboro, C. H., Janowitz, L. K., Guptill-Yoran, L., et al. (1999). A clinical trial of intranasal and subcutaneous vaccines to prevent upper respiratory infection in cats at an animal shelter. *Feline Practice, 27*, 7–13.
13. Edinboro, C. H., Ward, M. P., & Glickman, L. T. (2004). A placebo-controlled trial of two intranasal vaccines to prevent tracheobronchitis

(kennel cough) in dogs entering a humane shelter. *Preventive Veterinary Medicine, 62*, 89–99.

14. Ellis, J. A., Haines, D. M., West, K. H., et al. (2001). Effect of vaccination on experimental infection with *Bordetella bronchiseptica* in dogs. *Journal of the American Veterinary Medical Association, 218*, 367–375.
15. Goldkamp, C. E., Levy, J. K., et al. (2008). Seroprevalences of feline leukemia virus and feline immunodeficiency virus in cats with abscesses or bite wounds and rate of veterinarian compliance with current guidelines for retrovirus testing. *Journal of the American Veterinary Medical Association, 232*(8), 1152–1158.
16. Greene, C. (1998). Immunoprophylaxis and immunotherapy. In *Infectious Diseases of the Dog and Cat* (2nd ed., pp. 717–750). Philadelphia: W. B. Saunders.
17. Lappin, M. R., Sebring, R. W., et al. (2006). Effects of a single dose of an intranasal feline herpesvirus 1, calicivirus, and panleukopenia vaccine on clinical signs and virus shedding after challenge with virulent feline herpesvirus 1. *Journal of Feline Medicine and Surgery, 8*(3), 158–163.
18. Larson, L. J., & Schultz, R. D. (1997). Comparison of selected canine vaccines for their ability to induce protective immunity against canine parvovirus infection. *American Journal of Veterinary Research, 58*, 360–363.
19. Larson, L. J., & Schultz, R. D. (2006). Effect of vaccination with recombinant canine distemper virus vaccine immediately before exposure under shelter-like conditions. *Veterinary Therapeutics, 7*(2), 113–118.
20. Larson, L. J., & Schultz, R. D. (2008). Do two current canine parvovirus type 2 and 2b vaccines provide protection against the new type 2c variant? *Veterinary Therapeutics, 9*, 94–101.
21. Larson, L. J., Hageny, T. L., et al. (2006). Effect of recombinant canine distemper vaccine on antibody titers in previously vaccinated dogs. *Veterinary Therapeutics, 7*(2), 107–112.
22. Lauritzen, A., Jarrett, O., & Sabara, M. (1997). Serological analysis of feline calicivirus isolates from the United States and United Kingdom. *Veterinary Microbiology, 56*, 55–63.
23. Levy, J. K., Crawford, P. C., et al. (2001). Use of adult cat serum to correct failure of passive transfer in kittens. *Journal of the American Veterinary Medical Association, 219*(10), 1401–1405.
24. Levy, J. K., Crawford, P. C., et al. (2008). Differentiation of feline immunodeficiency virus vaccination, infection, or vaccination and infection in cats. *Journal of Veterinary Internal Medicine, 22*(2), 330–334.

25. Martella, V., Cavalli, A., et al. (2004). A canine parvovirus mutant is spreading in Italy. *Journal of Clinical Microbiology, 42*, 1333–1336.
26. Miyamoto, T., Taura, Y., Une, S., et al. (1995). Immunological responses after vaccination pre- and post-surgery in dogs. *Journal of Veterinary Medical Science, 57*, 29–32.
27. Mouzin, D. E., Lorenzen, M. J., et al. (2004a). Duration of serologic responses to five viral antigens in dogs. *Journal of the American Veterinary Medical Association, 224*(1), 55–60.
28. Mouzin, D. E., Lorenzen, M. J., et al. (2004b). Duration of serologic response to three viral antigens in cats. *Journal of the American Veterinary Medical Association, 224*(1), 61–66.
29. Newbury, S. P., Page, J., et al. (2007). A placebo-controlled field trial of an intranasal vaccine for feline calicivirus and feline herpesvirus to prevent clinical signs of feline infectious respiratory disease complex in an animal shelter. In *Conference of Research Workers in Animal Diseases, Proceedings of the 88th Annual Meeting* (p. 140). December 2–4, 2007, Chicago, Illinois.
30. Noon, K. F., Rogul, M., et al. (1980). Enzyme-linked immunosorbent assay for evaluation of antibody to canine distemper virus. *American Journal of Veterinary Research, 41*(4), 605–609.
31. Olsen, L. B., Larson, L. J., & Schultz, R. D. (1998). Canine parvovirus (CPV-2b) infection in cats. In *Proceedings of Conference of Research Workers in Animal Disease*, Abstract P167.
32. Patterson, E. V., Reese, M. J., et al. (2007). Effect of vaccination on parvovirus antigen testing in kittens. *Journal of the American Veterinary Medical Association, 230*(3), 359–363.
33. Pedersen, N. C., & Hawkins, K. F. (1995). Mechanisms for persistence of acute and chronic feline calicivirus infections in the face of vaccination. *Veterinary Microbiology, 47*, 141–156.
34. Reed, T. L., von Messling, V., et al. (2003). A comparative study of canine distemper vaccines. In *Proceedings of Conference of Research Workers in Animal Disease*, Poster 87.
35. Richards, J. (2003). 2001 Report of the American Association of Feline Practitioners and Academy of Feline Medicine Advisory Panel on feline retrovirus testing and management. *Journal of Feline Medicine and Surgery, 5*(1), 3–10.
36. Richards, J. R., Elston, T. H., et al. (2006). The 2006 American Association of Feline Practitioners Feline Vaccine Advisory Panel report. *Journal of the American Veterinary Medical Association, 229*(9), 1405–1441.

37. Schroeder, J. P., Bordt, D. W., et al. (1967). Studies of canine distemper immunization of puppies in a canine distemper-contaminated environment. *Veterinary Medicine/Small Animal Clinician, 62*(8), 782–787.
38. Schultz, R. D. (1998). Current and future canine and feline vaccination programs. *Veterinary Medicine, 93*, 233–254.
39. Schultz, R. D. (1999). Canine and feline vaccines. In R. D. Schultz (Ed.), *Veterinary Vaccines and Diagnostics–Advances in Veterinary Medicine* (pp. 289–358). San Diego: Academic Press.
40. Schultz, R. D. (2000). Considerations in designing effective and safe vaccination programs for dogs. In L. E. Carmichael (Ed.), *Recent Advances in Canine Infectious Diseases*. International Veterinary Information Service. Retrieved from: www.ivis.org (accessed on 6 December 2024).
41. Schultz, R. D. (2009). A commentary on parvovirus vaccination. *Journal of Feline Medicine and Surgery, 11*(2), 163–164.
42. Schultz, R. D., & Conklin, S. (1998). The immune system and vaccines. *Compendium of Continuing Education for Practicing Veterinarians, 20*, 5–18.
43. Schultz, R. D., Appel, M. J., & Carmichael, L. E. (1977). Canine vaccines and immunity. In R. W. Kirk (Ed.), *Current Veterinary Therapy VI* (pp. 1271–1275). Philadelphia: W. B. Saunders.
44. Schultz, R. D., Larson, L. J., & Newbury, S. (2007). Surveys of shelter dogs and cats in multiple states. *University of Wisconsin–Madison*, unpublished data.
45. Scott, F. W., & Geisinger, C. M. (1999). Long-term immunity in cats vaccinated with an inactivated trivalent vaccine. *American Journal of Veterinary Research, 60*, 652–658.
46. Tizard, I., & Ni, Y. (1998). Use of serologic testing to assess immune status of companion animals. *Journal of the American Veterinary Medical Association, 213*(1), 54–60.
47. Tizzard, I. R. (2000). Vaccination and vaccines. In I. R. Tizzard (Ed.), *Veterinary Immunology* (6th ed., pp. 235–252). Philadelphia: W. B. Saunders.
48. Toshach, K., Jackson, M. W., & Dubielzig, R. R. (1997). Hepatocellular necrosis associated with the subcutaneous injection of an intranasal *Bordetella bronchiseptica*-canine parainfluenza vaccine. *Journal of the American Animal Hospital Association, 33*, 126–128.
49. Truyen, U. (2006). Evolution of canine parvovirus: A need for new vaccines? *Veterinary Microbiology, 117*, 9–13.

50. Truyen, U., Evermann, J. F., et al. (1996). Evolution of canine parvovirus involved loss and gain of feline host range. *Virology, 215*, 186–189.

51. Weigler, B. J., Guy, J. S., Nasisse, M. P., et al. (1997). Effect of a live attenuated intranasal vaccine on latency and shedding of feline herpesvirus 1 in domestic cats. *Archives of Virology, 142*, 2389–2400.

CHAPTER 5

FELINE UPPER RESPIRATORY DISEASE

Contents

Introduction.......... 176

5.1. Agent Characteristics and Epidemiology.......... 176

5.2. Clinical Signs.......... 181

5.3. Therapy.......... 183

5.4. Modes of Transmission.......... 185

5.5. Diagnosis.......... 188

5.6. Disease Control and Prevention.......... 188

5.7. Disease Surveillance.......... 201

5.8. Staff and Volunteer Education and Motivation.......... 202

5.9. Adopting a Systematic Approach to Management of URTD.......... 203

References.......... 207

INTRODUCTION

Veterinarians who oversee the health of cats in shelters frequently deal with the frustrating issue of upper respiratory tract disease (URTD). The agents responsible for URTD are widespread, highly contagious, and well-suited to spread illness, especially among populations of people living in extremely vulnerable shelters. There are vaccines for the agents that most frequently cause disease, but they are invariably ineffective at preventing infection and frequently only mitigate the severity of respiratory symptoms rather than eliminating them altogether. The transmission of URTD is facilitated by people and other fomites (items contaminated with respiratory secretions), which makes respiratory disease control extremely challenging. Since complete prevention is not possible due to the nature of the agents and shelter populations, the goal of URTD management is to minimize its occurrence.

5.1. AGENT CHARACTERISTICS AND EPIDEMIOLOGY

5.1.1. Agents

Cat URTD is most commonly caused by two viruses: feline calicivirus (FCV) and feline herpesvirus-1 (formerly known as feline rhinotracheitis virus, or FVR). Mycoplasma species, *Chlamydophila felis*, and *Bordetella bronchiseptica* are among the bacterial agents causing feline URTD (Gaskell et al., 2006).

Herpesvirus and calicivirus are known to be the main respiratory pathogens. There is only one serotype of the alpha herpesvirus known as Feline Herpesvirus-1 (FHV-1) that affects domestic cats worldwide. Although there are different isolates, neither their antigenic makeup nor the symptoms they cause differ significantly (Dawson et al., 2007).

The virus known as feline calicivirus, or FCV, belongs to the Caliciviridae family. The U.S. is home to a single genogroup, known as genotype I, of an RNA virus that depends on the error-prone replication linked to RNA-dependent RNA polymerases. s. quite a few isolates and strains. Strains do not seem to differ by geographic location, except for one additional Japanese genotype (genotype II). The FCV genome's plasticity increases the virus's adaptability while also making the creation of vaccines that are effective against all strains more difficult. Additionally, due to this plasticity, strains of differing virulence have been produced, from those that cause inapparent infections to those that cause virulent systemic feline calicivirus disease (VS-FCV), which was formerly known as hemorrhagic-like fever and is also known as virulent systemic disease (VSD) in Great Britain (Pedersen et al., 2000).

The reason why the virulent strains are so concerning to shelters is not that they are frequently encountered, but rather that they seem to originate more frequently in rescue shelters and then spread from there to animals kept in veterinary offices and staff members' homes. Genetic differentiation between these strains and common field strains of FCV suggests that the VS-FCV isolates originated independently during each outbreak. As of the middle of 2008, certain virulent strains have not been protected against by current vaccines.

According to some research, caliciviruses are replicating at high rates in rescue shelter populations because the hosts have high levels of non-neutralizing immunity, which may lead to the emergence of hypervirulent strains in these populations. Theoretically, high viral concentrations and high transmission rates result from these quickly replicating strains. These populations of cats are thought to simultaneously heighten their immune response, though they may not show any symptoms themselves. Nevertheless, when diseased cats are added to populations of felines who have never before encountered these strains (e.g., veterinary clinics, and homes), the swiftly replicating viruses spread widely, overpower immune responses, and result in VS-FCV. Interestingly, these strains have mysteriously not survived longer than two or three months. This could be due to high mortality, prompt implementation of control measures, or ongoing viral mutations leading to less deadly forms (Hurley and Sykes, 2003).

It was once thought that *Chlamydophila felis*, formerly known as *Chlamydia psittaci* (var felis), frequently caused a wide range of respiratory symptoms. However, it is currently mainly known to cause conjunctivitis in cats. Cats with the infection rarely show systemic symptoms.

Bordetella bronchiseptica functions as a primary pathogen in laboratory settings. Some researchers consider this organism to be a primary pathogen in the field as well because of its capacity to paralyze the mucociliary apparatus. Its potential to function as a primary pathogen is further supported by the discovery that this organism can be recovered on its own from cats suffering from respiratory illnesses in the field (Binns et al., 1999; Dawson et al., 1991; Welsh, 1996).

Both Mycoplasma species and *Mycoplasma* spp. are cultivable. Cats with or without clinical signs of disease can have their DNA extracted from their nasal cavity or pharynx through the use of polymerase chain reaction (PCR). Although they have been recovered from cats with lower respiratory tract diseases, they are not present in the lungs of healthy cats. Though mainly thought to be opportunistic invaders, they may also function as primary pathogens in certain lower respiratory tract disease cases (Chandlers and Lappin 2002). Other secondary bacterial invaders include *Staphylococcus* spp., *Pasteurella multocida*, *Streptococcus spp.*, and *E. coli*.

5.1.2. Host Range and Zoonotic Potential

Domestic cats and a few other Felidae members are infected by the FCV and the feline herpesvirus-1. No zoonotic potential is known for these agents. At least one immunocompromised person with chronic conjunctivitis has been conclusively linked to infection with *Chlamydophila felis*, which is known to infect cats. 88% of Japanese small animal veterinarians who participated in a seroprevalence survey had antibodies to the feline strain of *Chlamydia psittaci* Fe/Pn1, although no information was given about clinical illness. *B. bronchiseptica* causes respiratory disease in dogs, cats, rabbits, horses, laboratory rodents, pigs, and in rare cases, humans. More than 35 cases of *B. bronchiseptica* have been documented in immunocompromised people, underscoring that a low zoonotic risk exists. In addition, at least one case of pertussis-like illness has been reported in a boy whose face was sprayed with an intransal *B. bronchiseptica*/parainfluenza vaccine while holding his dog for vaccination (Berkelman, 2003).

5.1.3. Distribution

Numerous reviews indicate that FCV and FHV cause about 80% of feline URTD cases, with each agent contributing roughly equally to the disease. The majority of research on the frequency of respiratory agents in cats is based on prevalence studies, which look for instances of virus shedding in cats owned by clients at a particular time.

Based on virus isolation, prevalence rates of calicivirus among clinically affected cats have varied from roughly 10 to 33%, while herpesvirus prevalence rates have ranged from 11 to 39%. The recovery ratio of calicivirus to herpesvirus has generally been higher in cats that show no symptoms. It has been documented that cats without clinical symptoms shed between 8 and 41% of their calicivirus and between 0.4 and 1.75% of their herpesvirus. In a recent study of 152 cats from 22 Swedish catteries, prevalence rates for herpesvirus and calicivirus in resident cats that appeared to be in good health were as low as 2.6% and 0%, respectively. Recovering calicivirus more frequently may not necessarily indicate a higher relative incidence of this virus because calicivirus shedding lasts longer than herpesvirus shedding on average.

Variations in the tests employed to recover and identify the agents make it difficult to compare the prevalence of infection with these agents across studies (e.g., viral isolation, PCR), the stage of the cats' illness at which they were sampled, the location of the sample collection (e.g., conjunctival sac, oropharynx), as well as the makeup of the populations that were sampled (e.g., purebred catteries, rescue cats, or household cats). According to some researchers, the proportion of herpesvirus-shedding cats has decreased despite widespread vaccination, but not the frequency of virus shedding. In what way do these prevalence estimates relate to animal shelters in the United States is unknown, but in a recent study of cats in seven out of eight animal shelters in California, calicivirus recovery outpaced herpesvirus recovery. Two Colorado shelters found an 85 point 2% prevalence of FHV-1 positive cats, which is significantly higher than previously reported (Veir et al., 2004).

To precisely determine the risk that cats in shelters face of contracting these infections, studies of the incidence of infection with each respiratory agent are required. Veterinarians could better target their preventive care with the help of such data. Pedersen et al. 2004 carried out a prevalence study in two shelters in California, collecting data on the occurrence of herpesvirus and calicivirus infections over time. This study indirectly provides information on the relative incidence of infections caused by these agents. When these cats entered the study shelters, 43.3% of them were shedding herpesvirus, and 10.5% were shedding calicivirus. After a week, the herpesvirus shedding rate had increased by over 10 times, to 51.7%, while the calicivirus shedding rate had increased slightly, to 16.7%. 40% of cats were still shedding FHV-1 at two weeks, and 20% were shedding calicivirus, despite the fact that the number of samples studied had significantly decreased by the second week. Sadly, the clinical status of these cats was not disclosed; however, during the first week in these shelters, the estimated incidence of infection was roughly 6% for calicivirus and 45% to 47% for herpesvirus (Pedersen et al., 2004). Upto 40% of cats in shelters may exhibit URTD symptoms during their first week there, and a higher percentage may show symptoms during the second week, according to incidence estimates based on clinical cases (which do not take into account the causative agent). Although more research is needed, it appears likely that the prevalence and incidence of respiratory agent infection differs by shelter depending on the community's exposure to these agents, the time of year, the virulence of certain strains (especially calicivirus), management, and other factors.

The prevalence of recovery of *Bordetella bronchiseptica* using oropharyngeal or nasal swabs from cats varies considerably across shelters. In the U.K. prevalence estimates ranged from 0% to 19.5% among cats from rescue catteries. In the U.S. a study of four shelters in southern Louisiana found 3.1% of cats were culture positive, and in a shelter in Colorado, 4.9% of cats with signs of acute respiratory disease were culture positive. Unfortunately, in the Louisiana study, prevalence estimates were calculated by lumping cats from all rescue catteries or shelters together, which obscured the variability of prevalence among the shelters studied. Similarly, the variability of *B. bronchiseptica* recovery among shelters within the same region and across regions is unknown.

Studies of seroprevalence of antibodies against Bordetella suggest that infections are common during the lifetime of cats. Seroprevalence estimates were as high as 83% among rescue cats in the U.K. (McArdle et al., 1994), but appeared to be lower (24.1%) in four U.S. shelters (Hoskins et al., 1998). Anecdotally, some shelters in the U.S. report endemic Bordetella associated with URTD, while others do not. Factors (e.g., strain differences, presence of other agents, management) that influence the persistence of Bordetella in shelters are poorly understood.

Evidence strongly suggests that cats and dogs infect each other with *B. bronchiseptica*. Molecular typing of Bordetella isolates from dogs and cats using pulsed–field gel electrophoresis (PFGE) supports the hypothesis that transmission between these species is likely to occur. Dawson et al. (2000) reported that the PFGE patterns of Bordetella isolates were more similar among dogs and cats living together, than from two cats (or two dogs) living separately. Rand et al. (2002) examining isolates from dogs and cats from two U.S. shelters, reported similar PFGE and antibiotic resistance patterns from affected dogs and cats. Using a different methodology, Binns, and coworkers (1998) found that living in a household with dogs with respiratory disease was a risk factor for infection of cats with Bordetella. Estimates from surveys in the U.S. suggest that *C. felis* infections may account for 5% to 10% of feline respiratory disease. Prevalence of *C. felis* infections based on PCR or culture among laboratory and privately owned cats with conjunctivitis ranged from 13% to 34%, depending on country of residence, origin of cats (e.g., shelter, cattery), and the assay used. Using PCR, Italian researchers found prevalence estimates in three colonies of stray cats to range from 13.9% to 27.3%. The organism has been isolated from 4% and 6% of clinically normal cats as well (from rectal and conjunctival swabs, respectively). Fukushi et al. (2002) demonstrated that prevalence estimates were higher using PCR than bacterial culture in Japan, but what proportion of PCR-positive cats are capable of transmitting agent is unknown. Colonies of cats with endemic *C. felis* infections have been identified, suggesting that shelters (particularly those housing cats long term) are at risk. *Chlamydophila felis* infections occur worldwide and do not have known geographic or environmental predilections (Ramsey, 2000).

Around the world, cats' oro-pharynxes have been found to contain Mycoplasma organisms. These organisms were recovered from the upper respiratory tract of clinically healthy cats at rates ranging from 39% to over 80%, with no discernible regional preference. According to a study conducted in a Colorado shelter, *Mycoplasma* spp., out of 60.7% of cats with respiratory symptoms (37 out of 61). All of the infected cats had co-infections with different organisms. Accurately assessing trends in prevalence over time is challenging for the reasons mentioned above. Geographical location determines seasonal variations, which typically correspond with shifts in the quantity of vulnerable kittens and the hours when cats are most likely to be outside socializing with other cats.

Two studies provide some insight into long-term trends. Evaluating submissions of samples to a diagnostic service in the U.K. from 1980 to 1989, Harbour et al. (1991) failed to demonstrate significant trends in viral recovery rates among cats with clinical signs in that decade. Also, in 1992, another study of clinically healthy cats attending cat shows in the U.K. demonstrated that the recovery rates of calicivirus and herpesvirus were not significantly different compared to those of a similar survey conducted before the introduction of widespread vaccination. These data suggest no effect of vaccination on the relative frequency of calici and herpesvirus infections over time.

5.2. CLINICAL SIGNS

Clinical signs alone cannot be used to diagnose the causative agent, even though some may be more frequently linked to a particular pathogen than another (Figure 5.1). Sneezing and discharges from the nose and eyes are linked to FCV, FHV-1, and *B. bronchiseptica* infections. For viral pathogens, feline upper respiratory infections typically take one to six days to incubate; for primary bacterial infections, this time may extend.

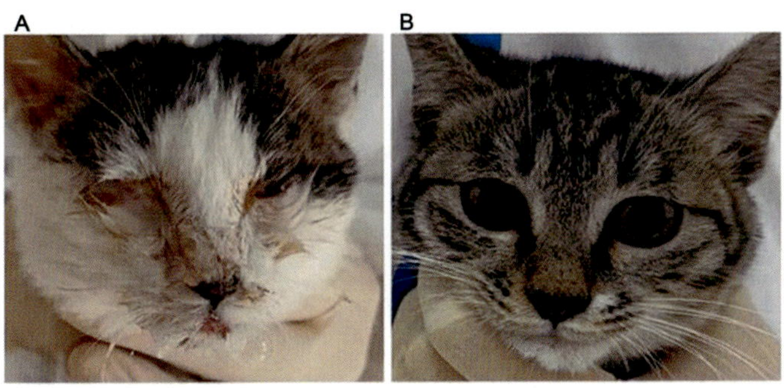

Figure 5.1. *Typical clinical signs after FHV infection. (**A**) FHV-A1 isolate; (**B**) negative control group.*

Source: Deng M, Liang H, Xu Y, Shi Q, Bao F, Mei C, Dai Z, Huang X. Identification, Genetic Characterization, and Pathogenicity of Three Feline Herpesvirus Type 1 Isolates from Domestic Cats in China. Veterinary Sciences. *2024; 11(7):285. https://doi.org/10.3390/vetsci11070285.*

Although it can also happen with a herpesvirus infection, oral ulceration (mainly on the tongue but also on other parts of the mouth, lips, or nose) is frequently linked to FCV infection. The virus has been linked to the genesis of lymphoplasmacytic gingivitis stomatitis (LPGS) complex, and lameness and skin lesions are also observed in some FCV-infected cats (possibly due to variations in tissue tropism among strains). Between 33 and 50% of people have died in outbreaks brought on by VS-FCV strains. Signs of widespread lesions in cats infected with hypervirulent strains include fever, anorexia, subcutaneous edema (often of limbs and face), and ulceration of the mouth and skin (especially on the paw pads, pinnae, and nares). Affected cats may have pancreatic, liver, and spleen necrosis, as well as bronchointerstitial pneumonia at necropsy. Herpesvirus infections, with the possible exception of hypervirulent FCV strains, typically result in more severe disease than calicivirus infections, particularly in young kittens where mortality rates may be elevated. Herpesvirus symptoms can manifest early as fever, depression, inappetance, and sneezing, which can be followed by serous nasal and ocular discharges. Drooling, a rare oral ulcer, viral pneumonia, and neurological symptoms are possible additional symptoms. While conjunctivitis, keratoconjunctivitis sicca, eosinophilic keratitis, stromal keratitis, symblepharon, and corneal sequestrum are common acute, mild ocular disease manifestations caused by FHV-1, it can also result in more chronic manifestations. Similarly, cats with herpetic dermatitis have been diagnosed thanks to the advancement of diagnostic PCR. Additionally, FHV-1 has the potential to cause osteolytic alterations in the nasal turbinate bones and could play a role in the development of chronic rhinosinusitis in cats (Coyne et al., 2007).

Signs of infection with *B. bronchiseptica* may include oculonasal discharges, sneezing, conjunctivitis, submandibular lymphadenopathy, fever, and lethargy. Coughing may also occur, but coughing in cats is much less common than in similarly infected dogs. Subclinical infections are common. Disease is generally more severe in kittens, sometimes producing pneumonia and death. The virulence among Bordetella strains varies, potentially explaining the wide variability among shelters reporting this organism as a significant problem.

C. felis is primarily linked to conjunctivitis, which typically affects one eye but can also affect the other. Conjunctival hyperemia, blepharospasm, chemosis, and serous to mucopurulent ocular discharge are examples of clinical signs. High dose exposure has been shown in experiments to cause systemic symptoms like fever, depression, nasal discharge, lameness, and stunted growth, as well as possibly in shelter settings.

Mycoplasma spp. infections in cats that are already compromised by other respiratory agents cause diseases of the lower respiratory tract, such as pyothorax, pneumonia, and bronchopneumonia. Growing evidence suggests that *Mycoplasma* spp. might also be the main respiratory pathogens in cats. Reinfections with these agents are not uncommon, even though cats recuperating from respiratory infections may be resistant to the agent causing the initial disease for several months to years after infection. Every respiratory agent has the potential to induce subclinical infection.

5.3. THERAPY

The way upper respiratory tract infections are treated in shelter cats is not the same as it is in owned animals. Due to the concentration of these bacteria in shelter environments, cats living there have an increased risk of contracting various secondary bacterial infections as well as respiratory agents like *Bordetella bronchiseptica* and *Chlamydophila felis*. Cats in shelters are frequently under a great deal of stress, which makes it harder for them to fight off infections from outside sources and endogenous bacterial overgrowth.

Without conducting antibiotic sensitivity testing or culturing, respiratory infections are typically treated empirically. Antimicrobial therapy should only be used for cats that have obvious signs of a secondary bacterial infection (e.g., purulent nasal or ocular discharges), since the majority of respiratory infections are viral in nature and because of growing concerns about the emergence of antibiotic-resistant bacteria. Rather than administering antibiotics to every cat exhibiting respiratory symptoms that is only mildly ill, steps that lower the chance of developing secondary bacterial infections should be taken. These include maintaining proper ventilation in spaces where ill cats are kept, making sure the mucociliary system gets enough water, lowering anxiety levels in the isolation unit, doing a thorough cleaning, and limiting the spread of fomite.

Broad-spectrum antibiotics with efficacy against anaerobes are recommended for bacterial infections secondary to normal flora overgrowth. Empirically, shelter veterinarians report better response to treatment of cats (with evidence of bacterial infections) when therapy is started with doxycycline (5–0 mg/kg q 12 h PO). This response to doxycycline probably reflects the higher frequency of Bordetella and Chlamydophila infections in shelter cats and the higher likelihood of secondary infections with *Mycoplasma* spps. normally resident in the upper respiratory tracts of cats (against which doxycycline is usually effective). Doxycycline is also relatively inexpensive and can be administered once daily. Caution should be exercised when administering doxycycline tablets, however, because of the risk of esophagitis and esophageal strictures. Doxycycline can be easily compounded in liquid form and treatment should be continued for a minimum of 10 days, regardless of remission of signs. Other antibiotics useful in treatment of URTD include amoxicillin-clavulanic acid (15 mg/kg orally every 12 hours) and first-generation cephalosporins (Lappin, 2003).

In shelters, cats with chronic rhinitis that may be secondary to osteochondritis from herpesvirus infections are sometimes seen by veterinarians. Because clindamycin is effective against gram-positive and anaerobic bacteria and can pierce bone and cartilage, it is frequently used to treat affected cats. The cats should receive treatment in shelters equipped to handle their needs for a minimum of four to six weeks, or until the respiratory symptoms have subsided after two weeks. Overall, to relieve dehydration, promote appetite, and increase comfort, afflicted cats should receive supportive therapies akin to those given to owned cats.

Mycoplasma spp. and *Chlamydophila felis* are usually isolated from cats that have conjunctivitis, and infected cats should apply topical tetracycline ointment three to four times a day. In some shelters where frequent treatment is not an option, a once-daily oral doxycycline regimen may be a more sensible option. It is also effective. To totally eradicate the Chlamydophila infection, treatment might need to be continued for three to six weeks. If the treatment is not given long enough, symptoms, and shedding that go away at first could come back after the treatment is stopped.

Ocular diseases associated with the herpesvirus can present in various ways. Because the disease can spread quickly, affected eyes need to be checked every day. Topical antibiotic ointments that lubricate the eye and treat secondary bacterial infections are effective in treating the majority of cases. The best medications are ointments with erythromycin or tetracycline, as they work well against Chlamydophila. The course of the disease will determine how long the therapy lasts. Topical antiviral products and antibiotics constitute the minimal effective treatment regimen for cats suffering from herpetic keratitis. The antibiotic's goal is to stop the viral corneal lesions from becoming infected with bacteria, which could cause serious, sight-threatening consequences. It is advised to apply NeoPolyBact or erythromycin ointment, both of which are less than $10 per tube, two to three times a day. Cats should be careful not to contaminate the tip of the tube if they are sharing it with one another, as this is a great way to spread FHV.

Right now, several strategies are both clinically and financially successful (unfortunately, the majority of other antiviral treatments are still highly expensive or necessitate frequent dosing). The shelter's resources will determine which approach is best to use. Using idoxuridine is one choice. This drug, which is older, has strong anti-FHV properties. Veterinary professionals need to compare prices because it needs to be compounded. Idoxuridine's main disadvantage is that it needs to be taken at least four times a day to be effective.

Cidofovir 0.5% solution is the second antiviral choice. Cidofovir has anecdotally been effective when used in shelters. This antiviral only needs to be taken twice a day to be highly effective and has good anti-FHV activity.

The drug is also not widely available and needs to be compounded. It costs roughly twice as much as idoxuridine, but because it needs to be taken less frequently, the bottle lasts a lot longer. An antiviral solution of betadine diluted 1:30 with sterile saline, applied one drop in the affected eye twice a day, has shown an encouraging effect, according to empirical reports from some shelter veterinarians. It is best to continue antiviral and antibiotic treatment for two weeks, or until the cornea is fluorescein negative. Culture and antimicrobial susceptibility testing are necessary to treat cases of severe lower respiratory infections or to contain respiratory disease outbreaks. While bronchoalveolar lavage (or necropsy, in the event that affected cats pass away or are put to sleep) is the recommended method for obtaining diagnostic samples for lower respiratory diseases, conjunctival, and oropharyngeal samples can be helpful in isolating upper respiratory pathogens.

5.4. MODES OF TRANSMISSION

URTD agents are primarily spread through oral, nasal, and ocular secretions and are extremely contagious. Cats' tidal volume is believed to be too small for effective aerosol, although when they sneeze, cats can release viral macro droplets upto four feet in the air. Droplet transmission is most likely a concern in shelters that have group housing, where cats can sneeze near to one another, and in which the distance between facing cage banks is less than four feet. Direct cat-to-cat transmission or fomites are the main ways that URTD is spread in shelters.

Even in a dried state, calicivirus can endure in the environment for up to 28 days at room temperature. In animal shelters where cats are housed individually, contaminated feed is most likely the main way that FCV spreads. According to recent research, the survival time of FCV varies depending on the surface, but it can persist for hours on improbable surfaces like computer keyboards and phone buttons. Herpesvirus-1 is more easily spread by fomites, despite being more delicate outside the body, lasting only 18 hours at 15□ in a moist environment and up to 12 hours in a dry one. This is likely due to frequent handling of infected cats and inadequate cleaning and disinfection.

Chlamydophila organisms have relatively poor survival outside the host, but Bordetella organisms may survive outside of the body in low-nutrient fluids. These agents are transmitted by direct cat-to-cat contact and indirectly, by fomites. *Chlamydophila felis* is also shed from the reproductive tract and young kittens can be infected during parturition. Conjunctivitis caused by *C. felis* can persist or recur, and shedding can continue for several months following apparent recovery unless effective antibiotic therapy of adequate duration is administered (Gaskell, 1992).

Cats with ocular or nasal discharges (serious or mucopurulent), sneezing, or nasal congestion are considered to have URTD since the agent causing URTD in cats in shelters cannot be reliably identified based solely on clinical signs. They need to be kept apart from healthy cats because they are the most effective shedders of the agent. For these agents, the carrier state is well-established. *Herpesvirus, calicivirus, Bordetella bronchiseptica*, and *Chlamydophila felis* infected cats shed the agent (and are contagious) during the infection process as well as during the incubation phase, during clinical signs, and after clinical disease recovery. Cats who never show any clinical symptoms are also susceptible to infection and shedding of any of these agents. However, the frequency and volume of shedding differ depending on the agent and infection stage. Herpesvirus shedding can happen as soon as 24 hours after infection and typically lasts for one to three weeks. Cats may still have FHV-1 DNA after three weeks, but it's unclear if they are actively shedding the virus at this point. 80–100% of cats become latent carriers after recovering from the initial infection; they are infected but do not shed the virus. Even in the presence of maternal antibodies, kittens may become latently infected after being exposed to the queen, guaranteeing the virus's spread to succeeding generations.

The trigeminal ganglia are the main location of latency. After a stressful event (e.g., being taken to a shelter, giving birth, and being rehoused inside the shelter), but in one study, about 45% of the cats started to reseed the virus four to 11 days (mean 7.2 days) after the stress and kept on shedding for 1 to 13 days (mean 6.5 days). In shelters, the percentage of infected cats whose infection relapses after a stressful event might even be higher. As previously mentioned, Pedersen et al. (2004) found that within the first seven days of entering two California shelters, the percentage of cats shedding the herpesvirus increased tenfold. About 24 hours after infection, infectious calicivirus also shows up in respiratory secretions, and it keeps shedding long after clinical recovery (August, 1984).

According to a study, 50% of cats were still shedding virus at 75 days after infection, and the majority of cats were still shedding 30 days after infection. Some cats continue to shed all their lives. Several studies indicate that different FCV strains may shed for varying lengths of time. The persistence of shedding is probably unrelated to stress, unlike FHV-1, regardless of the strain. One study involved *B bronchiseptica* -infected kittens, bacteria were seen in the oropharynx as long as 19 weeks after infection. It was unknown whether the organism concentrations were high enough to spread to other cats. One of the two pregnant queens experienced shedding due to parturition, but the young did not have the infection (Dawson et al., 1996).

Similarly, *C. felis* organisms can persist in the conjunctiva for two months or longer in naturally and experimentally infected cats, and the organism has been recovered from 4% to 6% of clinically normal cats. The transmission risk these cats pose to others is unknown. It is not known whether Chlamydophila infections remain latent, being reactivated when cats are stressed, or whether some infected cats are persistently infected with slowly replicating organisms. Regardless of the causative agent, it is not practical or necessary to routinely identify subclinical shedders in shelters. In light of the prolonged shedding period of calicivirus, recommendations to isolate recently recovered cats from healthy cats in purebred catteries have been made (Knowles and Gaskell, 1991).

The majority of shelters find it impractical to implement similar recommendations due to the high demand for cage space and the highly variable shedding period among cats. Thankfully, they do not shed the virus as effectively as cats that exhibit symptoms. However, extra care may be necessary when handling unusually virulent viral strains, like virulent systemic feline calicivirus. Reintroducing recently recovered cats into a vulnerable population or adoptive home should be done with caution in cases where severe or fatal disease has been documented, especially in otherwise healthy, vaccinated cats. Cats with virulent FCV who have recently become infected and are subclinically affected can spread the fatal disease.

5.4.1. Risk Factors for URTD

Feline URTD is common in cat populations, and it is most likely to affect cats living in groups, like shelters. Based on the findings of two studies and anecdotal reports, the duration of stay in the shelter seems to be a major risk factor for the development of URTD. In these studies, the risk of developing URTD increased between 5% and 13% per day after staying in the shelter for approximately five days. Although the frequency of the different agents may differ depending on the shelter, there is no established geographic pattern to their occurrence. In kittens, young cats, and rescue catteries, the common upper respiratory agents are more likely to cause severe illness on a population-level. Numerous URTD epidemiologic studies have also identified additional risk factors. Aged cats (11 years and older) were significantly more likely to develop URTD (compared to cats 7 to 11 months of age) in a cohort study of cats in a northeastern U.S. shelter). Increasing age, residence in a rescue cattery, and less than excellent hygiene increased the risk of *B. bronchiseptica* infection in European catteries. In a similar study of risk factors associated with *B. bronchiseptica* infection in rescue catteries in the U.K., only the presence of respiratory signs in dogs was significantly associated with Bordetella infections in cats. In studies of risk factors for *C. felis* infections, males, the Birman breed, and young age were associated with significantly higher risk. Other important risk factors include stress and increasing numbers of cats in shelters (Binns et al., 1999; Lait et al., 2005; Pedersen, 1991; Speakman et al., 1999).

5.5. DIAGNOSIS

Diagnostic testing is rarely used by shelters to confirm URTD or pinpoint the precise agent or agents causing illness in their cats. When an outbreak occurs, meaning that the frequency or severity of the disease is genuinely higher than what is typically observed, additional diagnostic testing is warranted to inform prevention and treatment plans. For instance, with the implementation of suitable preventive and control measures, the identification of cats exhibiting symptoms consistent with virulent calicivirus infection and subsequent confirmation with diagnostic testing could prevent transmission to other animals.

The diagnostic laboratory should be consulted prior to gathering and submitting samples for diagnosis. Swabs from the conjunctiva and oropharynx are most frequently used to cultivate respiratory agents. Viral transport media should be used to transfer samples intended for viral isolation to a diagnostic lab. Samples for chlamydial culture should be placed in a chlamydial transport media (e.g., 0.2 m sucrose and 0.02 m phosphate), and those to be cultured for *B. bronchiseptica* should be placed in charcoal Amies transport medium before transport to the laboratory. When isolating Bordetella organisms from cats, particularly those exhibiting symptoms of lower respiratory disorders, requires the collection of bronchoalveolar lavage samples. Ascertaining the objective of the examination, additional diagnostic methods (e.g., PCR). It is important to proceed with caution when interpreting PCR results, especially if you are trying to determine what is causing your current respiratory symptoms. Respiratory pathogen subclinical carriers are frequent, and a positive PCR result does not always indicate a cause-and-effect link.

5.6. DISEASE CONTROL AND PREVENTION

Due to carrier states, the agents' high infectivity, their ease of transmission (particularly on fomites), the incapacity of vaccination to prevent infection and shedding, and the susceptibility of cats to reinfection, the elimination of upper respiratory tract infections in cats in animal shelters is practically impossible. However, management strategies can significantly lower the incidence and severity of illness. A wide range of control measures must be used to minimize the occurrence of feline respiratory agents due to the numerous challenges posed by their biology.

5.6.1. Minimize Agent Concentration in the Environment

5.6.1.1. Avoid Overcrowding

Overcrowding is one of the biggest risks to the health of cats in many shelters nowadays. Overcrowding a facility with more animals than it can adequately care for results in higher levels of stress for caregivers and the animals themselves, worse sanitation, and more agents in the environment. Cats and other animals that have high rates of respiratory infections are at risk due to overcrowding. A shelter's capacity to house cats will be determined by a number of factors, such as the layout and size of the building, the kind of housing, the ventilation and air quality, the number of employees, and the efficiency of the cleaning and disinfection program. It's likely that shelters with high respiratory disease rates have too many cats to properly care for them.

5.6.1.2. Cleaning and Disinfection

Effective cleaning and disinfection protocols are crucial to minimizing the occurrence of URTD because respiratory agents can survive for variable lengths of time outside of the host's body and because fomites are crucial to the transmission of URTD. The nonenveloped calicivirus is the most challenging respiratory agent to eradicate from shelter surfaces and fomites among the five previously discussed respiratory agents. Whatever the product or agent, organic matter must be thoroughly cleaned and removed before disinfection can begin. This is an important step that shouldn't be skipped because most disinfectants become inactive when organic matter is present. Cage surfaces, especially the hard-to-clean cage doors, are frequently contaminated by nasal and ocular discharges; therefore, staff members must first scrub these surfaces with hot water and detergents before rinsing and disinfecting (Figure 5.2).

Calicivirus and other respiratory agents in the environment can be effectively killed by using 5% sodium hypochlorite (bleach) diluted 1:32 (1 part bleach to 32 parts water or ½ cup bleach per gallon of water) and allowed to sit on the surface for at least 10 minutes after further washing. Higher concentrations are irritating to both animal and human respiratory tracts, possibly making them more susceptible to respiratory infections, but they are not required to efficiently eradicate calicivirus. Higher concentrations should therefore not be regularly used in shelters (apart from during periodic "deep" cleanings when animals are temporarily removed from their surroundings or in response to ringworm outbreaks). To ensure that staff members prepare the concentration correctly each day, veterinarians should insist that instructions for making bleach solutions be posted close to the mixing area. To ensure that the concentration (when mixed) of bleach is still high enough to kill calicivirus and other respiratory agents, it must be kept out of direct sunlight and stored in either its original container or one that has been darkened or opaque. Before putting the animal back in its cage, the surface needs to be dried and the bleach solution should be rinsed off (Boothe, 1998; Lawler, 1989).

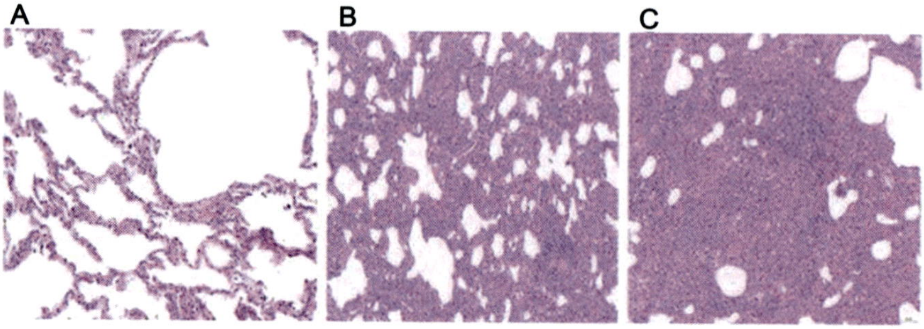

Figure 5.2. *Microscopic pathological changes of the lung after FHV infection. (**A**) Control; (**B**) mild interstitial pneumonia; (**C**) severe interstitial pneumonia. Scale bar = 50 μm.*

Source: Deng M, Liang H, Xu Y, Shi Q, Bao F, Mei C, Dai Z, Huang X. Identification, Genetic Characterization, and Pathogenicity of Three Feline Herpesvirus Type 1 Isolates from Domestic Cats in China. Veterinary Sciences. 2024; 11(7):285. https://doi.org/10.3390/vetsci11070285.

Quaternary ammonium disinfectants do not consistently inactivate calicivirus, but they are effective against the other respiratory agents. For the purpose of evaluating disinfectants, FCV has been widely used as a stand-in for human noroviruses, which are also members of the Caliciviridae family. There has been at least one published report on a novel quaternary ammonium product that effectively combats FCV. However, only goods with established efficacy ought to be utilized until further independent testing is conducted. Hard water, soap, and low pH decrease the effectiveness of quaternary ammonium disinfectants against bacterial and herpesvirus agents (Lawler, 1989; Boothe, 1998).

Although potassium peroxymonosulfate is more costly than bleach, it is also more effective against calicivirus and has other advantages over bleach, such as being less corrosive, less irritating to the respiratory system, and working better against organic matter and carpets. The disinfection protocol must include bleach or potassium peroxymonosulfate, though these don't have to be applied every day. Some shelters employ tactics like switching these compounds out for quaternary ammonium products to lessen the environmental corrosiveness of bleach and, if they're using Trifectant ®, to lower the cost of disinfection (Gilman, 2004). Veterinarians should track disease rates before and after protocol modifications to assess the efficacy of disinfectants in reducing URTD in their shelter, regardless of the approach they take.

It is imperative to wash and disinfect water and food dishes, toys, and other equipment that comes into contact with cats in shelters due to the significant importance of fomite transmission of feline respiratory agents. Similar to smocks, pens, and clipboards, these items should either stay in their assigned locations or be cleaned and sanitized before being moved between areas. Every residential area should ideally have access to both hot and cold water. Cleaning areas containing possibly exposed, sick, and healthy animals requires different personnel and equipment, as is the case with the control of all infectious diseases. Washable aprons or smocks with specific areas for use are ideal for staff members working in shelters where they must cover multiple tasks. If it isn't feasible, cleaning should start with areas that house high-risk residents (e.g., kittens) to the sickest ones at the end. Meeting and greet spaces need to be cleaned and disinfected as frequently as possible during the hectic adoption hours.

Many methods of cleaning have been suggested for shelters that house individual cat cages. Although methods that avoid or minimize handling the cats (thereby reducing stress and possible agent transmission by the handlers) are theoretically desirable, no research has shown which approach is better than the other. It has been shown that in a significant number of latently infected cats, simply moving the cats around in their cages can trigger herpesvirus shedding again. Therefore, it might be better to scoop the litter box and add fresh water and food every day in individual cat cages if the cage stays clean rather than handling (and stressing out) each cat during cleaning every day. In a similar vein, cats kept in cages with multiple compartments might be moved into one while the other is being cleaned, saving them from stress and possible contact with contaminated handlers.

5.6.1.3. Hand Disinfectants

Hand washing or hand sanitizing should be done frequently because people handle cats that shed respiratory agents. People can reduce the risk of spreading agents from one cat to another by washing or dilution of bacteria and viruses from their hands. Hand disinfectants with an alcohol base (70% alcohol is advised) work well against enveloped viruses and vegetative bacteria, but not nonenveloped viruses. They may need to be applied more than once because of their rapid evaporation, which depends on adequate contact time. In recent tests using FCV as a surrogate for human noroviruses, Gehrke et al. (2004) found isopropanol to be more effective at reducing FCV concentrations on finger tips than ethanol. Since there is strong interest in finding products effective against human noroviruses (closely related to FCV), other hand disinfectant products are being developed that may work well against FCV (as well as the other respiratory agents) (Galabov et al., 2006). Requiring the use of hand sanitizers between cats may be more practical because it is frequently impractical for staff, volunteers, or the general public to properly wash their hands after handling individual cats (despite their questionable efficacy against calicivirus). If there are known or suspected calicivirus outbreaks, people should wash their hands thoroughly or change their gloves frequently.

Gilman (2004) wrote, " While there are modern and effective disinfectants and cleaning tools, it is the animal care staff's attention to detail that will make a critical difference in the health of shelter animals."

Employees need to be inspired to follow these cleaning and disinfection procedures and made aware of how important they are. They might not be aware of the significance of their actions in promoting or preventing the spread of URTD since they are ignorant of the etiology and transmission of the disease. Veterinarians should make sure that written documentation of a shelter's sanitation procedures is available and that volunteer and new employee orientations include them. Furthermore, even with excellent training programs, resourceful employees may find ways to cut corners or overlook the significance of specific protocols. Because proper URTD management depends so heavily on good sanitation, veterinarians should routinely check on staff cleaning and take appropriate action when staff members deviate from protocol.

5.6.1.5. Ventilation

It is not necessary for veterinarians to be ventilation specialists to help shelters enhance the ventilation in their buildings. While it is unlikely that aerosolization of respiratory agents will be the main mode of transmission, adequate ventilation can help control humidity levels and lower ammonia and particle fumes that raise the risk of URTD (Johnson, 2004). Reducing the risk of secondary bacterial infections is likely another benefit of good ventilation. To prevent hair and other debris from obstructing air flow, veterinarians can advise shelters on the significance of routinely cleaning grills, ventilation ducts, and air filters. It is easy to determine whether and how much force air is moving by doing simple tasks like holding a piece of chapter up to an intake register. To accelerate drying after cleaning and lower humidity, veterinarians can also advise turning on fans. This will prevent mold from growing and bacteria from surviving in the environment (Gilman, 2004). Fans ought to be positioned so that the animals are not directly exposed to airflow.

Veterinarians should confer with heating, ventilation, and air conditioning (HVAC) specialists in shelters with high rates of unit recovery due to thermal shock to ensure that the HVAC system is operating at peak efficiency. HVAC specialists might not be aware of comparable guidelines for animal facilities, despite the fact that many of them are familiar with human ventilation optimization guidelines. Veterinarians need to know what is ideal in terms of disease control to consult effectively. Animal facilities should ideally have 10 to 12 full air changes per hour as opposed to recycling air inside the building, unlike human housing facilities. Even in the presence of adequate air circulation in the surrounding environment, cages with three walls may restrict air flow and result in less than optimal air quality for individual cats because it is likely that relevant air quality is at the level of a cat's nose. Separate ventilation has always been advised, especially for cat isolation rooms in shelters. Investing in pricey isolation rooms with separate ventilation may not be cost-effective, as it is unlikely that aerosolization of feline respiratory agents is a significant mode of transmission. There is no evidence of aerosol transmission between dogs and cats in shelters, however there might be an exception in places where kennel cough is endemic and cats have been found to have Bordetella infections. The ideal temperature range for the shelter is 65 to 75□, which is the lower end for adults and the upper end for kittens (Sinclair, 1997).

There may be temperature-extreme microenvironments in the shelter, such as cages beneath windows in the winter, which, if left unchecked, could put some cats at high risk of URTD. Although it is ideal to continuously bring in fresh air and exhaust circulated air, outside air frequently needs to be heated or cooled for the comfort of both humans and animals. Ventilation systems should ideally enable control of the recirculation ratio to reduce costs. This would involve raising the amount of outside air during respiratory disease outbreaks and lowering it during periods of low rates. Although ideal humidity levels are most likely closer to 35 to 45%, relative humidity levels of between 30 and 79% are still desirable.

When it comes to purchasing cages, when veterinarians have the luxury of offering advice, cages that allow air flow (e.g., compared to those enclosed on three sides, open fronts with back vents or two open sides) offer better air flow. As was previously mentioned, cages with shelves or connecting portals can reduce handling that causes stress when cleaning.

5.6.1.6. HEPA Filters, Ultraviolet Light, and Ozone Generators

UV light and HEPA (high-efficiency particulate air) filter units are frequently used by human health care and research facilities to lower the concentrations of microbes in their surroundings. Because costly filters need to be changed regularly due to clogging from animal dander and fur, stand-alone HEPA filter units are typically not feasible to run in animal facilities. They can remove all but micron-sized particles when installed correctly in central systems behind filters meant to remove hair and larger particles. HEPA filters must only be installed by professionals who are aware of their uses, placement restrictions, and maintenance requirements (Johnson, 2004). The mechanism of ultraviolet light disinfection involves breaking down cellular DNA, but the microbes must be exposed to the UV rays. Although they have numerous drawbacks, ultraviolet light units are utilized in medical facilities and research facilities to eradicate microorganisms from surfaces and air streams. Because UV light is harmful to skin and eyes, it must be used after cleaning to remove organic matter from surfaces. Additionally, people, and animals must leave the area before disinfection. Areas that are shaded from the UV light are not disinfected, necessitating the use of other methods to sanitize them. Consistency in surface disinfection is crucial. To disinfect air streams in ventilation systems, UV disinfection lights can also be installed. These devices seem ill-advised in shelters right now because UV light disinfection of air streams has a history of variable success, the aerosol transmission of feline respiratory agents is unlikely to play a major role in URTD transmission, and UV light does nothing to remove particulate matter and other respiratory irritants from the air.

Ozone is produced by ozone generators, and it irritates the respiratory system. Ozone generators are not advised for use in shelters because ozone can weaken the body's defenses against respiratory infections and exacerbate long-term respiratory conditions like asthma. It is unknown how much of a difference UV light disinfection units or HEPA filters make in terms of the prevalence of feline respiratory diseases in shelters. Installing HEPA filters and UV light disinfection units won't seem to have the same impact on URTD rates as good old-fashioned cleaning and disinfecting protocols using old methods, along with enhancements to staff training, stress reduction, efforts to reduce fomite transmission, and appropriate vaccination protocols. Probably the most effective way to improve air quality is source reduction, e. g. reduce the amount of airborne pollutants produced in cat housing areas by preventing overcrowding and improving cleaning (EPA, 1990).

5.6.2. Enhancing Host Resistance

5.6.2.1. Reduce Stress

Any source of stress(e.g., emotional, physical, environmental) can potentially reduce immune function (including responsiveness to vaccination) and increase shelter cats' susceptibility to the development of URTD or prolong their recovery. Stress (e.g., entering a shelter) is a documented trigger for the recrudescence of herpesvirus shedding and disease. Therefore, a consistently high plane of nutrition (appropriate for age and physiologic status), environment enrichment, appropriate handling, a comfortable environment (e.g., away from loud noises, barking dogs), and other stress-reducing strategies in the shelter such as turning off the lights at night are essential to an effective URTD management program (Case and Fahey 2004; Goldman et al., 2004; Griffin and Hume, 2006; Lawler

1998). Staff stress is one risk factor for URTD that isn't as well-known. Elevated levels of staff stress are almost certainly linked to a higher risk of URTD in cats, most likely via multiple pathways. Depending on where the stress is coming from, high staff stress can lead to things like handling cats roughly or paying insufficient attention to details in preventive medicine protocols. Effective URTD control programs depend on the staff and volunteers in shelters being managed properly.

5.6.2.2. Vaccination

Merely depending on vaccinations will not be sufficient to lower the incidence of URTD in shelters. The respiratory vaccinations do not shield against infection, agent shedding, or carrier states that develop after infection. Instead, the goal of these vaccinations is to decrease the severity of the illness. Before entering the shelter, all cats (or only those deemed adoptable, for long-term care, and in court cases, if cost prohibitive) should receive a modified live vaccination against FCV and FHV-1. The American Association of Feline Practitioners (AAFP) lists these as essential shelter vaccinations. It is not advised for most shelters to use killed vaccines because the goal is for cats to become protected quickly. Achieving a quick immune response ensures that cats in the shelter are as protected as possible from other infectious agents in that setting. Shelters in the U.S. are required to select between parenteral modified live virus (MLV) and intranasal (IN) vaccinations against FCV and FHV-1. In addition to having the advantage of inducing mucosal immunity where respiratory agents enter the body, IN vaccines can be used in the face of maternal antibodies in young kittens and provide faster immunity than their parenteral counterparts. It goes without saying that shelter staff members need to be qualified to give IN vaccinations.

In cats vaccinated four or six days prior to challenge with FHV-1, there was a noteworthy decrease in respiratory disease scores following the challenge, as evidenced by a recent study on a commercial IN vaccine for panleukopenia, FCV, and FHV-1. In cases where it is not feasible to foster kittens outside of the shelter, IN vaccinations against respiratory viruses may be administered to kittens as early as three weeks of age. One 0.5 ml IN dose may be split among two or three kittens (three to five weeks old), even though it is not mentioned on the label and has not been scientifically tested, to offer early protection. 4–6 days after vaccination, cats receiving IN vaccines may experience brief, mild clinical symptoms. It is necessary to allow for the possibility of mild URTD symptoms emerging during this time in shelters where cats are put to death due to URTD symptoms. Six-week-old kittens vaccinated parenterally with MLV vaccination will develop significantly higher antibody titers than their littermates who are not vaccinated in shelters that do not use IN vaccines. Kittens older than six weeks should be vaccinated every two weeks until they are 16 weeks old if they stay in the shelter for longer than two weeks. The AAFP report suggests that cats be vaccinated against FCV and FHV-1 using the schedule of every three years used for pet cats in sanctuaries where they may live for years. However, more research is required to support this recommendation. Parenteral vaccination is advised for the best defense against panleukopenia. The use of an IN vaccine, with or without calicivirus and herpes components, should be viewed as an addition to, not a replacement for, parenterally modified live vaccinations.

Due to its high genetic plasticity, FCV is able to adapt to the selection pressures in its environment, which makes it more difficult to develop vaccines that offer equal protection against all isolates. The majority of the vaccines on the market today were developed after the frequency distribution of FCV field isolates was altered. The majority of current vaccines are based on a single strain, either FCV 255 or FCV-F9. Current vaccinations frequently do not offer protection against an increasing number of novel isolates, even though these strains are still in circulation. Vaccinated cats have not shown protection against virus isolates that cause virulent systemic disease, and vaccination failures have been linked to infection with FCV strains that are antigenically different from vaccine strains. Research is currently underway to develop a new generation of FCV vaccines that are effective against a wider range of FCV isolates. Europe has at least one vaccine that is effective against two strains of FCV, and a manufacturer in the U.S. has unveiled a vaccine to protect against the strain that causes virulent systemic (VS) illness. Given that isolates linked to VS-FCV outbreaks have proven to be distinct from both the current vaccine strains and one another, it appears highly improbable that the novel vaccine will offer defense against VS-FCV outbreaks in the future.

Bordetella bronchiseptica and *Chlamydophila felis* vaccines are not core vaccines for cats in shelters. These bacterins, like the URTD viral vaccines, do not prevent infection or shedding of agents. Their use may be warranted in the face of outbreaks where these organisms have been identified, or in shelters with endemic infections caused by these agents. When used, IN vaccination for *B. bronchiseptica* induces a nonspecific immunity that protects temporarily against other respiratory pathogens. When necessary, kittens may be vaccinated as early as four weeks of age for *B. bronchiseptica* and at nine weeks of age for *C. felis*. Anecdotally, some shelters experiencing *B. bronchiseptica* outbreaks in their cats report significant declines in clinical signs and in kitten mortality after the introduction of *B. bronchiseptica* vaccine. On the other hand, Schultz, and Ford report that in their experience this has not been true.

The above-mentioned recommendations are primarily grounded in immunological principles and research conducted outside of animal shelters. There aren't many studies on the effectiveness of vaccinations in shelters. The combination of an MLV intranasal and killed parenteral vaccination protocol reduced respiratory signs approximately 66% more than the parenteral vaccine alone, according to a vaccine trial conducted in a California shelter with 57 cats. This suggests that concurrent vaccination with parenteral and IN vaccines is preferable to either mode of vaccination alone. It's unclear, though, if this advantage was linked to better protection from a modified live vaccine or the mode of administration. IN vaccination reduced signs in litters of kittens and individual kittens compared with unvaccinated kittens during the first five days of shelter residence, but not in adults during the same period. This observational (and nonrandomized) study included 701 litters of kittens, 531 kittens, and 2203 adult cats in a Massachusetts shelter over a 50 week period. When compared to cats receiving the IN vaccine during the same period, parenteral vaccination with an MLV vaccine, however, seemed to reduce the frequency of URTD by about 60% in cats who had spent more than five days in the shelter. To determine the best vaccination schedules for shelters, more research is required. Shelter veterinarians should base their recommendations until then on current knowledge and the application of different protocols within their own shelter.

5.6.2.3. Drug Prophylaxis

Shelters have tried to prevent infection in cats by using L-lysine and interferon, which lessen the incidence, severity, and amount of virus shed (s). It has been demonstrated that the amino acid L-lysine decreases the recurrence rate, lesion severity, and healing time in people with herpes simplex virus type 1 (HSV-1)-related recurrent clinical lesions in clinical trials. L-lysine is believed to prevent viral replication by competing with arginine, which is necessary for the replication of herpesviruses. Lysine has been demonstrated in cell culture to impede FHV-1 viral replication, most likely through this mechanism. A few trials in cats are emerging. Stiles, Townsend et al. (2002) demonstrated a significant reduction in the severity of signs of FHV-1 associated conjunctivitis in a blinded, randomized trial of eight cats; there were four cats receiving 500 mg of lysine twice daily compared with four cats receiving a lactose placebo. The incidence of disease and frequency of virus isolation were not significantly different between the groups (although the statistical power to find differences between the groups was low).

In a 2003 study, cats were randomly assigned to receive either a placebo or 400 mg of oral lysine daily, and they were immediately stressed by a change in housing. A period of 15 days was observed before the administration of intramuscular methylprednisolone acetate to all cats. After that, observations were made for a further 15 days. After the shift in housing, the number of virus-shedding episodes in the treated group was significantly lower than in the placebo group, but not after prednisolone-induced immunosuppression. Unfortunately, the statistical power to identify differences in the incidence of signs or severity of disease was low because few cats in either group developed clinical signs.

In a subsequent clinical trial, Maggs et al. (2007) evaluated the safety and efficacy of oral lysine supplementation in diminishing the severity of respiratory disease in cats and measured plasma lysine and arginine concentrations. Trial cats were selected from a colony of cats that had recently experienced an outbreak of URTD. Designed to more closely simulate populations in shelters, Maggs et al., found that cats in the supplemented group had significantly more severe signs than those not supplemented. Male cats in the supplemented group also demonstrated more aggressive behavior towards each other than observed in the control cats. Since the arginine concentrations of supplemented cats were reduced in the supplemented group and failed to return to basal concentrations after treatment, Maggs et al. (2007) recommended more research to demonstrate the safety of supplementing with lysine. In light of these results, it seems prudent to avoid the use of lysine until its safety and efficacy have been more extensively evaluated in shelters.

Proteins called interferons, which are derived from cells, prevent a variety of DNA and RNA viruses from being synthesized, assembled, and released. Human and feline interferons have been demonstrated to exhibit antiviral activity *in vitro*; however, *in vivo*, this activity was only noticeable at high doses. In 1999, URTD was found to be less severe in cats treated with 25 U of natural human interferon alpha early in the disease than in control cats. There was no decrease in viral shed. To determine whether administering interferon to shelter cats can lessen the occurrence or severity of URTD, more research is required.

5.6.3. Minimizing Exposure

5.6.3.1. Quarantine

Quarantine is a highly recommended strategy in preventive medicine to lessen the spread of infectious agents among populations. In certain adoption guarantee facilities and sanctuaries, the application of quarantine (10–14 days) may be necessary. However, quarantine is probably ineffective in most shelters given the strong correlation between the length of stay and the risk of URTD and the need to keep days of residence as short as possible (to maximize the number of cats saved). Once examined, vaccinated, and neutered, cats that appear healthy should be admitted to the shelter. These animals are less likely to get sick in the shelter if they are placed on the adoption floor and taken out of the building as soon as possible. Prospective owners should be informed (ideally verbally and through a handout) that any cat adopted from a shelter may get an upper respiratory tract infection or "cold" (just like their children when they start day care). Some cats will develop URTD after adoption. Given this risk, the handout ought to outline the symptoms of URTD, explain why prevention is not possible, and advise new owners to keep their new cat alone for about two weeks. Adopters should also be counseled to seek veterinary assistance immediately if their cat develops mucopurulent ocular or nasal discharge or systemic signs (e.g., anorexia).

Euthanizing cats with URTD should happen as soon as possible in isolation shelters to stop the disease from spreading from these cats. Affected cats in shelters that treat respiratory infections should be quickly isolated in isolation rooms from healthy cats after identification (s). It is ideal for the room to have adequate ventilation without requiring the doors to be kept open. The goal of an isolation room is to reduce the possibility that sick cats will infect healthy ones with infectious agents. Isolation is accomplished by physically dividing sick cats from healthy ones and restricting the flow of personnel and other items between "clean" (sick cat housing) and contaminated (sick cat housing) areas. To stop the spread of highly contagious respiratory agents, this is crucial. In a facility, isolation doesn't mean having a different set of cages in a hallway or on one side of a room. The isolation room should ideally be situated away from young, healthy animals that are more prone to illness and should be placed for adoption. The isolation chamber should only be used by a small number of well-trained individuals who are aware of the dangers involved in working with cats that have URTD. The door(s) to the isolated area(s) should always be closed, and the area(s) should be clearly marked as such. After employees have left, doors may be left open if ventilation is hampered by closed doors. Staff are reminded by the closed doors and signage that there are cats with respiratory diseases living there and that visitors should wear smocks (which stay in that room). Cleaning supplies and other items should only be used in that room, and hands must be completely cleaned after leaving the space or gloves should be put on and taken off before leaving. To reduce the need to transport sick cats through areas with healthy animals to examine them, an exam or treatment table should ideally be located in the isolation room or very close to it contaminated stethoscopes, smocks, and so on, must be stored in a designated area so that no one will forget to use them with healthy cats. The goal of isolation is undermined by shelters that host sick cats in the hallways or permit unrestricted access to the isolation room.

Cats should be taken out of isolation once they have fully recovered, even though they may continue to shed organisms for variable and frequently extended periods of time after clinical signs have disappeared. After being taken out of isolation, these cats should be put up for adoption as soon as possible because it is not feasible for most shelters to keep them in isolation. Certain shelters have set aside spaces or group rooms for cats who have just recovered from illness as opposed to cats who have never been sick. Cleaning and care can then move from unaffected to recently recovered cats, which could be useful. Employees occasionally neglect to report sick cats promptly or even try to "hide" them with signs in the hopes that they will find homes instead of being put to death or kept in isolation. The most effective way to spread these highly contagious agents is through sick cats, so leaving sick cats among healthy cats puts the healthy cats at risk of contracting the disease as well. Employees need to be taught that prompt action can lessen animal suffering, stop symptoms from getting worse, or even save the lives of sick kittens. Similarly, staff members frequently want to let the isolation ward cats use the floor for cleaning purposes, citing the fact that the cats are rarely let out of their cages. Although the risks of this practice are unknown, it is known that cats can have multiple respiratory agent coinfections. An infection with a second agent is likely to prolong the recovery process for affected cats because they are already severely disabled. Mild exercise and socialization may be beneficial for sick cats, but these activities should be done in a location far from the floor, as this can be contaminated by debris that drifts out of cages during the day and by other cats. In between cats, this area needs to be cleaned.

5.6.3.2. Foster Care

Using a well-planned and run foster care program is one of the best ways to lower exposure among kittens and other high-risk cats. As soon as possible, kittens under the age of eight weeks should be placed in foster care. If they can stay out of the shelter or only go somewhere where there aren't any other cats, that would be great. Additionally, it would be great if these kittens could be adopted straight from the shelter, avoiding contact with agents there. Should kittens need to be kept in the shelter, they should have a space apart from the adult cats and away from dogs that bark. Due to declining mother immunity, eight week old kittens returned from foster care are particularly vulnerable to respiratory infections. It is crucial to house these kittens apart from adult cats and to market them as soon as possible. Similar to this, some shelters place ill cats that require a long recovery period in foster homes, where they frequently recover quickly after being taken out of the stressful shelter setting.

5.6.4. Minimizing Transmission

5.6.4.1. Housing and Traffic Patterns

Employees cleaning or tending to dogs with canine respiratory disease complex or cats with URTD should refrain from handling or passing by areas housing healthy cats, at least until they have changed into clean clothes and thoroughly cleaned their hands (because *B. bronchiseptica* is most likely spread by cats and dogs. Older cats should have quarters apart from younger ones, and kittens should be kept apart from adult cats in the house. Cats should ideally not live near dogs that bark or have their areas patrolled by dogs. Double-tiered cages are preferable to triple-tiered ones when used for single-cat housing because the latter is more difficult to maintain and clean. In a similar vein, it is challenging for staff members or prospective adopters to get a good look at these animals.

All cat cages should be free of drafts and other extremes in the microenvironment, and cages facing each other should be spaced at least four feet apart to prevent respiratory agent droplet transmission during sneezes. Even though overcrowded shelters occasionally temporarily position banks of cages in hallways, this should only be done in an emergency and not in heavily trafficked hallways or places where dogs may walk right by the cages.

Although data are not yet available, studies are being conducted to compare the URTD rates of cats kept in communal housing versus those kept in single cages. Based on the author's personal experience, anecdotally, most shelters in the U.S. who have experience housing cats in both types of settings, think that cats living in communal housing have lower rates of respiratory illnesses. Some studies in the U.K. show that cats living in groups experience higher levels of stress than cats living alone. However, it is unclear which type of housing is best for promoting low URTD rates until comparative studies are published. Colony housing is becoming more and more common, so it is imperative to conduct a thorough scientific analysis of its effects on URTD in shelters.

Human trafficking should ideally move from the most disease-susceptible animals to the most disease-resistant ones in all animal areas. Therefore, kittens should be seen by the public before adults to support URTD. As with many circumstances in shelters, adoption objectives might not align with the best practices in preventive medicine. To promote the adoption of adults over kittens, many shelters require prospective adopters to meet their adults first. Although they think that this tactic boosts adult cat adoptions, it might put kittens in danger.

5.7. DISEASE SURVEILLANCE

The majority of veterinarians that treat small animals haven't given population disease surveillance any thought. Given that in the U.S. disease surveillance is not particularly necessary for small animal veterinarians, who typically practice in homes with three or fewer animals. But in animal shelters with a large number of animals, staff members, and volunteers, veterinarians have to depend on others to quickly recognize and report sick animals. In a similar vein, they have to determine the frequency of disease occurrence, monitor changes in incidence over time, and assess the impact of modifications to preventive medicine protocols.

In order for a surveillance system to be effective, it must be able to identify diseases quickly and accurately, report diseases consistently and quickly, provide laboratory support for diagnostic testing when needed, enter data into a data capture system, summarize, and interpret data, and report diseases to the appropriate parties, including management. Lack of a thermometer and stethoscope to evaluate an individual animal's health is comparable to not having a disease surveillance system to keep an eye on the health of the population. It is hard to evaluate the success of preventive medicine protocols, spot mild to moderate outbreaks, and track the reduction of disease if case counts and disease rates cannot be determined.

Programs for surveillance can be basic or complex. In an ideal world, disease rates would be determined taking into consideration variations in the number of cats in the shelter (e.g., by season), but merely tally the URTD cases to get a sense of the disease's occurrence over time. The typical seasonal variation in cases will become visible after a few years of data collection, and significant changes can be evaluated by comparing the number of cases observed in the current year with those from previous years. Regardless of how the cases are tracked, staff must be trained to identify signs of URTD (including those associated with VS-FCV) and motivated to report infected cats promptly to a designated person (e.g., veterinarian, veterinary technician, manager). Employees need to be aware that boarding sick cats with well-behaved cats can spread infections, cause more cases, contaminate the environment, and increase the prevalence of URTD. Medical records for impacted animals in the shelter improve the standard of care provided to individual cats and are a crucial part of the surveillance system. Caregivers can view each cat's progress and treatments on any given day thanks to their medical records. They can be utilized to track cases over time as well.

5.8. STAFF AND VOLUNTEER EDUCATION AND MOTIVATION

Most of the preventive and control measures that shelter veterinarians have to put in place are carried out by volunteers and staff. Veterinarians should be involved in the creation of written protocols, provide justifications for their significance, and, when feasible, aid in staff motivation to attain optimal compliance. The veterinarian can be a strong motivator to get volunteers and staff to comply, especially with the backing of the shelter management. Effective URTD management programs necessitate a team approach because highly infectious respiratory agents contaminate so many areas and items in the shelter, including staff. Employees need to be aware of the overall strategy for URTD control as well as their critical role in it. They should receive regular updates on disease incidence data so they feel invested in the advancement and inspired to keep up the good work.

5.9. ADOPTING A SYSTEMATIC APPROACH TO MANAGEMENT OF URTD

Veterinarians need to take a methodical approach when looking into an outbreak or examining an animal to make sure they don't miss anything crucial. In a similar vein, it's critical to create a methodical strategy for disease control, which should include URTD management in shelters. This makes it more likely that crucial intervention techniques will be remembered and that a management plan will be developed and distributed to all managers and caregivers. Everyone has to be aware of and accountable for their specific roles within the larger URTD program.

The management program needs to be multifaceted to minimize the amount of contamination in the environment, maximize host resistance, and reduce the transmission of the several respiratory agents involved due to their complex epidemiology. The systematic approach should involve:

1. A disease surveillance system, including detailed medical records for sick cats.
2. Specification of goals, both short and long term.
3. An assessment of the total facility from the perspective of how the facility enhances or diminishes the likelihood of upper respiratory disease in cats.
4. The identification of impediments to achieving the URTD goals and strategies to overcome them.
5. Plans to achieve the goals, both long and short term.
6. Plans to evaluate the data from the surveillance system to evaluate whether the goals are being met.
7. Mechanisms to report results to everyone who has participated in the URTD program, as well as to the executive director and the board of directors.
8. Plans to celebrate and move on to new goals if the old ones have been met; if the goals have not been met, time to reflect on why and to revise the plans (or change the goals) and begin the process again.

Making time to reflect, plan, and build a team is arguably the most crucial aspect of this strategy. This is a challenging task in the hectic, demanding, and frequently overwhelming shelter environment. Still, many shelters probably prioritize planning time as their top priority. When planning is put into practice, it will help to reduce crises, enable a collaborative approach to the numerous issues that come up every day, and develop a successful program for managing diseases.

1. **Surveillance System:** A disease surveillance system may vary greatly in complexity based on the software and level of experience in the shelter. The objective is to monitor the number of cats in the population that contract URTD and observe changes in that number over time, relative to the size of the population. The most effective way to do this is with a medical records system that improves the care of individual animals. The health management team can then determine whether URTD incidence is rising, falling, or staying the same. This count should ideally be done about the total number of cats housed at the shelter (i.e., rates are determined). But it can be useful to start by defining what a case is and counting and monitoring the number of sick cats. If the total number of cats entering the shelter each year stays relatively constant over time, the number of cats who develop URTD at a particular point in time (e.g., this July) can be contrasted with the quantity the subsequent year (e.g., July of next year). Enhanced surveillance using a paid software program can yield even more precise data for figuring out rates and classifying cases according to age, other attributes, or place of origin (stray or owner surrender). Once a system is in place, it is possible to track how management changes affect the incidence of URTD.
2. **Setting Goals***:* There should be a time limit and a set of clear, achievable goals. One objective could be to document the disinfection procedure before the subsequent staff gathering. Alternatively, if it is found that one in three cats who enter the shelter go on to develop URTD, a target could be set to lower that percentage to one in four cats who enter the shelter during the course of the upcoming year. Establishing objectives gives.
3. **Complete Review of the Facilities:** A comprehensive review and discussion of the facilities (such as traffic flow, temperature control throughout the shelter, and ventilation) and current management practices that impact URTD are necessary. Usually, this review recommends reasonably simple adjustments to current practices (or even reasonably priced facility modifications) that could immediately lower the rate of illness. For instance, one shelter

discovered that their garage had devolved into a chaotic catch-all area after conducting an inventory of their physical facilities. The garage had not held animal control vehicles in years. This space was divided into two cat holding areas for a minimal cost to improve isolation for kittens coming back from foster care.

4. **Explicitly Identifying Those Factors, Policies, and Concerns That Serve As Impediments to Achieving The Goals:** If potential impediments is identified, preemptive strategies can be developed to minimize or eliminate their impact. For example, if there is an insufficient number of staff members to finish cleaning all cat cages before the shelter opens and allowing the disinfectant to work for 10 minutes, then additional staff may need to be added or current staff responsibilities rearranged.
5. **Plans to Achieve the Goals Must Be Developed:** To accomplish the objectives, plans need to be created. The plans may not look all that different from what is currently being done in some shelters. For instance, it might include an assessment of the URTD program's development and a strategy for resolving any issues. Plans in other shelters might entail setting up a system for medical records. Plans for controlling URTD should provide access to a diagnostic lab where samples can be sent as needed. Similarly, shelter staff should have easy access to protocols for gathering and sending frequently needed samples for URTD diagnosis.
6. **Data Analysis and Review:** At first, it might be challenging to determine progress with accuracy because numerous factors can alter the number of sick cats in shelters, including the time of year, shifts in population dynamics (e.g., percentage of kittens), and so forth. It is possible to graph numbers, or better yet, incidence rates, and talk about possible causes for their fluctuations. In due course, evaluations can be conducted to compare the URTD rates for various sections of the shelter between this year and last year.
7. **Feedback Provided to Staff, Volunteers, and Board Members:** Reactions provide even more motivation to carry out the plan. Volunteers and employees at shelters prefer to believe that their work is improving the animals' well-being. For them to remain motivated to follow preventative procedures, they require feedback. Notifying employees that URTD cases are declining can be a potent incentive. Comparably, discovering stable or rising rates of URTD incidence can start conversations about why advancements have not been made. (e.g., communal housing as opposed to private cage housing), and so on.

8. **Reassess Goals and Draft New Ones:** When objectives are achieved, it is appropriate to rejoice and reaffirm efforts to establish and accomplish new objectives. If objectives are not reached, then the causes can be addressed. Data generated by a surveillance system can be used to track advancement or support the case for more funding to accelerate it (August, 1990).

REFERENCES

1. Abd-Eldaim, M., Potgeiter, L., et al. (2005). Genetic analysis of feline caliciviruses associated with a hemorrhagic-like disease. *Journal of Veterinary Diagnostic Investigation, 17*, 420–429.
2. Andersen, B. M., Banrud, H., et al. (2006). Comparison of UV-C light and chemicals for disinfection of surfaces in hospital isolation units. *Infectious Control and Hospital Epidemiology, 27*, 729–734.
3. August, J. R. (1984). Feline viral respiratory disease: The carrier state, vaccination and control. *Veterinary Clinics of North America: Small Animal Practice, 14*, 1159–1171.
4. August, J. R. (1990). The control and eradication of feline upper respiratory infections in cluster populations. *Veterinary Medicine, 85*, 1002–1006.
5. Bannasch, M. J., & Foley, J. E. (2005). Epidemiologic evaluation of multiple respiratory pathogens in cats in animal shelters. *Journal of Feline Medicine and Surgery, 7*, 109–119.
6. Baulch-Brown, C., Love, D. N., et al. (1997). Feline caliciviruses: A need for vaccine modification. *Australian Veterinary Journal, 75*, 209–213.
7. Bemis, D. A. (1992). Bordetella and Mycoplasma respiratory infections in dogs and cats. *Veterinary Clinics of North America: Small Animal Practice, 22*(5), 1173–1186.
8. Berkelman, R. L. (2003). Human illness associated with use of veterinary vaccines. *Clinical Infectious Diseases, 37*, 407–414.
9. Binns, S. H., Dawson, S., et al. (1999). Prevalence and risk factors for feline *Bordetella bronchiseptica* infection. *Veterinary Record, 144*, 575–580.
10. Binns, S. H., Dawson, S., et al. (2000). A study of feline upper respiratory tract disease with reference to prevalence and risk factors for infection with feline calicivirus and feline herpesvirus. *Journal of Feline Medicine and Surgery, 2*, 123–133.
11. Binns, S. H., Speakman, A. J., et al. (1998). The use of pulsed-field gel electrophoresis to examine the epidemiology of *Bordetella bronchiseptica* isolated from cats and other species. *Epidemiology and Infection, 120*, 201–208.
12. Blackmore, D. K., Hill, A., et al. (1971). The incidence of mycoplasma in pet and colony-maintained cats. *Journal of Small Animal Practice, 12*, 207–216.
13. Boothe, H. W. (1998). Antiseptics and disinfectants. *Veterinary Clinics of North America: Small Animal Practice, 28*, 233–248.

14. Brogden, K. A., Lehmkuhl, H. D., et al. (1998). *Pasteurella* hemolytica complicated respiratory infections in sheep and goats. *Veterinary Research, 29*, 233–254.
15. Browning, G. F. (2004). Is *Chlamydophila felis* a significant zoonotic pathogen? *Australian Veterinary Journal, 82*, 695–696.
16. Cai, Y., Fukushi, H., et al. (2002). An etiological investigation of domestic cats with conjunctivitis and upper respiratory tract disease in Japan. *Journal of Veterinary Medical Science, 64*, 215–219.
17. Case, L. P., & Fahey, G. C. (2004). Nutritional challenges for shelter animals. In L. Miller & S. Zawistowski (Eds.), *Shelter Medicine for Veterinarians and Staff* (pp. 79–93). Ames, IA: Blackwell Publishing.
18. Chandler, J. C., & Lappin, M. R. (2002). Mycoplasmal respiratory infections in small animals: 17 cases (1988–1999). *Journal of the American Animal Hospital Association, 38*, 111–119.
19. Clay, S., Maherchandani, S., et al. (2006). Survival on uncommon fomites of feline calicivirus, a surrogate of noroviruses. *American Journal of Infection Control, 34*(1), 41–43.
20. Cocker, F. M., Howard, P. E., et al. (1987). Effect of human alpha-hybrid interferon on the course of feline viral rhinotracheitis. *Veterinary Record, 120*(16), 391–393.
21. Coutts, A. J., Dawson, S., et al. (1994). Isolation of feline respiratory viruses from clinically healthy cats at UK cat shows. *Veterinary Record, 135*, 555–556.
22. Coutts, A. J., Dawson, S., et al. (1996). Studies on natural transmission of *Bordetella bronchiseptica* in cats. *Veterinary Microbiology, 48*, 19–27.
23. Coyne, K. P., Jones, B. R., et al. (2006). Lethal outbreak of disease associated with feline calicivirus infection in cats. *Veterinary Record, 158*(16), 544–550.
24. Datz, C. (2003). *Bordetella* infections in dogs and cats: Pathogenesis, clinical signs and diagnosis. *Compendium on Continuing Education for the Practicing Veterinarian, 25*(12), 896–900.
25. Dawson, S., Jones, D., et al. (2000). *Bordetella bronchiseptica* infection in cats following contact with infected dogs. *Veterinary Record, 146*, 46–48.
26. Dawson, S., McArdle, H. C., et al. (1993). Investigation of vaccine reactions and breakdowns after feline calicivirus vaccination. *Veterinary Record, 132*, 346–350.
27. DiFrancesco, A., Carelle, M. S., et al. (2003). Feline chlamydiosis in Italian stray cat homes. *Veterinary Record, 153*, 244–245.

28. Dinnage, J. D., Scarlett, J. M., & Richards, J. R. (2008). Descriptive epidemiology of feline upper respiratory tract disease in an animal shelter. *Journal of Feline Medicine and Surgery*. Accepted for publication.
29. Doultree, J. C., Druce, J. D., et al. (1999). Inactivation of feline calicivirus, a Norwalk virus surrogate. *Journal of Hospital Infection, 41*(1), 51–57.
30. Dworkin, M. S., Sullivan, P. S., et al. (1999). *Bordetella bronchiseptica* infection in human immunodeficiency virus-infected patients. *Clinical Infectious Diseases, 28*, 1095–1099.
31. Edinboro, C. H., Janowitz, L. K., et al. (1999). A clinical trial of intranasal and subcutaneous vaccines to prevent upper respiratory infection in cats at an animal shelter. *Feline Practice, 27*, 7–13.
32. Eleraky, N. Z., Potgieter, L. N. D., et al. (2002). Virucidal efficacy of four new disinfectants. *Journal of the American Animal Hospital Association, 38*, 231–234.
33. Elliot, H. (1991). *Bordetella bronchiseptica* in a closed cat colony. *Veterinary Record, 129*, 474–475.
34. Evermann, J. F., Laurenson, M., et al. (1993). Infectious disease surveillance in captive and free-living cheetahs–An integral part of the species survival plan. *Zoo Biology, 12*, 125–133.
35. Foley, J. E., Rand, C., et al. (2002). Molecular epidemiology of feline bordetellosis in two animal shelters in California, USA. *Preventive Veterinary Medicine, 54*, 141–156.
36. Ford, R. B. (1993). Role of infectious agents in respiratory disease. In *Feline Infectious Diseases* (pp. 17–35). *Veterinary Clinics of North America: Small Animal Practice, 23*.
37. Ford, R. B. (2004). Vaccination strategies in the animal shelter environment. In L. Miller & S. Zawistowski (Eds.), *Shelter Medicine for Veterinarians and Staff* (pp. 285–305). Ames, IA: Blackwell Publishing.
38. Foster, S. F., Barrs, V. R., et al. (1998). Pneumonia associated with *Mycoplasma* spp. in three cats. *Australian Veterinary Journal, 76*(7), 460–464.
39. Fulton, R. W., & Burge, L. J. (1985). Susceptibility of feline herpesvirus-1 and feline calicivirus to feline interferon and recombinant human leukocyte interferons. *Antimicrobial Agents and Chemotherapy, 28*, 698–699.
40. Gaskell, R. M. (1992). Upper respiratory disease in the cat (including chlamydia): Control and prevention. *Feline Practice, 20*, 7–12.
41. Gaskell, R. M., & Povey, R. C. (1977). Experimental induction of feline viral rhinotracheitis virus re-excretion in FVR-recovered cats. *Veterinary Record, 100*, 128–133.

42. Gaskell, R. M., & Povey, R. C. (1982). Transmission of feline viral rhinotracheitis. *Veterinary Record, 111*, 359–362.

43. Gaskell, R. M., & Wardley, R. C. (1977). Feline viral respiratory disease: A review with particular reference to its epizootiology and control. *Journal of Small Animal Practice, 19*, 1–16.

44. Gaskell, R., & Dawson, S. (1998). Feline respiratory disease. In C. E. Greene (Ed.), *Infectious Diseases of the Dog and Cat* (pp. 97–106). Philadelphia: W. B. Saunders Co.

45. Gaskell, R., Dawson, S., et al. (2006). Feline respiratory disease. In C. E. Greene (Ed.), *Infectious Diseases of the Dog and Cat* (pp. 145–154). St. Louis: Saunders Elsevier.

46. Gaskell, R., Dawson, S., et al. (2007). Feline herpesvirus. *Veterinary Research, 38*, 337–354.

47. Gehrke, C., Steinmann, J., et al. (2004). Inactivation of feline calicivirus, a surrogate of norovirus (formerly Norwalk-like viruses), by different types of alcohol *in vitro* and *in vivo*. *Journal of Hospital Infection, 56*, 49–55.

48. Gilman, N. (2004). Sanitation in the animal shelter. In L. Miller & S. Zawistowski (Eds.), *Shelter Medicine for Veterinarians and Staff* (pp. 67–78). Ames, IA: Blackwell Publishing.

49. Greene, C. E. (2006a). Chlamydial infections. In *Infectious Diseases of the Dog and Cat* (pp. 245–252). St. Louis: Saunders Elsevier.

50. Greene, C. E. (2006b). Environmental factors in infectious disease. In *Infectious Diseases of the Dog and Cat* (pp. 991–1013). St. Louis: Saunders Elsevier.

51. Greene, C. E. (2006c). Immunoprophylaxis. In *Infectious Diseases of the Dog and Cat* (pp. 1069–1119). St. Louis: Saunders Elsevier.

52. Griffin, B., & Hume, K. R. (2006). Recognition and management of stress in housed cats. In J. R. August (Ed.), *Consultations in Feline Medicine* (Vol. 5, pp. 717–734). St. Louis, MO: Elsevier Saunders.

53. Griffith, R. S., DeLong, D. C., et al. (1981). Relation of arginine-lysine antagonism to herpes simplex growth in tissue culture. *Chemotherapy, 27*, 209–213.

54. Griffith, R. S., Norins, A. L., et al. (1978). A multicentered study of lysine therapy in herpes simplex infection. *Dermatologica, 156*, 257–267.

55. Griffith, R. S., Walsh, D. E., et al. (1987). Success of L-lysine therapy in frequently recurrent herpes simplex infection: Treatment and prophylaxis. *Dermatologica, 175*, 183–190.

56. Gruffydd-Jones, T. J., Jones, B. R., et al. (1995). Chlamydia infection in cats in New Zealand. *New Zealand Veterinary Journal, 43*, 201–203.

57. Harbour, D. A., Howard, P. E., et al. (1991). Isolation of feline calicivirus and feline herpesvirus from domestic cats 1980 to 1989. *Veterinary Record, 128*, 77–80.

58. Hargis, A. M., Ginn, P. E., et al. (1999a). Feline herpesvirus 1-associated facial and nasal dermatitis and stomatitis in domestic cats. *Veterinary Clinics of North America: Small Animal Practice, 29*, 1281–1290.

59. Hargis, A. M., Ginn, P. E., et al. (1999b). Ulcerative facial and nasal dermatitis and stomatitis in cats associated with feline herpesvirus 1. *Veterinary Dermatology, 10*, 267–274.

60. Helps, C. R., Lait, P., et al. (2005). Factors associated with upper respiratory tract disease caused by feline herpesvirus, feline calicivirus, *Chlamydophila felis*, and *Bordetella bronchiseptica* in cats: Experience from 218 European catteries. *Veterinary Record, 156*(21), 669–673.

61. Holst, B. S., Berndtsson, L. T., et al. (2005). Isolation of feline herpesvirus-1 and feline calicivirus from healthy cats in Swedish breeding catteries. *Journal of Feline Medicine and Surgery, 7*(6), 325–331.

62. Hoover, E. A., & Kahn, D. E. (1987). Viral respiratory diseases and chlamydiosis. In J. Holzworth (Ed.), *Diseases of the Cat: Medicine and Surgery* (pp. 214–237). Philadelphia: WB Saunders Co.

63. Hoskins, J. D., Williams, J., et al. (1998). Isolation and characterization of *Bordetella bronchiseptica* from cats in southern Louisiana. *Veterinary Immunology and Immunopathology, 65*, 173–176.

64. Hurley, K. F. (2004). Implementing a population health plan in an animal shelter: Goal setting, data collection and monitoring, and policy development. In L. Miller & S. Zawistowski (Eds.), *Shelter Medicine for Veterinarians and Staff* (pp. 211–234). Ames, IA: Blackwell Publishing.

65. Hurley, K. F., & Sykes, J. E. (2003). Update on feline calicivirus: New trends. *Veterinary Clinics of North America: Small Animal Practice, 33*, 759–772.

66. Hurley, K. F., Pesavento, P. A., et al. (2004). An outbreak of virulent systemic feline calicivirus disease. *Journal of the American Veterinary Medical Association, 224*, 241–249.

67. Jacobs, A. A., Chalmers, W. S., et al. (1993). Feline bordetellosis: Challenge and vaccine studies. *Veterinary Record, 133*, 260–263.

68. Jimenez, L., & Chiang, M. (2006). Virucidal activity of a quaternary ammonium compound disinfectant against feline calicivirus: A surrogate for norovirus. *American Journal of Infection Control, 34*(5), 269–273.

69. Johnson, T. (2004). The animal shelter building: Design and maintenance of a healthy and efficient facility. In L. Miller & S. Zawistowski (Eds.),

Shelter Medicine for Veterinarians and Staff (pp. 55–66). Ames, IA: Blackwell Publishing.

70. Kennedy, M. A., Mellon, V. S., et al. (1995). Virucidal efficacy of the newer quaternary ammonium compounds. *Journal of the American Animal Hospital Association, 31*, 254–258.
71. Knowles, J. O., & Gaskell, R. M. (1991). Control of upper respiratory diseases in multiple cat households and catteries. In J. R. August (Ed.), *Consultations in Feline Internal Medicine* (pp. 563–569). Philadelphia: WB Saunders Co.
72. Knowles, J. O., Gaskell, R. M., et al. (1989). Prevalence of feline calicivirus, feline leukemia virus, and antibodies to FIV in cats with chronic stomatitis. *Veterinary Record, 124*, 336–338.
73. Kowalski, W. J., & Bahnfleth, W. P. (2000). UVGI design basics. *Heating/Piping/Air Conditioning, January*, 100–110.
74. Kramer, A., Galabov, A. S., et al. (2006). Virucidal activity of a new hand disinfectant with reduced ethanol content: Comparison with other alcohol-based formulations. *Journal of Hospital Infection, 62*(1), 98–106.
75. Lappin, M. (1999). Roundtable discussion: *Bordetella bronchiseptica*. *Feline Practice, 27*, 10–12.
76. Lappin, M. R. (2003). Practical antimicrobial therapy. In R. W. Nelson & C. G. Couto (Eds.), *Small Animal Internal Medicine* (pp. 228–232). St. Louis, MO: Mosby.
77. Lappin, M. R., Sebring, R. W., et al. (2006). Effects of a single dose of an intranasal feline herpesvirus 1, calicivirus, and panleukopenia vaccine on clinical signs and virus shedding after challenge with virulent feline herpesvirus 1. *Journal of Feline Medicine and Surgery, 8*, 158–163.
78. Lawler, D. F. (1989). Disinfection of animal environments. In R. W. Kirk & J. D. Bonagura (Eds.), *Current Veterinary Therapy X: Small Animal Practice* (pp. 90–95). Philadelphia: WB Saunders Co.
79. Lawler, D. F. (1998). Prevention and management of infection in catteries. In C. E. Greene (Ed.), *Infectious Diseases of the Dog and Cat* (pp. 701–706). Philadelphia: WB Saunders Co.
80. Lommer, M. J., & Verstrakte, F. J. (2003). Concurrent shedding of feline calicivirus and feline herpesvirus 1 in cats with chronic gingivostomatitis. *Oral Microbiology and Immunology, 18*, 131–134.
81. Maggs, D. J., Collins, B. K., et al. (2000). Effects of L-lysine and L-arginine on *in-vitro* replication of feline herpesvirus type-1. *American Journal of Veterinary Research, 61*, 1474–1478.
82. Maggs, D. J., Nasisse, M. P., et al. (2003). Efficacy of oral supplementation

with L-lysine in cats latently infected with feline herpesvirus. *American Journal of Veterinary Research, 64*, 37–42.

83. Maggs, D. J., Sykes, J. E., et al. (2007). Effects of dietary lysine supplementation in cats with enzootic upper respiratory disease. *Journal of Feline Medicine and Surgery, 9*(2), 97–108.
84. McArdle, H. C., Dawson, S., et al. (1994). Seroprevalence and isolation rate of *Bordetella bronchiseptica* in cats in the UK. *Veterinary Record, 135*, 506–507.
85. Miller, L. (2004). Dog and cat care in the animal shelter. In L. Miller & S. Zawistowski (Eds.), *Shelter Medicine for Veterinarians and Staff* (pp. 95–123). Ames, IA: Blackwell Publishing.
86. Nasisse, M. P., Halenda, R. M., et al. (1996). Efficacy of low dose oral, natural human interferon alpha in acute feline herpesvirus-1 (FHV-1) infection: A preliminary dose determination trial. *Proceedings of the 27th Annual Meeting of the American College of Veterinary Ophthalmologists*, 79.
87. Ossiboff, R. J., Sheh, A., et al. (2007). Feline caliciviruses (FCVs) isolated from cats with virulent systemic disease possess *in vitro* phenotypes distinct from those of other FCV isolates. *Journal of General Virology, 88*, 506–517.
88. Osterholm, M. T., Hedberg, C. W., et al. (2000). Epidemiology of infectious diseases. In G. L. Mandell, J. E. Bennett, & R. Dolin (Eds.), *Principles and Practice of Infectious Diseases* (pp. 156–167). Philadelphia: Churchill and Livingstone.
89. Pedersen, N. C. (1988). Chlamydiosis. In *Feline Infectious Diseases* (pp. 231–236). Goleta, CA: American Veterinary Publications, Inc.
90. Pedersen, N. C. (1991). Common infectious diseases of multiple-cat environments. In N. Pedersen (Ed.), *Feline Husbandry* (pp. 163–176). Goleta, CA: American Veterinary Publications.
91. Pedersen, N. C., Elliot, J. B., et al. (2000). An isolated epizootic of hemorrhagic-like fever in cats caused by a novel and highly virulent strain of feline calicivirus. *Veterinary Microbiology, 73*, 281–300.
92. Pedersen, N. C., Sato, R., et al. (2004). Common virus infections in cats, before and after being placed in shelters, with emphasis on feline enteric coronavirus. *Journal of Feline Medicine and Surgery, 6*, 83–88.
93. Pesavento, P. A., MacLachlan, N. J., et al. (2004). Pathologic, immunohistochemical, and electron microscopic findings in naturally occurring virulent systemic feline calicivirus infection in cats. *Veterinary Pathology, 41*, 257–263.

94. Porter, J. F., Parton, R., et al. (1991). Growth and survival of *Bordetella bronchiseptica* in natural waters and in buffered saline without added nutrients. *Applied and Environmental Microbiology, 57*, 1202–1206.

95. Poulet, H., Brunet, S., et al. (2005). Immunization with a combination of two complementary feline calicivirus strains induces a broad cross-protection against heterologous challenges. *Veterinary Microbiology, 106*(1–2), 17–31. E-pub January 21, 2005.

96. Povey, R. C. (1979). A review of feline viral rhinotracheitis (feline herpesvirus-1 infection). *Comparative Immunology, Microbiology and Infectious Diseases, 2*, 373–387.

97. Povey, R. C., & Johnson, R. H. (1970). Observations on the epidemiology and control of viral respiratory disease in cats. *Journal of Small Animal Practice, 11*, 485–494.

98. Radford, A. D., Coyne, K. P., et al. (2007). Feline calicivirus. *Veterinary Research, 38*, 319–335.

99. Radford, A. D., Dawson, S., et al. (2006). The challenge for the next generation of feline calicivirus vaccines. *Veterinary Microbiology, 117*, 14–18.

100. Radostits, O. M. (2001). Principles of health management of food-producing animals. In O. M. Radostits (Ed.), *Herd Health: Food Animal Production* (pp. 1–45). Philadelphia: WB Saunders Co.

101. Rampazzo, A., Appino, S., et al. (2003). Prevalence of *Chlamydophila felis* and feline herpesvirus-1 in cats with conjunctivitis in Northern Italy. *Journal of Veterinary Internal Medicine, 17*, 799–807.

102. Ramsey, D. T. (2000). Feline chlamydia and calicivirus infections. *Veterinary Clinics of North America: Small Animal Practice, 30*, 1015–1028.

103. Randolph, J. F., Moise, N. S., et al. (1993). Prevalence of mycoplasma and ureaplasma recovery from tracheobronchial lavages and prevalence of mycoplasma recovery from pharyngeal swab specimens in cats with or without pulmonary disease. *American Journal of Veterinary Research, 54*, 897–900.

104. Richards, J. R., Elston, T. H., et al. (2006). The 2006 American Association of Feline Practitioners Feline Vaccine Advisory Panel report. *Journal of the American Veterinary Medical Association, 229*(9), 1405–1441.

105. Shewen, P. E., Povey, R. C., et al. (1980). A survey of the conjunctival flora of clinically normal cats and cats with conjunctivitis. *Canadian Veterinary Journal, 21*, 231–233.

106. TerWee, J., Sabara, M., et al. (1998). Characterization of the systemic disease and ocular signs induced by experimental infection with *Chlamydia psittaci* in cats. *Veterinary Microbiology, 59*(4), 259–281.

107. Wardley, R. C., & Povey, R. C. (1977). Aerosol transmission of feline caliciviruses: An assessment of its epidemiological importance. *British Veterinary Journal, 133*, 504–508.

CHAPTER 6

AVIAN INFECTIONS DISEASE CONTROL AND CARE IN THE ANIMAL SHELTER

Contents

Introduction 218

6.1. Species 218

6.2. Husbandry 220

6.3. History and Physical Examination 224

6.4. Infectious Disease 227

6.5. Nutritional/Metabolic Disorders 247

6.6. Behavior 252

6.7. Control of Avian Infectious Diseases 260

References 266

INTRODUCTION

The prevention and treatment of avian infectious diseases in animal shelters is essential to preserving the health and welfare of bird populations there. Similar to other creatures, birds can contract a range of infectious diseases, some of which can spread quickly in a shelter's confined spaces. If left untreated, these infections–which can be brought on by bacteria, fungi, viruses, and parasites–can result in serious health problems or even death.

Comprehensive disease control in avian populations necessitates a multifaceted strategy that includes stringent biosecurity protocols, frequent health evaluations, and quick isolation of carriers. To stop the introduction and spread of pathogens, biosecurity measures include things like good sanitation, restricted access to bird enclosures, and the use of personal protective equipment (PPE) by shelter employees. Frequent health evaluations aid in the early identification of illnesses, enabling prompt intervention and treatment. Furthermore, it is essential to isolate sick birds to stop the spread of infectious agents to healthy people.

Providing a suitable environment that satisfies the physical, behavioral, and nutritional needs of birds in shelters is another aspect of providing care for them. One important factor that can weaken a bird's immune system and increase their susceptibility to infections is stress. Therefore, it is essential for their general health and resistance to disease to make sure that birds are kept in suitable housing, have access to clean water, a balanced diet, and environmental enrichment.

6.1. SPECIES

The majority of pet birds found in shelters are Psittaciformes, which includes African grays, macaws, lories/lorikeets, Amazons, cockatoos, conures, cockatiels, and parakeets. Canaries, finches, mynahs, and other pass arises are among the animals that are kept as pets and may be given to shelters (AWC, API, and ASPCA, 2009). Birds are a diverse group of animals, unlike dogs and cats, which are distinct species with numerous breeds within the species. By itself, parrot species number 352. These species have a variety of needs that must be satisfied in captivity because they are found in a variety of ecosystems throughout the world in the wild. It is crucial to accurately identify any new species that is brought to a shelter and to learn about each individual pet's unique requirements (Figures 6.1 and 6.2).

Figure 6.1. *Grey African parrot.*

Source: By Ssemmanda will-Own work, CC0, https://commons.wikimedia.org/w/index.php?curid=151191029.

Figure 6.2. *Indian ringneck parakeet (Psittacula krameri).*

Source: By MagnaErucae-Own work, CC BY-SA 4.0, https://commons.wikimedia.org/w/index.php?curid=105455559.

6.2. HUSBANDRY

6.2.1. Housing

A quiet, separate space away from other animals, such as dogs, cats, reptiles, and so on., is the best place to keep birds. The bird room needs to have good ventilation and be draft-free. Because Old World and New World parrots have never been exposed to each other, many experts advise keeping them apart out of concern that viruses that the birds may be more vulnerable to could spread. Standard quarantine procedures should allay worries, though, as these viruses are rarely found in shelter environments. It is possible to pair different species together, but caution must be used. It is challenging to develop guidelines for cohousing these birds because some species are known to be aggressive, and some individual birds also exhibit aggressive behavior. It is ideal to locate birds that can live in a group so that they can become socialized. If a breeding pair is bonded to one another rather than laying eggs, it may be okay to keep them together; however, this may occasionally make them less sociable with people and less likely to be adopted.

The majority of birds can tolerate temperatures between 60 and 80°F, but healthy, acclimated, and feathered birds can tolerate far greater temperatures. Birds should not be exposed to harsh chemicals like bleach or ammonia, cooking odors, Teflon, smoke (including second-hand smoke), household paint, candles, insecticides, and so on, because they are typically very sensitive to a wide range of different environmental toxins. Odor-producing chemicals shouldn't be kept in the bird room. If air filtration or purification systems are used, they must strictly follow the manufacturer's recommendations for regular maintenance and cannot use ionization technology or produce ozone.

Aviaries and cages for birds should have strong doors that cannot be unlocked to allow escape. Wood cannot be properly disinfected and is easily chewed through by most parrots, so it should not be used to build cages. Because of their toxicity and propensity to rust, iron, lead, and zinc shouldn't be present in the cage. Best is stainless steel. News chapter (apart from glossy sheets), plain chapter, or waxed chapter can be used to line the bottom of the cage; however, the news chapter needs to be taken out every day utilizing woodchips, sawdust, shavings, and so on. is not advised since food and bird droppings can promote the growth of bacteria. Cages need to be at least roomy enough for birds to spread their wings, flap them in all directions, climb the bars, jump from one perch to another, and play with toys.

For birds kept in captivity, minimum requirements for cage design have been established. The minimum cage size for small birds, such as parakeets, finches, and lovebirds, is one square foot, with cage bars spaced no more than

0.5 inches apart. The cage size of medium-sized birds, such as cockatiels, small conures, and caciques, should be at least 1.5 square feet and the cage bars should not be wider than three-quarters of an inch. To prevent their heads from fitting through the bars, larger birds such as macaws, cockatoos, African grays, and Amazons should have cage bars spaced appropriately for their size. According to the (AWC, API, and ASPCA 2009), the cage itself should have enough room inside for the bird to spread its wings in all directions, plus extra room so that the bird's tail doesn't poke through the bars.

There are various options for perch thickness and material; birds should have perches that fit their feet. Perches that are both big enough for the bird to sit with his or her feet flat and small enough for the bird's feet to surround them should be present. To prevent birds from dropping their waste into food dishes or water, perches shouldn't be positioned above them. Enough perches should be available so that birds can coexist peacefully while perching in their natural positions at the same time. For birds to produce enough vitamin D, they require full-spectrum light. If birds are housed indoors, full-spectrum light bulbs should be changed regularly. It is advised that birds have daily access to natural sunlight (ideally five hours), as it is far more efficient than any type of bulb. However, they also need to be able to move out of direct sunlight. For them to be able to sleep, they also need access to darkness for about 10 hours.

News chapter is frequently placed at the bottom of bird cages, ideally on a tray beneath the cage's barring so the bird cannot step in the excrement. Everyday news chapter changes ought to occur. Fresh water and a fresh bowl should be added to water bowls every day. Every two weeks, all non-cloth toys, bowls, and perches can be washed on high in the dishwasher. You can use the washer and dryer for toys made of cloth. The cage should be thoroughly cleaned with a disinfectant once a month. While other disinfectants may be safe to use, this author uses either Trifectant or chlorhexidine. The cage can be cleaned with water after it has been scrubbed thoroughly. For this task, power washers are ideal. To ensure that all the fumes have dissipated, the cage should be left outside for three to four hours after cleaning is finished, before the bird is put back inside. Diluted bleach can also be used to disinfect the cage in case of an infectious disease outbreak. The cage should then be left outside for a full day to air out before being brought back inside.

Although these are minimum housing standards, birds should be able to fly and typically spend most of their days foraging and feeding in the wild (Echols, 2004). For this reason, providing a bird with food and water in a cage is insufficient to meet all of its needs. Certain parrots are extremely social and intelligent. Having an emotionally and physically healthy bird will require enrichment. Sturdy toys that fit their size and are designed specifically for birds should be provided. The aviary or cage for the bird should ideally be big enough for it to be able to fly. As an alternative, birds should be allowed to spend time outside of their cages in a secure setting under the guidance of a person with experience with birds. Unwanted behaviors like biting, screaming, feather plucking, aggression, and so on, are signs that a bird's needs are not being satisfied. For an illustration of a correctly set-up bird cage, refer to Figure 6.3.

Figure 6.3. *Cage setup with perching options and foraging toys.*

Source: By Chris Murray-originally posted to Flickr as DSC_5260, CC BY 2.0, https://commons.wikimedia.org/w/index.php?curid=5666967.

6.2.2. Diet

For exotic birds, there is no diet designed specifically for that species. Generally speaking, a diet consisting primarily of 60–75% pellets and minimal seeds is advised. It is known that psittacines given a diet high in seeds can experience health issues like fatty liver, atherosclerosis, vitamin A deficiency, and so on. (Fiskett and Reavill, 1904). The pellet brands Harrisons, Lafeber, Roudybush, and Zupreem are a few that are suggested. To help ensure that all birds can be switched from seeds to pellets, it is a good idea to have multiple brands of pellets. The majority of psittacines prefer nuts and seeds over foods that are good for them because they adore them. Since most psittacines are seed motivated, nuts, and seeds are acceptable in moderation for healthy birds and can be used in foraging toys and behavioral training. A lot of birds will be food-motivated when given nutritious treats. Whole grains, fresh veggies, and modest amounts of berries and other fruits should make up 25–40% of the remaining diet. Whole grains such as spelt, millet, flax seed, barley, and oat grouts are acceptable. Broccoli, cauliflower, brussels sprouts, beans, sweet potatoes, peppers, and dark green vegetables are nutritious veggies to feed. To prevent the birds from being contaminated with fungus spores and other toxins, all food provided to them should be of human-grade quality. It's not advisable to leave fresh food in the cage overnight.

Since many species in captivity have diverse diets in the wild, it is challenging to recommend a single standard protocol for all species. It's crucial to investigate the dietary needs of the particular species.

The following are some general dietary considerations for various species:

1. Granivorous budgies and cockatiels should have 50% whole grains and 50% pellets, with leafy vegetables and minimal fruit.
2. Cockatiels fed an all-pellet diet have been known to have problems with hypercalcemia leading to renal mineralization (Fiskett and Reavill, 2004).
3. Macaws require a higher fat content in their diet, so a higher percentage of nuts–about 10%–is recommended.
4. Never give seed to lories or lorikeets. Soft leafy vegetables are added to fruits in their diet of ground pellets or specially prepared lorie food. Instead of using fresh produce, you can use pureed fruits and vegetables, natural juices, and commercial human baby foods.

90% of birds can usually be easily transitioned from a seed diet to a pellet diet. It will require extra effort from a tiny portion of the birds to complete the transition. Paying close attention to what the birds are eating is necessary to prevent starvation during the process. On the other hand, losing too much weight too soon could be harmful. The majority of "seed junkies" are overweight and may have fatty liver disease. The secret is to not allow them to lose too much weight too quickly, but to keep them sufficiently hungry to want to try new foods. To successfully transition birds that only desire seeds to a nutritious diet, regular weighing and observation are required.

There should always be fresh, clean drinking water available to all birds. It is important to firmly position water dishes against the cage's sides to keep birds from tipping them over. Since some birds enjoy taking baths in the water, it needs to be watched over and changed twice a day. It is recommended to use bottled or filtered water or to let tap water run for a few minutes before filling the dish because birds are sensitive to contaminants in the water that may not harm humans (AWC, API, and ASPCA, 2009).

Water and food dishes should be simple to sanitize and clean. Dry food can be arranged in ceramic or stainless steel dishes. Water bottles are not recommended for use in shelters due to their increased likelihood of containing mold and harmful bacteria, as well as their difficulty in being swiftly cleaned and disinfected.

6.3. HISTORY AND PHYSICAL EXAMINATION

Prey species include passerines and psittacines. Prey species conceal obvious symptoms of disease to evade being devoured by predators. Owners of birds frequently find them dead and discover they had no idea something was wrong. The shelter needs to take a thorough history and examine the bird physically to provide the right care. The author suggests inserting the microchip while the bird is sedated if it is to be used for identification, something that many shelters would likely find difficult or impossible to accomplish. Leg bands are a useful tool for identity, though the author is hesitant to use them in a shelter. The main reason for band identification is regulatory, and attempts to remove them have unintentionally resulted in some injuries.

6.3.1. The History

The history should make an effort to get the following information from the owner or whoever seized or relinquished the bird, even though it is not always possible to learn much about a bird that has been turned in to the shelter or seized for various reasons:

1. The bird's origin.
2. Where the bird was kept in the house?
3. What other species the bird was in contact with?
4. Whether any other animals or humans in the household were sick?
5. What the cage was made of?
6. What was in the cage?
7. What diet was provided and what the bird actually ate?
8. When the bird began exhibiting any abnormal behaviors such as plucking feathers?
9. Whether there were any recent environmental changes?

6.3.2. Physical Examination

The physical examination should be methodical and comprehensive, taking into account the bird's health and stress level, just like with any other animal examination. Every bird should have a personalized medical file that contains information about their species, a precise physical description, the full findings of the physical examination, a history of behavior and health issues, information about their diet, past, and current medical conditions, and more. In certain cases, particularly when the bird is sick, the examination and diagnostic process should move slowly and in stages, contingent on the patient's stability. While this textbook cannot provide comprehensive instructions on how to examine a bird, it will provide some key points. To guarantee that birds are examined appropriately, shelters that accept animals should have a copy of one of the many avian textbooks on hand for basic knowledge. They should also collaborate closely with a veterinarian who has experience treating birds. Examining the bird from head to feet should come after assessing its general appearance and behavior. The bird should be bright and alert, responsive to outside stimuli, and show signs of normal behavior, such as vocalizing, eating normally, and grooming itself.

The bird will need to be caught for additional inspection after being watched in the cage. When confining birds, it is crucial to minimize their stress levels. For handlers who might not have experience or who are uncomfortable, there are plenty of references available to explain this procedure. Remove the bowl, perks, and any other obstructions from the cage. The handler can cover the bird with a towel in a calm and gentle manner. The handler can then stabilize the bird's head with one hand while supporting the bird's body in the towel with the other. To ensure complete control over the bird's head, the handler should gently grasp it with the thumb and middle finger beneath the mandible and the index finger on top. After the bird's head is securely in place, the handler can continue the examination without attempting to remove the bird from its cage.

Examine the mouth, ears, nares, eyes, cere (a soft, fleshy swelling on the beak), and nares for any swellings, masses, discharge, and so on. Birds that have upper respiratory infections may develop paranasal sinus swellings surrounding their eyes. They might also breathe through their mouths, wheeze, or sneeze. When a bird is deficient in vitamin A, as is often the case with birds fed only seeds, the papillae on its hard palate, or choanal slit, will be blunted or absent. Since papillae do not regrow, their absence merely indicates a nutritional shortfall that the bird experienced at some point in its life. It is important to auscultate the bird's heart, lungs, and air sacs to check for any anomalies. The bird's physical state can be assessed by palpating its pectoral muscle mass. A patient who has severe obesity would be indicated by palpating no keel at all, while a patient with a sharp, prominent keel (breastbone) would indicate very poor body condition.

A complete examination of the feathers and skin is necessary because they may show clinical indications of malnutrition, metabolic disorders, and infectious diseases. Smooth, bright, glossy feathers that rest flat against the body are ideal. In order for the bird to properly eat and preen, its beak needs to be free of discharges, swellings, chips, or dents, and it should also be aligned when closed. Reduced feather maintenance and preening could indicate a medical condition. Examining the bird's cloaca, which is a vent or region similar to the anus in mammals, involves everting the tissue to check for papillomas or other anomalies in the proctodeum, the cloaca's most distal chamber.

The feet should be flexible and dexterous, and they should be checked for growths, sores, and indications of swelling or inflammation. While it's common for birds to stand with one leg tucked under them, it's still advisable to check for any issues with both legs. During daily rounds, birds should be visually inspected to determine their current state of health. The best approach to prevent serious disease and treat health issues is early detection. Changes in weight, appetite, or activity (preening, vocalizing, playing, etc.) are some indicators of illness frequent scratching or rubbing, a change in the consistency, color, or frequency of bird droppings, vomiting, a limping gait, difficulty perching, swollen joints, and so on.

To prevent the bird from escaping and injuring itself, the primary flight feathers that are still intact should be trimmed by a veterinarian or another expert in this field. Unless a bird is misaligned, a veterinarian with experience working with beaks should be consulted before trimming a beak. Two people can do the task of trimming nails if necessary. While the other person trims the bird, one person must hold the bird in place. Because black nails are more likely to bleed, Kwik Stop is usually enough to control and stop the blood flow. Using a Dremel., which produces a smoother surface and reduces the chance of scratches, or the conventional trimmers used on dogs and cats, are the two methods used for trimming nails.

Observing bird droppings is important when making rounds. The three components of normal droppings are urine, urates, and feces. Normal droppings do not smell. The actual feces are dark and worm-like, but depending on what else the bird is eating, they can also be green, brown, pink, purple, or red. This can happen if the bird is mainly eating pellets. According to (AWC, API, and ASPCA 2009), the urine is the clear portion of the mixture and shouldn't be confused with diarrhea, whereas the urates are typically off-white or cream-colored.

6.3.3. Guidelines for Disease Management

In addition to being carriers of diseases that humans and other birds can contract, birds can also conceal the symptoms of a variety of illnesses. Before being introduced to other birds in the shelter, any recently admitted birds should ideally undergo a screening process to check for infectious diseases and be quarantined for a minimum of 30 days. At Best Friend's Animal Sanctuary, a paired sample (choanal and/or cloacal swab and whole blood) is submitted for PCR testing to test for chlamydia. It is impossible to definitively rule out the possibility that a bird carries the zoonotic disease, meaning that it can infect humans, even if the bird's test results are negative for chlamydia. Additionally, even after the quarantine period has ended, the bird may still be diseased. Macaws, Amazons, and conures are tested for Pacheco's disease; birds from Africa, Asia, and Australia are tested for Psittacine beak and feather disease. Upon intake of a new bird, complete blood counts and blood chemistries, including bile acids, are also carried out. Since radiographs are reasonably priced, they are often taken at the time of intake for new birds. Heart disease, proventricular dilatory disease, neoplasia, and other conditions are frequently found on radiographs that are not found on physical examination or blood work. It is also advised to perform complete fecal examinations and gram stains to identify parasites, bacterial infections, Candida overgrowth, and so on. Much of the history is frequently absent when a bird is brought to a shelter or sanctuary, making a comprehensive workup even more crucial to ascertain the bird's overall health.

6.4. INFECTIOUS DISEASE

The diseases for which the author feels that shelters should be most concerned or most frequently encountered are included in these brief, basic disease descriptions. It is not meant to take the place of veterinary professionals' or avian experts' care and guidance. The shelter veterinarian or a veterinarian with experience treating birds should be consulted for advice and input on all diagnoses, treatments, and adoption placements.

6.4.1. Viral Diseases

6.4.1.1. Psittacine Beak and Feather Disease (PBFD)

A circovirus known to infect birds from Asia, Australia, and Africa is responsible for Psittacine Beak and Feather Disease (PBFD) (Phalen, 2006a). The disease most frequently affects African gray parrots, lovebirds, eclectus parrots, budgerigars, lories, and lorikeets. Although PBFD is rare and manifests in a less severe form in New World parrots, it is still possible. It is possible for lovebirds to be infected without exhibiting any symptoms of illness. The infection's acute form is linked to feather lesions, pneumonia, enteritis, and even death. According to Fitzpatrick and Reavill (2004), cockatoos with beak pathology frequently exhibit progressive dystrophic feather lesions in the chronic form.

When feather dust is used as a medium of transmission, PBFD can be extremely contagious. A carrier state exists for it. Testing and a 30 day quarantine are advised to screen for the illness. The birds' whole blood and PCR (Polymerase Chain Reaction) swabs, environment PCR probe testing, and biopsy of the afflicted feathers are used to make the diagnosis. When birds arrive at shelters, many of them will have plucked feathers or poorly condition feathers. Excluding PBFD is crucial, as is not assuming that the bird's lack of feathers is due to a dietary shortfall or behavioral problem. The only kind of treatment is supportive, and recovering birds should only live in homes without other vulnerable birds. Depending on the species and age of the bird, the disease's severity varies; birds that are severely affected and for whom there are no suitable placement options may need to be put to death. Keeping up with hygiene is a key management strategy for preventing the spread of feather dust.

6.4.1.2. Herpes Viruses

Psittacine herpes viruses are multivalent. Birds that show clinical signs of Pacheco's disease usually die from the disease, which is highly contagious. Macaws, Amazons, and some species of conures are most likely carriers of the virus (Phalen, 2006a). Although many bird species are susceptible, the disease is more severe in New World parrots. Large-scale liver necrosis usually results in bird death. The virus takes three to 14 days to incubate; the disease is spread by ingestion, inhalation, and contact with the conjunctiva through fomites (i.e., hands, clothing, and inanimate objects). It is recommended to submit a paired sample consisting of a combined cloaca/choanal swab and blood for PCR testing to identify the serotypes causing Pacheco's disease (Phalen, 2006a).

Any bird that recovers from the illness needs to be identified as a carrier. Acyclovir is a treatment that can reduce the number of people who die during an outbreak. Although a vaccine exists, it's unclear if it offers protection against every herpes virus serotype. Before considering vaccination, it would be important to know the specific herpes serotype involved.

A white-gray growth on mucous membranes in the mouth cavity, upper gastrointestinal tract, cloaca, bile, and pancreatic ducts, is papillomatosis, another condition caused by a herpes virus. The lesions, which are most frequently observed in macaws, hawk-headed parrots, and Amazons, are associated with the herpes virus and Pacheco's disease (Fiskett and Reavill, 2004). Acetic acid treatment will cause cloacal lesions to turn white, and histology and PCR probes will help with the diagnosis. Treatment depends on how severe the illness is; many birds can live with it for years. Additionally, pancreatic duct and bile carcinomas are associated with papillomatosis (Phalen, 2006a).

6.4.1.3. Polyomavirus

All psittacine birds are susceptible to polyomavirus disease, although macaws, cockatiels, and lovebirds are frequently afflicted. When the virus started causing explosive outbreaks in commercial aviaries, it was initially identified in budgerigars (Phalen, 1997). Young canaries and finches have a high rate of morbidity and mortality; nestlings and juveniles are most frequently affected. Certain juveniles and nestlings will have concomitant PBFD infection. The age, species, and health of the exposed bird all affect the likelihood of a clinical disease developing. Anorexia, crop stasis, subcutaneous hemorrhage, tremors, ataxia, and decreased down and contour feather formation are examples of clinical signs. The virus can propagate both horizontally and vertically and is released in feather dander and bird droppings. There is a carrier state in which the virus resides in the kidney and spreads sporadically, particularly during stressful events. The most effective method for making a diagnosis is to use DNA-PCR probes on samples of blood, tissue, or ambient material.

There is a vaccine that can be helpful in the event of an outbreak (Fiskett and Reavill, 2004; Phalen, 2006a).

Usually administered as a preventive measure when the nestlings are about four weeks old, the vaccine is followed by a booster in another four weeks. Given that birds typically catch the disease before they turn eight weeks old, there is debate over how effective the vaccination is. Because of this, giving or carrying the vaccine in a shelter is not feasible. Obtaining a good history and routinely quarantining all incoming birds for 30 days is highly recommended, especially if the bird is from a breeding situation. Positive offspring ought to be put to sleep or separated. It is important to perform thorough cleaning and disinfection to stop the disease from infecting other young birds.

6.4.1.4. Paramyxovirus

Birds can contract paramyxovirus from a variety of serotypes. Parmyxovirus-1, the cause of exotic Newcastle disease (END), is a serious concern for any birds being imported into or exported from the United States and is regarded as a very high risk to the poultry industry. There are laws governing the transportation of birds into and out of the nation as well as between states, but birds that are brought into the country illegally may be carrying END. All species of birds are susceptible to the deadly and highly contagious END disease. Two to 15 days are needed for incubation. Parrots infected with END may exhibit respiratory, gastrointestinal, or neurological system symptoms, or they may show no symptoms at all (USDA and APHIS, 2003). Whether or not chickens are vaccinated, END can infect them and result in death. Because END is so severe, many birds pass away without exhibiting any symptoms. Notify federal and state animal health officials and isolate the bird right away if it shows clinical signs of endometriosis (END) or if it appears that it may have been exposed to other infected birds (USDA and APHIS, 2003).

6.4.1.5. Adenovirus

Adenovirus can occasionally result in disease outbreaks in African grays, pionus species, lovebirds, budgies, parrotlets, and cockatiels. Anemia is one of the clinical signs, which can also affect the central nervous system, gastrointestinal tract, kidneys, and lungs. Additionally, diseases can strike birds suddenly and without warning symptoms. Although renal enlargement can be detected radiographically, necropsies are typically used to make the diagnosis (Fiskett and Reavill, 2004). Although it is not a very common disease, other birds who were exposed to the infected bird would be at risk if it were to be identified.

6.4.1.6. Poxvirus

There are numerous pox viruses; the most well-known ones are the canary/finch, lovebird, and amazona pox, which can affect a variety of bird species. The apteric (featherless) areas of the body are affected by the dry cutaneous form of pox, which causes papules or elevated scab lesions to appear around the eyes, beak, nares, tibiotarsus, and feet. The wet diphtheroid form has a higher death rate and damages mucous membranes. There are lesions on the trachea, bronchi, oral membranes, and conjunctiva. Lesions should be surgically removed if they are causing clinical signs or obstructing airways.

Finches and canaries are common species with the septicemic form. It starts suddenly and results in cyanosis, anorexia, and ruffled feathers. The septicemic form rarely has cutaneous lesions. Typically, birds will pass away in three days. Histology of the lesions is used to make the diagnosis (Fiskett and Reavill, 2004; Phalen, 2006a). The only ways to treat the poxvirus are by treating secondary infections and offering supportive care. Whether treatment in a shelter setting should be attempted would depend on the severity of the disease. Controlling insects–mosquitoes can transmit the disease and avoiding contact with other birds that have lesions are two ways to stop the disease from spreading.

6.4.1.7. Proventricular Dilatation Disease (PDD)

Lymphoplasmacytic inflammatory infiltrates within the peripheral and/or central nervous systems are the cause of proventricular dilatation disease (PDD). It has been established that the borna virus causes the illness, but it remains unclear whether other viruses or medical conditions also play a role in the disease process (Rinder et al., 2009). Commonly affected parrot species include macaws, African grays, conures, cockatoos, and eclectus parrots. Central nervous system symptoms are common in African gray parrots. Clinical symptoms include ataxia, seizures, weight loss, persistent, and sporadic regurgitation, undigested seed in feces, depression, and deficits in proprioception and motor function. A combination of radiographs demonstrating a dilated proventriculus, a slow gastrointestinal emptying time, PCR testing for the borna virus, a biopsy of the crop's highly vascularized areas, and clinical signs are used to make the diagnosis. This disease has no known cure and is difficult to identify with certainty in a live bird. Treatments for secondary infections are frequently required, and nonsteroidal anti-inflammatory drugs (NSAIDs) like meloxicam can be useful in easing the clinical signs (Fiskett and Reavill, 2004; Phalen, 2006a). When a bird exhibits clinical symptoms, it should be kept apart from other birds and kept under strict isolation guidelines.

6.4.1.8. Avian Influenza

Although infections in other gallinaceous species can be fatal, domestic poultry are most frequently affected by the highly pathogenic avian influenza (HPAI) virus, which causes a serious and frequently fatal infection. Approximately 17 million birds were destroyed during an outbreak in the United States in 1983–1984, which cost close to $65 million. Psittacines and other species may be infected by HPAI, but it should be emphasized that companion birds rarely contract HPAI and that poultry are more likely to exhibit its typical symptoms (CDC, 2008; CFSPH, 2009b).

An influenza virus of type A causes HPAI. Though type A influenza viruses are similar in all species, they usually have a higher specificity for a single species or related animal groups. Although human subtype A influenzas typically only infect humans, human cases of avian influenza subtypes have been documented, usually as a result of human contact with infected poultry. Since the influenza A virus is ever-evolving, it's possible that birds are the source of some human subtypes, including H_1N_1, H_1N_2, and H_3N_2. When avian influenza strikes companion birds, it can cause symptoms that are vague and even central nervous system-related. It is important to report the disease if a tracheal or cloacal swab is used to identify it.

6.4.2. Bacteria and Other Infectious Causes of Disease

6.4.2.1. Chlamydia Psittaci

Chlamydia is a zoonotic intracellular bacterial parasite that causes pisittacosis. It is also referred to as ornithosis or parrot fever on occasion. Numerous bird species have been identified as having *C. psittaci* like pigeons and other companion bird species pose the biggest threat to public health. Clinical symptoms such as sinusitis, dyspnea, conjunctivitis, polyuria, diarrhea, lethargy, anorexia, yellow to dark-green droppings, poor feathering, and persistent weight loss are typical of the classical form of the disease in amazons. The diagnosis of psittacosis is aided by hematology, radiography, biochemistry, PCR probes, antigen capture tests, antibody tests, cytology, and cultures. Acute infections frequently cause leukocytosis. The disease is hard to diagnose because the bacterium is intracellular and sheds periodically. Additionally, if the bird has taken antibiotics, this will result in false negative results. The bird cannot be proven to be disease-free by a conclusive test.

A five-day to several-week incubation period is observed. The most frequent way of transmission is by inhaling dried dust contaminated with bird droppings; ingestion is less common. The illness can also be dispersed by insects and termites. Subclinical infections happen frequently. Stressed-out carrier birds frequently release the virus. The illness is zoonotic, meaning it infects humans' respiratory systems. If caught early and treated effectively, chlamydia usually does not pose a serious threat to humans or birds. It also reacts well to tetracyclines. If psittacosis is suspected in birds, they should be quarantined for 30 days. Wearing a surgical cap, mask, gloves, and protective clothes is recommended for handlers. To stop the spread of the infectious feather dust, necropsies should be carried out under a hood by personnel dressed in the protective gear mentioned above. The bird should also be submerged in a disinfectant. Since the waste material in the cage can spread disease for weeks, it is imperative to maintain proper sanitation to stop the spread of illness. According to several studies, quaternary ammonium compounds, chlorophenols, iodophor disinfectants, formaldehyde, and 80% isopropyl alcohol can all be effective against Chlamydia (CFSPH, 2009a ; Fiskett and Reavill, 2004; Gerlack, 1994; Tully, 2006).

6.4.2.2. Salmonella

A zoonotic bacterial disease called salmonella can cause African gray parrots, lories, and young birds to die at a high rate. Salmonella can infect most vertebrates, although host susceptibility and carrier state development differ significantly between species.

Salmonella is spread by eating contaminated food or water or by direct contact with aerosolized feces or feather dust. It is excreted in feces as well as nasal, ocular, and other secretions. Insects, rodents, flies, and other vermin can also spread it. The most frequent causes are stress, crowded conditions, and inadequate ventilation. The disease has a carrier state, and the most common way to get infected is through the droppings of chronically infected birds. Although the sporadic shedding of the bacteria makes diagnosis challenging, it can be achieved by PCR or sequence assays, bacterium isolation, and identification.

Clinical signs of the disease include depression, lethargy, diarrhea, pasty vents, lameness, abscesses, convulsions, dehydration, crop stasis, meningitis, osteoarthritis, and sudden death. African gray parrots are especially prone to chronic illness, which manifests as dermatitis, arthritis, excessive thirst, and nares discharge. Lories may pass away unexpectedly due to an acute illness. The outcomes of bacterial cultures and sensitivity testing determine the course of treatment (Fiskett and Reavill, 2004). Many antibiotics work well against salmonella; sensitivity and culture should be taken into consideration when choosing a course of treatment.

The bacteria can survive for extended periods of time outside the bird's body. When dry, contaminated droppings resemble dust. Sanitation inside cages is essential to stopping the spread of disease. Several disinfectants that shelters frequently use, like quaternary ammonium products, inactivate salmonella. When handling birds suspected of having salmonella, as limited personnel as possible should wear protective clothing as outlined in the Chlamydia section. The birds should also be placed in quarantine for observation and receive a complete course of treatment before release or transfer. Three to six weeks after treatment, a culture should be carried out to ensure the bird is disease-free and ready for adoption without any cautions. It should be mentioned that birds can contract salmonella from humans as well.

6.4.2.3. Mycobacterium

Although mycobacterium infections in psittacines are uncommon, they typically involve *M. genavense* or *M. avium.* Although it is uncommon, these two mycobacterium species do have the ability to spread to other animals. Although it is more frequently reported that a bird contracts tuberculosis from a human, it has been reported that psittacines have *Mycobacterium tuberculosis.* The most frequently impacted birds are finches/canaries, amazons, and pionus. The most common way that mycobacterium from other animals' excrement contaminates soil or water is through ingestion or inhalation. Individual differences in susceptibility to the disease are significant.

The range of lesions is from multiple granulomas that are easily mistaken for tumors to diffuse granulomatous inflammation, which typically occurs within the gastrointestinal tract. Skin conditions, ocular lesions, and respiratory granulomas are less common. A patient's leukogram may be noticeably elevated or it may be perfectly normal. Aspiration or biopsy of the mass is used for diagnosis; cytology and acid-fast stains are used to assess intestinal thickenings and the mass. Everyone who comes into contact with the bird must see a doctor if the illness is found. Granulomas should be surgically removed if at all possible. Multidrug and other effective therapies based on sensitivity and culture should be used if the disease spreads (Pollock, 2006). A bird with a confirmed mycobacterium by biopsy will probably have to be put down in a shelter because of the poor prognosis, considerable medical and surgical care required, and potential for zoonotic disease.

6.4.2.4. Aspergillosis

The most common cause of respiratory illnesses in companion birds is aspergillosis. This fungus, which is frequently present in the environment, has been exposed to the majority of birds. Although nearly all avian species are affected, it is said that the African gray parrot and pionus parrot are more susceptible to it. Usually brought on by immunosuppression or exposure to a high concentration of spores, aspergillosis develops as a consequence. It can manifest as a localized mass, a disseminated systemic infection, or both in acute and chronic forms. To find out if the environment is a source of fungal spores, samples such as wood shavings, corncob bedding, and seed hulls should be tested. Foods unfit for human consumption, like peanuts, can also harbor aspergillosis. Should food be suspected of harboring aspergillums, it ought to be thrown out. It is important to take precautions against mold growth in the enclosure and to guarantee that only fresh food is served. Unless the bird is immunocompromised or exposed to extremely unhygienic conditions, most exposed birds do not get the disease.

Significant heterophilic leukocytosis is often brought on by aspergillosis and is linked to both the acute and chronic forms of the illness. For the diagnosis of aspergillosis, radiography, endoscopy, and cytologic evaluation are all helpful. Treatment options for aspergillosis include antifungal medications, surgical excision, nebulization, and supportive care for any underlying compromised states. The disease is difficult to diagnose and treat (Dahlhausen, 2006). Depending on the severity of the condition, euthanasia might need to be considered if systemic aspergillosis in a bird shelter setting is confirmed.

6.4.2.5. Macrorhabdus–Avian Gastric Yeast

It is now known that Macrorhabdus ornithogaster, commonly referred to as megabacteria or avian gastric yeast, is a yeast rather than a bacteria. It is frequently reported that avian gastric yeast affects budgies, cockatiels, finches, canaries, lovebirds, neophema parakeets, rosellas, lesser sulfur-crested cockatoos, and chickens. Signs of depression and weight loss are evident in affected birds. In canaries and budgies, the upper digestive tract may contain avian gastric yeast. Within the superficial mucosal glands of the isthmus, the organism resides in the lower part of the proventriculus. It is unknown what is being transmitted. Wet mounts, gram stains, and PCR probe tests can all be used to identify the yeast. Nonspecific symptoms of anemia, leukocytosis, heterophilia, monocytosis, lymphocytosis, and basophilia can be seen in budgies. Although amphotericin B and nystatin may be useful in treating the disease's clinical symptoms, it is unlikely that they will totally cure birds of it (Fiskett and Reavill, 2004; Phalen, 2006b).

6.4.2.6. Candida Albicans

Changes in the typical bacterial flora, the breakdown of the alimentary mucosa, or immunological incompetence can allow resident yeast–most likely *Candida albicans*–to proliferate and possibly spread. Lovebirds, budgies, and cockatiels are particularly vulnerable. Young birds are more likely to contract diseases, which can develop into fatal systemic diseases. Predisposing factors for this disease may include antibiotics, hypovitaminosis A, spoiled foods, an unhygienic and stressful environment, malnourishment, and bacterial or viral infections. Babies that are not handled carefully will often experience crop stasis, which can lead to Candida overgrowth. Clinical signs, history, lesions, and laboratory samples are used to make the diagnosis. A yeast infection cannot be diagnosed by culture alone. To effectively treat this condition, underlying causes of the infection must be found and addressed (Fiskett and Reavill, 2004).

6.4.2.7. Sarcocystis

A protozoan infection called sarcocystis leads to a high rate of morbidity and mortality in psittacine birds from Australia, Asia, and Africa, as well as occasionally in exotic columbiformes. The most frequently impacted species are lories and lorikeets, which are followed by cockatoos, African gray and eclectus parrots, and lorries.

Transmission occurs via mechanical vectors like cockroaches as well as the definitive host, the opossum's excrement. Two examples of intermediate hosts are grackles and cowbirds. Because of the protozoans' asexual reproduction in the lung, which causes deadly respiratory collapse, death happens quickly. Perhaps the only identifiable lesions are pulmonary congestion and hemorrhage. It is crucial to control cockroaches and opossums, and pyrimethamine may be helpful if the disease is identified and treated early (Fiskett and Reavill, 2004).

6.4.3. Parasites

6.4.3.1. Knemidokoptes

A common mite found in canaries and budgies is called klemidokoptes. Typically, apteric skin areas such as the beak, cere, vent, legs, and feet are home to mites. Nail and beak growth can become malformed as a result of hyperkeratosis. Mites are spread horizontally from bird to bird, but they can also spread vertically to featherless progeny. Reduced immune function may lead to severe secondary infections. The diagnosis is made by microscopically observing the mites on the hyperkeratotic skin. Treatment with ivermectin is common (Fiskett and Reavill, 2004). All toys, bowls, and detachable items should be completely hand-washed or put through a high-heat dishwasher after the cage has been cleaned and disinfected.

6.4.3.2. Giardia

Giardia can cause a range of clinical signs, from diarrhea and weight loss to failure to thrive, and has been found in psittacines, poultry, waterfowl, finches, and toucans. Giardia infections in cockatiels have been reported to cause severe pruritus and feather picking. *Hexamita/Spironucleus* spp. are now thought to be the primary cause of the disease in cockatiels. *A fecal* PCR test or a direct wet mount can be used for diagnosis. Although metronidazole and rovidazole are frequently used to treat giardia, *Hexamita/Spironucleus* spp. is more challenging to successfully clear (Gelis, 2006). Table 6.1 provides an overview of several significant bird viral diseases.

Table 6.1. *Virus Names, Abbreviations, and Genus/Species Designations are Listed According to the Taxonomy and Nomenclature Approved by the International Committee on Taxonomy of Viruses (ICTV)*

Viral Family	Genus	Viral Taxa	Abbreviations	Nomenclature	Bird Family	Clinical Signs	Mortality Rate
dsDNA Viruses							
Herpesviridae	*Iltovirus*	*Gallid herpesvirus 1*	GaHV-1	*Iltovirus gallidalpha1*	Turdidae, Estrildidae, Sturnidae	Subclinical infection/ Respiratory signs	High
		Psittacid herpesvirus 1	PsHV-1	*Iltovirus psittacidalpha1*	Sturnidae		
		passerid herpesvirus 1	–	not recognized	Fringillidae		
	Mardivirus	*Columbid herpesvirus 1*	CoHV-1	*Mardivirus columbidalpha1*	Corvidae, Turdidae	Subclinical infection	Probably not
Poxviridae	*Avipoxvirus*	*Fowlpox virus* (Lineage A1)	FWPV	Same	Fringillidae, Passeridae	Cutaneous lesions/ Upper respiratory and digestive tract lesions	High
		Pigeonpox virus (Lineage A2)	PGPV	Same	Fringillidae		High/ Medium
		Accipitriformespox virus (Lineage A7)	–	Same	Lanidae		High/ Medium
		Canarypox virus (Lineage B1)	CNPV	Same	Fringillidae, Paridae, Corvidae (and 11 other families)		High/ Medium
		Starlingpox virus (Lineage B2)	–	Same	Sturnidae, Passeridae		High/ Medium

		Lineage B3 poxvirus	–	Not recognized	Corvidae (and eight other mostly Passeriformes families)		High
Adenoviridae	*Atadenovirus*	European robin adenovirus	–	Not recognized	Meliphagidae, Fringillidae, Ploeceidae, Estrildidae, Turdidae, Ptilonorhynchidae, Sturnidae, Zosteropidae, Maluridae, Petroicidae	Subclinical infection/ Not known	Probably not
		European greenfinch adenovirus	–	Not recognized			
		Eurasian siskin adenovirus	–	Not recognized			
		Eurasian bullfinch adenovirus	–	Not recognized			
		common chaffinch adenovirus	–	Not recognized			
		passerine adenovirus 1	–	Not recognized			
		white plumed honeyeater adenovirus 1, 2	–	Not recognized			

		vitelline masked weaver adenovirus strains 37869 and 38132	–	Not recognized			
	Aviadenovirus	great tit adenovirus strain 47292	–	Not recognized	Fringillidae, Estrildidae, Paridae, Ploceidae, Sylvidae, Falcunculidae, Petroicidae	Subclinical infection/ Not known	Probably not
		european greenfinch adenovirus	–	Not recognized			
		goldfinch adenovirus	–	Not recognized			
		vitelline masked weaver adenovirus strain 39658	–	Not recognized			
	Siadenovirus	double-barred finch adenovirus	–	Not recognized	Estrildidae, Paridae, Sylvidae, Thraupidae, Meliphagidae, Pardalotidae, Rhipiduridae, Sturnidae	Subclinical infection/ Probable renal lesions	Probably not
		eurasian blackcap adenovirus	–	Not recognized			
		gouldian finch adenovirus 1	GFAdV-1	Not recognized			

		great tit adenovirus strain 47292	GTAdV-1	*Great tit siadenovirus A*			
		zebra finch adenovirus strain 47535	–	Not recognized			
Papillomaviridae	*Etapap illomavirus*	*Fringilla coelebs Papillomavirus 1*	FcPV1	*Etapapillomavirus 1*	Fringillidae	Cutaneous lesions	Low
		Serinus canaria papillomavirus 1	ScPV1	Not recognized	Fringillidae	Not known	Not known
ssDNA VIRUSES							
Circoviridae	*Circovirus*	*Beak and feather disease virus*	BFDV	*Circovirus parrot*	Artamidae, Corvidae, Estrildidae, Fringillidae, Petoicidae, Sturnidae, Thamnophilidae	Feather lesions	Probably not
		Circovirus canary	CaCV	Same		Feather diseases and immunosupression/ Subclinical infection	Yes/Not known
		Circovirus finch	FiCV	Same			
		Circovirus raven	RaCV	Same			
		Circovirus starling	StCV	Same			
		Circovirus zebrafinch	ZfiCV	Same			
	Cyclovirus	*Robinz virus RP_736*	RobinzV736	*Cyclovirus cervienka*	Petroicidae	Not known	Not known
		Robinz virus RP_1170	RobinzV1170	*Cyclovirus liepsnele*			

		Robinz virus RP_493	RobinzV493	*Cyclovirus pettirosso*			
		Robinz virus RP_620	RobinzV620	*Cyclovirus prihor*			
		Robinz virus RP_526	RobinzV526	*Cyclovirus punarinta*			
		Robinz virus RP_584	RobinzV584	*Cyclovirus rudzik*			
		Robinz virus RP_259	RobinzV259	*Cyclovirus totoi*			
	Probably new genus	Gyrovirus 11	–	Not recognized	Thamnophilidae	Not known	Not known
Genomoviridae	*Gemycir culavirus*	new species	–	Not recognized	Petroicidae, Aegithalidae, Corvidae, Emberizidae, Fringillidae, Muscicapidae, Paridae, Sylviidae, Turdidae, Zosteropidae	Not known	Not known
		Chickadee associated gemycircularvirus 1	–	*Gemycircularvirus chickad1*	Paridae		
		Finch associated genomovirus 5	–	*Gemycircularvirus haeme1*	Fringillidae		
		Finch associated genomovirus 6	–	*Gemycircularvirus haeme2*			
	Gemykibivirus	*Blackbird associated gemykibivirus 1*	–	*Gemykibivirus blabi1*	Turdidae		

		Black robin associated gemykibivirus 1	–	*Gemykibivirus blaro1*	Petroicidae		
		Finch associated genomovirus 3	–	*Gemykibivirus haeme1*	Fringillidae		
		Finch associated genomovirus 2	–	*Gemykibivirus haeme3*			
		Finch associated genomovirus 4	–	*Gemykibivirus haeme4*			
		Finch associated genomovirus 1	–	*Gemykibivirus haeme5*			
	Gemykroznavirus	Finch associated genomovirus 8	–	*Gemykronzavirus haeme1*	Fringillidae		
	Gemygorvirus	*Starling associated gemygorvirus 1*	–	*Gemygorvirus stara1*	Sturnidae		
Parvoviridae	*Aveparvovirus*	*Aveparvovirus passeriform1*	PfPV	Same	Thraupidae	Not known	Not known
ssRNA VIRUSES							
Orthomyxoviridae	*Alphainfluenzavirus* 1	Highly Pathogenic Avian Influenza H5N1	FLUAV/ (H5N1)	*Alphainfluenzavirus influenzae*	Passeridae, Turdidae, Ploceidae, Corvidae, Hirundidae, Icteridae (and many other families)	Mild/Severe disease	High

Paramyxoviridae	*Orthoavulavirus*	*Avian paramyxovirus 1* (NDV, *Newcastle Disease virus*)	APMV-1	*Orthoavulavirus javaense*	Passeridae, Motacillidae, Passarellidae	Not known/ Subclinical disease	High/Not known
	Metaavulavirus	Avian paramyxovirus 2	APMV-2	*Metaavulavirus yucaipaense*	Corvidae, Motacillidae, Troglodytidae, Muscicapidae, Emberizidae, Estrildidae	Mild/Severe disease	
	Paraavulavirus	Avian paramyxovirus 3	APMV-3	*Paraavulavirus wisconsinense*			
	Paraavulavirus	Avian paramyxovirus 4	APMV-4	*Paraavulavirus hongkongense*			
Pneumoviridae	*Metapneumovirus*	*Avian metapneumovirus*	AMPV	*Metapneumovirus avis*	Passeridae, Sturnidae	Not known	Not known
Bornaviridae	*Orthobornavirus*	*Estrildid finch bornavirus 1*	EsBV-1	*Orthobornavirus estrildidae*	Estrildidae, Corvidae, Fringillidae	Not known	Not known
		Canary bornavirus 1	CnBV-1	*Orthobornavirus serini*			
		Canary bornavirus 2	CnBV-2				
		Canary bornavirus 3	CnBV-3				
Flaviviridae	*Orthoflavivirus* [1]	*Japanese encephalitis virus*	JEV	*Orthoflavivirus japonicum*	Passeridae, Corvidae, Fringillidae, Turdidae (and >23 more families)	Not known	High

		Saint Louis encephalitis virus	SLEV	*Orthoflavivirus louisense*			
		Murray Valley Encephalitis virus	MVEV	*Orthoflavivirus murrayense*			
		Usutu virus	USUV	*Orthoflavivirus usutuense*		Severe disease	
		West Nile virus	WNV	*Orthoflavivirus nilense*		Mild/Severe disease	
Togaviridae	*Alphavirus* [1]	*Eastern equine encephalitis virus*	EEEV	Same	Turdidae, Hirundidae, Icteridae, Petroicidae, Estrildidae, Monarchidae	Neurological disease	Yes
		Highlands J virus	HJV	Same		Mild disease	
		Sindbis virus	SINV	Same		Severe disease	
		Ross River virus	RRV	Same		Not known	
		Mayaro virus	MAYV	Same			
		Buggy Creek virus	BUCGV	*Fort Morgan virus*		Neurological disease	
	Deltacoronavirus [2]	*Bulbul coronavirus HKU11*	BulCV_HKU11	Same	Turdidae, Estrildidae, Pycnonotidae, Passeridae, Zosteropidae, Fringillidae, Emeberizidae	Severe disease	Probably high

		Munia coronavirus HKU13	MunCV_HKU13	Same			
		Thrush coronavirus	–	Not recognized			
		White-eye coronavirus HKU16	WECoV_HKU16	Same			
		Coronavirus HKU15	PoCoV_HKU15	Same			
Picornaviridae	*Oscivirus*	Oscivirus A1 and A2 (turdivirus two and 3)	OsV-A1/OsV-A2	*Oscivirus A*	Turdidae, Muscicapidae	Severe disease	Not known
	Passerivirus	*Passerivirus A and B*	PasV-A/PasV-B	Same	Turdidae, Estrildidae		High
	Poecivirus	*Poecivirus A1*	PoeV-A	*Poecivirus A*	Paridae		Yes
	Probably new genus	French Guiana Picornavirus	FGPV	Not recognized	Thamnophilidae	Not known	Not known
Astroviridae	*Avastovirus*	novel avastrovirus clades four and 5	–	Not recognized	Fringillidae, Monarchidae Parulidae, Passeridae	Subclinical infectionMild disease	Not known
Caliciviridae	Probably new genera	new calicivirus	–	Not recognized	Prunellidae, Turdidae, Emberizidae, Fringillidae, Muscicapidae, Petroicidae	Subclinical infection	Not known
Hepeviridae	Probably new genera	new hepe-like viruses	–	Not recognized	Thamnophilidae, Turdidae, Zosteropidae	Not known	Not known
dsRNA viruses							

Reoviridae	*Orthoreovirus* [2]	new orthoreovirus	–	Not recognized	Emberizidae, Turdidae, Fringillidae	Not known	High mortality
RNA reverse transcribing virus							
Retroviridae	*Alpharetrovirus*	*Avian leukosis virus*	ALV	Same	Emberizidae, Paridae, Corvidae, Muscicapidae, Thraupidae, Phylloscopidae	Subclinical infection	Possibly

All ICTV Accepted Viral Taxa are in Italics.

Source: Williams RAJ, Sánchez-Llatas CJ, Doménech A, Madrid R, Fandiño S, Cea-Callejo P, Gomez-Lucia E, Benítez L. Emerging and Novel Viruses in Passerine Birds. Microorganisms. *2023; 11(9):2355. https://doi.org/10.3390/microorganisms11092355.*

6.5. NUTRITIONAL/METABOLIC DISORDERS

6.5.1. Chronic Liver Disease

Psittacines often suffer from chronic liver disease. Hepatotoxins, severe vacuolar hepatopathy, fatty liver, chronic congestion from heart disease, and infectious inflammatory hepatitis are a few examples of prior hepatic insults that can cause lesions. Psittacines fed all-seed diets and receiving little exercise are prone to developing fatty liver disease. Because they are unique to the liver, bile acids are a valuable tool for diagnosing reduced liver function. Radiographs aid in the detection of masses or an enlarged liver. A liver biopsy or aspirate is used to diagnose liver disease. Hepatic disease management involves the use of nutritional counseling, antibiotics, lactulose, milk thistle, colchicines, and other nutriceuticals in addition to the identification and correction of underlying and contributing disease conditions (Fiskett and Reavill, 2004; Hochleithner et al., 2006). Many birds in shelters will seem clinically normal; each bird needs to be evaluated to determine its quality of life and whether or not it can be placed in a home.

6.5.2. Renal Disease

Psittacines are prone to renal disease, though it can be challenging to identify. Elevated levels of protein and uric acid are indicators of disease, but serious illness can also coexist with perfectly normal blood work. Numerous other medical conditions can lead to renal disease. Cockatiels and budgies frequently develop gout, with many of these birds having underlying kidney disease that results in elevated blood levels of uric acid. In articular gout, urates are deposited in the joints, causing white swellings over the joints and tendons, particularly in the feet. Clinical signs of disease include nonspecific feather picking, shifting leg lameness, and an inability to bend the toes. Deposits on visceral organs are a symptom of visceral gout, which is secondary to elevated uric acid levels. Urate deposits on internal organs and in joints have been observed in birds with visceral and articular gout, although these conditions typically occur separately. Allopurinol has been used to reduce the production of uric acid; however, it does not affect tissue deposits. Treatment for renal disease includes dietary changes, omega-3 fats, and fluid therapy. Because articular gout causes significant pain and inflammation, NSAIDs (non-steroidal anti-inflammatory drugs) are recommended for managing the condition (Echols, 2006; Fiskett and Reavill, 2004).

6.5.3. Atherosclerosis

Plaque lesions of the elastic and muscular arteries' tunica media and tunica intima are known as atherosclerosis. Atypical vascular flow, endothelial integrity loss with reduced elasticity, and varying degrees of vessel narrowing are all brought on by the plaques. Thrombosis may result from modifications to the vessel walls. The disease most frequently affects African gray parrots, lovebirds, Amazon parrots, and cockatiels. Lesions typically affect the brachiocephalic and aortic arteries in birds.

Obesity, inactivity, and poor eating habits are the main causes of atherosclerosis. Psittacines may frequently pass away unexpectedly without exhibiting any symptoms due to this illness. Clinical signs that may be evident include progressive paresis of the legs, dyspnea, muscle wasting, and seizure activity. Treatment options include omega-3 fats, weight loss, dietary modifications, and heart medication if necessary (Fiskett and Reavill, 2004).

6.5.4. Hypovitaminosis A

Pet birds fed a diet high in seeds or cereal grains frequently develop hypovitaminosis A. Reduced vitamin A levels lead to epithelial squamous metaplasia, which causes the respiratory tract, conjunctiva, nasolacrimal duct, upper alimentary tract, and oral cavity to become hyperkeratotic. The submucosal glands will enlarge due to secondary infections caused by keratinizing epithelium blocking their ducts, resulting in large keratin granulomas inside the nasal or oral cavity. Additionally, this modifies the mucosal defenses, which increases the risk of bacterial and fungal infections, particularly in the respiratory tract. A history and physical examination are typically used to make a diagnosis. When the liver's vitamin A concentration falls below 50 IU/kg, epithelial changes take place. Treatment options include treating secondary infections, changing one's diet, and taking vitamin A supplements (Fiskett and Reavill, 2004).

6.5.5. Hypocalcemia

The most common association between hypocalcemia and tetany or seizures in African gray parrots is lower blood calcium levels. It is most likely a diet-related condition. Total calcium and albumin concentration in the plasma have been shown to be strongly correlated, so the calcium levels should be interpreted in conjunction with the albumin levels (Fiskett and Reavill, 2004).

6.5.6. Scissors Beak

A macaw's scissors beak is typical. It is the upper beak's lateral deviation over the lower beak. Although the exact cause is unknown, possible causes include trauma, viral infections, low or imbalanced calcium levels in the diet, improper hand-feeding techniques, and changes to the mandibular occlusal surface. If caught early, it can be treated by cutting off the lower beak, or gnathotheca, and using therapeutic massage techniques to reshape the upper and lower beak, or rhamphotheca, on its occlusal surface. If this approach is unsuccessful, pressure can be applied to correct the rhinotheca (upper beak) by attaching a ramp made of dental acrylic over a stainless steel mesh to the gnathotheca, taking care not to overcorrect (Martin and Ritchie, 1994). Care should be given by a veterinarian who has worked with birds before.

6.5.7. Other Conditions

6.5.7.1. Chronic Obstructive Pulmonary Disease (COPD)

In macaws, chronic obstructive pulmonary disease (COPD) is common, particularly in blue and gold macaws kept with African grays, cockatiels, and cockatoos. This type of pulmonary hypersensitivity is typically linked to polycythemia. Because of the reserve capacity of avian lungs and the relative inactivity of macaws kept in captivity, early diagnosis is challenging. Following the onset of polycythemia, lung lesions progress. The majority of these macaws have been housed in poorly ventilated aviaries. Better ventilation and isolation from cockatiels, cockatoos, and African gray species are the recommended treatments (Fiskett and Reavill, 2004).

6.5.7.2. Brown Hypertrophy of the Cere

It is typical for the cerebrum of female budgerigars to turn brown due to the production of estrogen. The cornified layer of the cere can experience extensive hyperplasia in older females, resulting in growths that resemble horns. Depending on the severity, this condition may or may not require surgical correction.

6.5.7.3. Millet Seed Aspiration

Canaries, budgies, and cockatiels frequently have seed aspirations. The bird will exhibit upper respiratory distress symptoms. It is frequently recommended to place an air sac breathing tube and remove seeds by flushing or aspirating. It is suggested that the flush be cytology. The best person to treat this would be a veterinarian with experience treating birds. (For the location of the air sac tube, *see* Harrison et al., 2006).

6.5.7.4. Choanal Atresia

A closed choanal slit is the result of a persistent membrane on the bony portion of the palate within the nasal cavity, which causes choanal atresia. This disorder obstructs the regular flow of nasal secretions into the oral cavity. Young birds may exhibit a persistent nasal and ocular discharge, as well as a distended infraorbital sinus cavity filled with a transparent fluid. Surgical correction is advised, depending on the severity.

6.5.7.5. Hematuria Syndrome of Neonates

Certain Amazon and pionus species, as well as young African gray parrots, have been reported to have reddish urine and urates. It can happen with some handfeeding recipes and is more noticeable with specific bedding types. A fecal occult blood test is used to make the diagnosis. There is no pathology or clinical sign associated with the condition, and treatment is not required.

6.5.7.6. Amazon Foot Skin Necrosis

In Amazons, especially the yellow-naped Amazon, skin necrosis of the Amazon foot occurs. Clinical lesions manifest as a brown to black discoloration of the apteric skin of the legs and feet, or as irregular patterns of erythema. Because birds are extremely prickly, they frequently injure themselves by attacking the skin's afflicted areas. The goal of treatment is to stop the mutilation while allowing the lesions to heal; the cause is unknown. Certain birds will experience recurrent episodes.

6.5.7.7. Feather Follicle Cysts

Although the exact mode of inheritance for feather follicle cysts is unknown, viruses could be the cause of the secondary folliculitis. Canaries of the new color that are dimorphic, crested, crest-bred, and have an extra down-type feather that makes them more prone to cyst formation. Cysts commonly occur in the feather tracts of the pectoral and scapular regions. Surgery is not a good option if the cysts are widespread throughout the body, but it is possible to remove them if they are solitary. The breeding of these birds ought to be prohibited.

6.5.7.8. Tail Avulsions

When wing-clipped fledglings fall and hyperextend their tails, they frequently sustain tail avulsions. The ventral tail base muscles are severed from the pygostyle, and a prolapsed cloaca may also be present. It might be necessary to close the skin surgically.

6.5.7.9. Self-Mutilation/Feather Picking

Psittacines frequently pick their feathers and self-mutilate. To ascertain whether the behavior is due to a primary behavior problem or an underlying condition, it is imperative that the clinician conducts a comprehensive medical workup. Along the point of the keel, sternum mutilation is common, and there may be an open wound involving the skin and muscle. Persistent self-harm can result in infection and necrosis, which can reach deeper muscles and the keel bone. It is common to find mixed bacteria of the species *Streptococcus*, *Staphylococcus aureus*, *Escherichia coli*, *Pseudomonas*, *Klebsiella*, and occasionally fungal organisms. Secondary infections are often the result of pruritus (Fiskett and Reavill, 2004) (Figure 6.4).

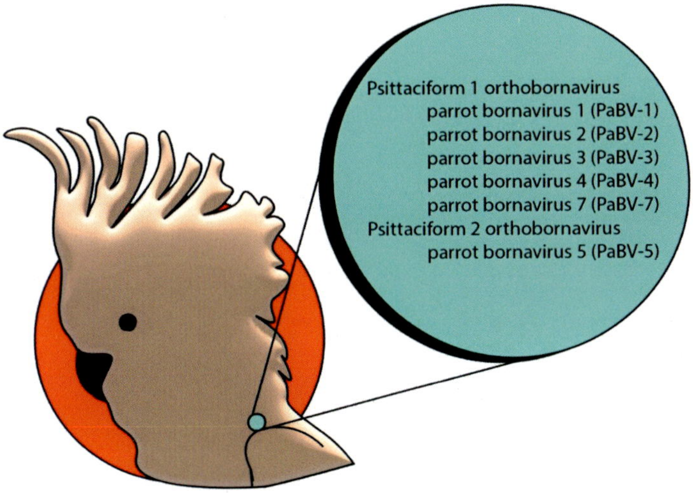

Figure 6.4. *Bornaviruses specific to psittacine birds.*

Source: Leal de Araújo J, Rech RR. Seeing beyond a Dilated Proventriculus: Diagnostic Tools for Proventricular Dilatation Disease in Psittacine Birds. Animals. *2021; 11(12):3558. https://doi.org/10.3390/ani11123558.*

In heavy-bodied birds such as Amazon parrots and African grays, self-mutilation can occur from improper wing trimming, which can cause trauma during landing and damage to the keel. Behavior changes must be addressed after all medical explanations for the mutilation have been exhausted. In these situations, a thorough history is highly beneficial in determining the behavior's cause and possible methods of behavior modification. Based on the author's observations, cases of self-mutilation involving Moluccan cockatoos are typically the most challenging to handle. When they start self-mutilating, it develops into a compulsive, repetitive disorder from which they are unable to stop. With enough time and care, these birds' energy can be channeled into constructive behavior. These are extremely time-consuming cases, so the bird should be placed with a facility or individual who can handle them if the shelter does not have the skilled staff or the time to handle them. These birds' conditions rapidly deteriorate and result in permanent emotional and medical harm if they are left in a cage without much care (Armstrong et al., 2001; Centers for Disease Control and Prevention, 2008) (Figure 6.5).

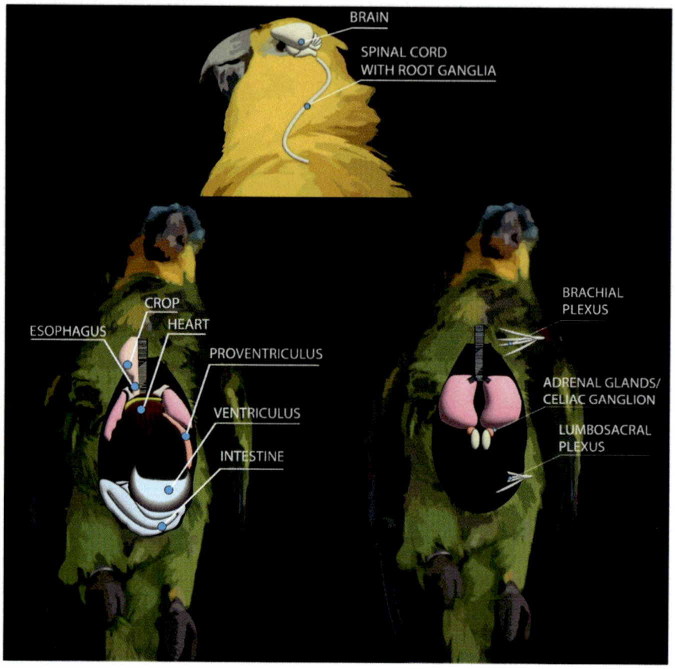

Figure 6.5. *These are the main organs that should be investigated using histopathology, immunohistochemistry, and RT-PCR for confirmation of PDD.*

Source: Leal de Araújo J, Rech RR. Seeing beyond a Dilated Proventriculus: Diagnostic Tools for Proventricular Dilatation Disease in Psittacine Birds. Animals. *2021; 11(12):3558. https://doi.org/10.3390/ani11123558.*

6.6. BEHAVIOR

As prey animals, psittacines react to stimuli considerably differently than do dogs and cats. Because parrots typically live in flocks in the wild, they might be less tolerant of isolated captivity settings than other species. Most of their time is spent searching for food, keeping an eye out for predators, and tending to their own needs. They have an enormous amount of extra time and energy in captivity because all of their basic needs–food, water, and protection from predators are satisfied. Obesity, boredom, and a host of behavioral issues result from this. For a few simple ways to offer the kind of stimulation that helps remove behavior issues linked to inactivity and purposelessness, see relevant research (Table 6.2).

Table 6.2. *Key Recommendations to Strengthen Surveillance of Wild Birds and Collaboration*

Knowledge gap or need	Recommendation
Capability to safely capture and handle wild birds for sampling	More wild bird capture and handling trainings for wildlife biologists and connection of ornithologists with infectious diseases specialists.
Biosafety in the field	Trainings for ornithologists for capturing of birds that have increased likelihood of infections or near outbreak areas.
Trust in multidisciplinary teams	Focus on trust building between laboratory specialist and ornithologists is imperative.
Hypothesis-testing to understand the ecology of infectious diseases in wild birds	Education on the gaps of understanding on the ecology of host/pathogen/vector systems and additional samples or data systematically collected from birds to answer key questions.
Sample and data sharing	Connection of laboratory and ornithologists for additional analysis of DNA/RNA or other analysis. Sharing of data and including more metadata in publications. Protocols for safe shipment of samples or extracted DNA/RNA.
Environmental and climate drivers of zoonotic diseases in wild bird populations	Connection between ornithologists, disease specialists, and climate scientists to use surveillance data and analysis to forecast future outbreaks and pathogen range shifts.
Microbial evolution	Connection with surveillance programs to viral and bacterial bioinformatic and evolution specialists and availability to sequencing.
Wild bird conservation	Inclusion and outreach to conservation organizations and governments to understand the impacts of diseases on wild bird populations globally and local populations.
Impact and risks to poultry	Better connection between wild bird surveillance and the poultry industry and government authorities to access and know of disease risks.
Education and outreach	Connection to public communication experts to design educational information for public on wild birds and ecology of disease, especially to correct mis/disinformation.
Communication with all stakeholders	Increased transparency with all government authorities, organizations, or entities that can assist in surveillance or have a need to know of results. Trainings for all personnel involved in surveillance to ensure rigorous scientific method and data quality to trust the results.

Source: Fair JM, Al-Hmoud N, Alrwashdeh M, Bartlow AW, Balkhamishvili S, Daraselia I, Elshoff A, Fakhouri L, Javakhishvili Z, Khoury F, Muzyka D, Ninua L, Tsao J, Urushadze L and Owen J (2024) Transboundary determinants of avian zoonotic infectious diseases: challenges for strengthening research capacity and connecting surveillance networks. Front. Microbiol. 15:1341842. doi: 10.3389/fmicb.2024.1341842.

Psittacines are nocturnal predators possessing keen senses of hearing and sight. Birds in homes and shelters are prone to disrupting their sleep, and if kept awake, they may roost, grind their beaks, or become very chatty. Although birds naturally breed when the days get longer in the spring, natural photoperiods are preferred. However, some birds with overactive breeding behaviors may need to have their exposure to light shortened. Psittacines may experience anxiety and fear due to improper cage placement. Certain animals will benefit from being in low-traffic areas if they are either timid or willfully defend their cage. It will be necessary to put additional gregarious, social birds in a busy area with lots of human interaction. Because of their flock behavior in the wild, most captive birds need a lot of interactions with humans and/or other birds, though this will vary depending on the individual bird. Research indicates that compared to their singly housed counterparts, orange-winged Amazon parrots kept in pairs spent less time preening, less time screaming, and more time being active (Meehan et al., 2003). When kept in pairs, parrots did not exhibit repetitive behaviors and showed less fear of human handlers (Hooijmeijer et al., 2004) (Figure 6.6).

Figure 6.6. *Outdoor enclosure for parrots with lots of space, and foraging toys.*

Source: Gupfinger R, Kaltenbrunner M. The Design of Musical Instruments for Grey Parrots: An Artistic Contribution toward Auditory Enrichment in the Context of ACI. Multimodal Technologies and Interaction. *2020; 4(2):16. https://doi.org/10.3390/mti4020016.*

6.6.1. Foraging Toys

Foraging toys are objects designed to conceal food, making it necessary for birds to search for it. There are easy and difficult toys. While some birds will have no trouble using the more challenging toys, others will need to start with the simpler ones and work their way up to the more complex ones. Treats can be first placed in small chapter cups that have been wadded up, or the food bowl can be wrapped with chapter or cardboard. You can create a foraging perch by hanging different foraging toys or by drilling holes where nuts, seeds, or other rewards can fit snugly. A foraging tree is an excellent feature for a shelter where various birds can visit throughout the day. It can be equipped with varying degrees of foraging stations to accommodate the diverse foraging needs of the birds. Given that pet trade has several aspects, Table 6.3 shows a study's result in this regard.

Table 6.3. *A List of Non-Native Bird Species Reported as Lost (Missing), Found, or Sighted by Pet Bird Breeders, Keepers, and Traders*

Species Name	Common Name	Family	NEM:BA A and IS Categories (NL = Not Listed)	Number of Reports		
				Lost	Found	Sighted
Acridotheres tristis †	Common mynah	Passeriformes	3	2	3	0
Agapornis fischeri †	Fischer's lovebird	Psittaculidae	NL	8	9	1
Agapornis lilianae	Lilian's lovebird	Psittaculidae	NL	1	3	0
Agapornis nigrigenis	Black-cheeked lovebird	Psittaculidae	NL	1	0	0
Agapornis personatus *†	Yellow-collared lovebird	Psittaculidae	NL	8	1	0
Agapornis roseicollis *†	Rosy-faced lovebird	Psittaculidae	NL	18	14	2
Amazona amazonica †	Orange-winged amazon	Psittacidae	NL	4	0	0
Amazona autumnalis	Red-lored amazon	Psittacidae	NL	1	0	0
Amazona oratrix	Yellow-headed amazon	Psittacidae	NL	0	3	0
Amazona aestiva	Blue-fronted amazon	Psittacidae	NL	4	3	0

Anas platyrhynchos domesticus †	Domestic mallard duck	Anatidae	3	1	0	0
Anodorhynchus leari	Lear's macaw	Psittacidae	NL	0	0	1
Ara ararauna †	Blue-and-yellow macaw	Psittacidae	NL	20	3	1
Ara chloropterus †	Red-and-green macaw	Psittacidae	NL	1	0	0
Ara macao †	Scarlet macaw	Psittacidae	NL	1	0	0
Ara severus †	Chestnut-fronted macaw	Psittacidae	NL	1	0	0
Aratinga auricapillus	Golden-capped parakeet	Psittacidae	NL	1	0	0
Aratinga solstitialis	Sun conure	Psittacidae	NL	10	10	1
Aratinga weddellii	Dusky-headed Parakeet	Psittacidae	NL	1	0	0
Cacatua alba *	White cockatoo	Cacatuidae	NL	5	4	0
Cacatua galerita †	Sulfur-crested cockatoo	Cacatuidae	NL	1	0	0
Cacatua galerita eleonora †	Eleonora cockatoo	Cacatuidae	NL	1	0	0
Cacatua goffiniana †	Tanimbar corella	Cacatuidae	NL	2	0	0
Cacatua leadbeateri	Major Mitchell's cockatoo	Cacatuidae	NL	2	0	0
Cacatua sanguinea	Little corella	Cacatuidae	NL	2	0	0
Cairina moschata †	Muscovy duck	Anatidae	NL	1	0	0
Chloebia gouldiae	Gouldian finch	Estrildidae	NL	1	1	0

Columba livia †	Rock dove	Columbidae	2 and 3	1	2	0
Cyanoliseus patagonus	Burrowing parrot	Psittacidae	NL	1	0	0
Cyanoramphus auriceps	Yellow-crowned parakeet	Psittaculidae	NL	0	1	0
Cyanoramphus novaezelandiae	Red-crowned parakeet	Psittaculidae	NL	6	4	0
Diopsittaca nobilis	Red-shouldered macaw	Psittacidae	NL	10	0	0
Eclectus roratus *	Moluccan eclectus	Psittaculidae	NL	17	8	1
Eolophus roseicapilla	Galah cockatoo	Cacatuidae	NL	11	4	3
Erythrura cyaneovirens	Red-headed parrotfinch	Estrildidae	NL	0	1	0
Eupsittula aurea	Peach-fronted parakeet	Psittacidae	NL	2	0	0
Eupsittula canicularis †	Orange-fronted parakeet	Psittacidae	NL	1	0	0
Forpus coelestis	Pacific parrotlet	Psittacidae	NL	7	0	0
Geopelia cuneata	Diamond dove	Columbidae	NL	0	1	0
Melopsittacus undulatus *†	Budgerigar	Psittaculidae	NL	26	42	5
Musophaga violacea †	Violet turaco	Musophagidae	NL	1	0	0
Myiopsitta monachus †	Monk parakeet	Psittacidae	NL	20	7	0
Neophema pulchella	Turquoise parrot	Psittaculidae	NL	0	2	0
Neopsephotus bourkii	Bourke's parrot	Psittaculidae	NL	0	3	0
Nymphicus hollandicus *†	Cockatiel	Cacatuidae	NL	220	50	7

Pavo cristatus *†	Indian peafowl	Phasianidae	NL	2	1	0
Pionus chalcopterus	Bronze-winged parrot	Psittacidae	NL	1	0	0
Pionus maximiliani	Scaly-headed parrot	Psittacidae	NL	5	1	0
Pionus senilis	White-crowned poinus	Psittacidae	NL	4	0	0
Platycercus elegans †	Crimson rosella	Psittaculidae	NL	4	2	0
Platycercus eximius †	Eastern rosella	Psittaculidae	NL	1	1	1
Poicephalus cryptoxanthus	Brown-headed parrot	Psittacidae	NL	0	2	0
Poicephalus fuscicollis †	Brown-necked parrot	Psittacidae	NL	1	0	0
Poicephalus gulielmi *	Red-fronted parrot	Psittacidae	NL	25	4	0
Poicephalus meyeri	Meyer's parrot	Psittacidae	NL	0	4	0
Poicephalus rufiventris	Red-bellied parrot	Psittacidae	NL	2	2	0
Poicephalus senegalus *†	Senegal parrot	Psittacidae	NL	31	7	0
Polytelis anthopeplus	Regent parrot	Psittaculidae	NL	0	1	0
Primolius auricollis	Golden-collared macaw	Psittacidae	NL	2	1	0
Primolius maracana	Blue-winged macaw	Psittacidae	NL	4	0	0
Psephotus haematonotus	Red-rumped parrot	Psittaculidae	NL	2	4	0
Psittacara erythrogenys †	Red-masked parakeet	Psittacidae	NL	0	2	0
Psittacara leucophthalmus	White-eyed parakeet	Psittacidae	NL	1	0	0

Psittacula alexandri †	Red-breasted parakeet	Psittaculidae	NL	3	1	0
Psittacula cyanocephala	Plum-headed parakeet	Psittaculidae	NL	0	1	0
Psittacula eupatria †	Alexandrine parakeet	Psittaculidae	NL	10	1	0
Psittacula krameri *†	Rose-ringed parakeet	Psittaculidae	2	149	82	12
Psittacus erithacus †	African grey	Psittacidae	NL	314	76	7
Pyrrhura lepida	Pearly parakeet	Psittacidae	NL	1	0	0
Pyrrhura molinae	Green-cheeked conure	Psittacidae	NL	41	18	3
Pyrrhura perlata	Crimson-bellied parakeet	Psittacidae	NL	1	0	0
Pyrrhura rupicola	Black-capped parakeet	Psittacidae	NL	0	1	0
Sturnus vulgaris *†	Common starling	Sturnidae	3	1	0	0
Taeniopygia castanotis †	Australian zebra finch	Estrildidae	NL	2	0	0
Treron calvus †	African green-pigeon	Columbidae	NL	0	1	0
Trichoglossus moluccanus	Rainbow lorikeet	Psittaculidae	NL	2	1	0

The National Environmental Management: Biodiversity Act (NEM:BA, Act 10 of 2004) alien and invasive species categories classifies species based on their ecological impact and management priority. An asterisk (*) indicates species with stolen reports, while those with a cross (†) are species with feral populations outside their native ranges.

NEM:BA categories: two = regulated species that may only be possessed, transported, and traded under a permit. three = species that are not prohibited require a permit for trade and must be managed to prevent them from becoming invasive.

Source: Shivambu TC, Shivambu N, Nelufule T, Moshobane MC, Seoraj-Pillai N, Nangammbi TC. Evaluating the Status of Lost, Found, and Sighted Non-Native Pet Bird Species in South Africa. Diversity. *2024; 16(5):283. https://doi.org/10.3390/d16050283.*

6.7. CONTROL OF AVIAN INFECTIOUS DISEASES

The variety of bird species, the range in susceptibilities to pathogens, and the close quarters in which birds are frequently housed create a special set of challenges for the control of avian infectious diseases in animal shelters. If not promptly and effectively managed, avian infectious diseases can spread quickly within a shelter environment, resulting in high rates of morbidity and mortality. For the purpose of safeguarding the health and welfare of avian populations housed in shelters, comprehensive disease control measures must be put into place (Armstrong et al., 2001; Centers for Disease Control and Prevention, 2008).

6.7.1. Biosecurity Measures for Avian Species

In settings like animal shelters, aviaries, and breeding facilities where birds are housed in close quarters, biosecurity measures are essential for halting the introduction and spread of infectious diseases among bird species. Strict biosecurity guidelines are put in place to guard against diseases that can seriously impair and kill bird populations, as well as to stop possible human-to-zoonotic disease transmission.

The following are key biosecurity measures tailored for avian species:

1. Quarantine Procedures
   - i. **Isolation of New Arrivals:** A minimum of 30 days should pass during the quarantine of any new birds brought into the facility. They should go through any necessary diagnostic testing and be watched for any indications of illness during this time. To avoid cross-contamination, physical separation between quarantine areas and the main bird population is necessary.
   - ii. **Dedicated Equipment and Staff:** Apply different equipment (e.g., perches, and feeding bowls) for birds housed in isolation. To reduce the chance of pathogens spreading to the broader bird population, designate specialized personnel to take care of birds in quarantine.
2. Disinfection and Sanitation
   - i. **Routine Cleaning:** To get rid of infections, common areas, feeding equipment, and bird enclosures must be cleaned and disinfected on a regular basis. Daily cleaning of enclosures and weekly or as needed, especially in the event of an illness discovery, should be followed by thorough disinfection.
   - ii. **Effective Disinfectants:** Make use of disinfectants that have

been shown to be successful in combating a variety of avian pathogens, such as fungi, viruses, and bacteria. Applying disinfectants correctly requires paying attention to contact time and concentration as well as following the manufacturer's instructions.

iii. **Footbaths and Hand Sanitizers:** Footbaths with disinfectant should be placed at the entrance to bird housing areas, and visitors and staff should use them before proceeding inside. Hand sanitizers ought to be accessible and used both prior to and following bird handling.

3. Control of Movement

i. **Restricted Access:** Limit access to bird housing areas to essential personnel only. Visitors should be restricted or required to follow strict biosecurity protocols, including wearing protective clothing and sanitizing hands and shoes.

ii. **Minimized Bird Transfers:** Avoid unnecessary movement of birds between different areas within the facility. If transfers are required, ensure that birds are healthy and that appropriate measures are taken to prevent the spread of pathogens.

4. Personal Protective Equipment (PPE)

i. **Protective Clothing:** PPE should be worn by staff members when working with birds or cleaning enclosures. Examples of PPE include gloves, masks, gowns, and shoe covers. This lowers the chance of zoonotic transmission and stops diseases from spreading from bird to bird.

ii. **Proper Disposal:** After use, PPE should be disposed of in designated containers to prevent contamination. Reusable protective clothing should be laundered regularly at high temperatures.

5. Environmental Control

i. **Ventilation:** To preserve air quality and lessen the accumulation of airborne pathogens, proper ventilation is crucial in bird housing areas. Make sure enclosures have air exchange systems that stop contaminated air from recirculating and adequate ventilation.

ii. **Temperature and Humidity Regulation:** Keep the temperature and humidity at ideal levels to help birds feel less stressed and become less susceptible to illness. Certain environmental needs may apply to different species and should be fulfilled.

6. Health Monitoring and Surveillance
   i. **Regular Health Checks:** Perform regular health evaluations on all birds, emphasizing physical examinations, behavioral observations, and appetite and weight tracking. The prevention of outbreaks depends on early illness detection.
   ii. **Disease Testing:** Conduct routine testing for common bird diseases, particularly when introducing new birds or in high-risk environments. Identification of infections prior to their spread can be facilitated by diagnostic tests like cultures, serology, and PCR.
   iii. **Record Keeping:** Keep thorough records of every bird's medical history, including immunization history, prescribed medications, and any symptoms noticed. Tracking disease trends and controlling outbreaks require this information.
7. Vector Control
   i. **Pest Management:** As pests can spread disease, use pest control methods to keep rodents, insects, and other animals out of areas used for housing birds. To get rid of possible entry points, regularly check and maintain enclosures.
   ii. **Wild Bird Exclusion:** Prevent contact between shelter birds and wild birds, which can carry and transmit diseases. Use netting or other barriers to keep wild birds away from feeding areas and enclosures.
8. Vaccination Programs
   i. **Species Specific Vaccinations:** Whenever possible, vaccinate birds against avian influenza, Newcastle Disease, and psittacosis. An avian veterinarian should be consulted when developing vaccination programs, and they should be customized for the particular species and disease risks found in the facility.
9. Emergency Response Planning
   i. **Outbreak Management:** Create a response plan that outlines procedures for containing afflicted birds, alerting pertinent authorities, and carrying out disease testing in the event of an outbreak. Communication strategies for staff, volunteers, and the general public should also be covered in the plan.
   ii. **Euthanasia and Disposal:** Euthanasia might be required to stop the spread of disease in situations where humane treatment options have run out or where disease control is not

practical. Establish procedures for the humane death of birds and the handling of their remains.

10. Staff Training and Education

 i. **Ongoing Training:** Staff members and volunteers should receive frequent training on biosecurity procedures, illness identification, and safe handling methods. Keeping personnel up to date on the most recent advancements in avian health is crucial to preserving a bio secure environment.

 ii. **Clear Communication:** Ensure that all staff members understand the importance of biosecurity measures and are committed to following protocols consistently. Clear communication and regular updates help reinforce these practices.

By implementing these biosecurity measures, facilities that house avian species can significantly reduce the risk of infectious disease outbreaks, protect the health of their bird populations, and contribute to overall public health safety.

6.7.2. Disease Surveillance and Early Detection

Preventing widespread outbreaks of infectious diseases in shelters requires early detection. A successful surveillance system must include routine health monitoring and timely reporting of any symptoms of illness.

1. **Health Assessments:** All birds should have routine health examinations performed upon intake and on a regular basis while they are in the shelter. A comprehensive physical examination, behavior observations, and weight and appetite tracking should all be part of these evaluations. Any bird exhibiting symptoms of disease should be quarantined right away and subjected to additional testing for diagnosis.
2. **Diagnostic Testing:** Diagnostic testing that is both quick and precise is essential when a disease is suspected. Tests like serology, polymerase chain reaction (PCR) for bacterial and viral pathogens, and cultures for fungal infections ought to be easily accessible to shelters. Working together with veterinary laboratories can help with prompt and precise diagnosis.
3. **Record Keeping:** It is essential to keep thorough records of every bird's health status, including any symptoms seen, treatments given, and results, to track disease trends and spot possible outbreaks. In the event that a zoonotic disease is discovered, this information may also be helpful for coordinating with public health authorities.

6.7.3. Treatment Protocols and Disease Management

To control an infectious disease and stop it from spreading to other birds, timely diagnosis and appropriate treatment are required. Treatment plans ought to be created with avian veterinarians in mind, taking into account the unique requirements of every species of bird.

1. **Medication Administration:** Treatment options for a given disease may include antiviral, antifungal, or antibiotic medication administration. Shelters must make sure that all of their staff members are properly trained in the dosage and administration of these drugs, whether they are to be taken orally, intravenously, or inhaled.
2. **Supportive Care:** Birds suffering from infectious diseases frequently need supportive care in addition to specialized treatments to help them heal. This can entail giving the birds a calm and stress-free environment, making sure they get enough food and water, and providing heat support for any that are frail or unable to control their body temperature.
3. **Euthanasia Considerations:** Euthanasia might be the most humane course of action when a bird's prognosis is bad or the illness poses a serious risk to the birds in the shelter. A veterinarian should be consulted before making this choice, which should take into account the welfare of each individual bird as well as the general health of the population housed in the shelter.

6.7.4. Staff Training and Education

Effective disease control requires educating volunteers and shelter staff about avian infectious diseases and the value of biosecurity. Topics including disease recognition, safe handling and quarantine protocols, disinfection methods, and PPE use should all be included in ongoing training programs.

1. **Workshops and Seminars:** Staff at shelters can stay up to date on the most recent advancements in the management and control of avian diseases by attending regular workshops and seminars conducted by specialists in avian health. These workshops ought to be customized to the unique difficulties that shelters encounter and ought to incorporate hands-on activities.
2. ***Protocols and Guidelines:*** It is recommended that shelters establish and uphold unambiguous procedures and policies concerning disease prevention and management, ensuring that all personnel have easy access to them. To take into account new information

and procedures, these protocols should be routinely reviewed and updated.

3. **Communication and Reporting:** It is crucial to promote candid communication and prompt reporting of any issues or observations pertaining to the health of birds. Since early intervention is essential to preventing disease outbreaks, staff members should feel free to report possible problems without worrying about facing consequences (Armstrong et al., 2001; Centers for Disease Control and Prevention, 2008).

REFERENCES

1. Armstrong, M. C., Farinato, R. H., & Telecky, T. M. (2001). A wing and a prayer: Birds and their protection under law. *Journal of Avian Medicine and Surgery, 15*, 310–315.
2. Avian Welfare Coalition, Animal Protection Institute, & American Society for the Prevention of Cruelty to Animals. (2009). *Captive Exotic Bird Care: A Guide for Shelters*. Retrieved from: www.avianwelfare.org (accessed on 6 December 2024).
3. Center for Food Security and Public Health (CFSPH), Iowa State University. (2009a). *Chlamydiosis (Avian), Technical Factsheet*, 1–4.
4. Center for Food Security and Public Health (CFSPH), Iowa State University. (2009b). *Highly Pathogenic Avian Influenza, Technical Factsheet*, 1–4.
5. Centers for Disease Control and Prevention. (2008). *Key Facts about Avian Influenza (bird flu) and Highly Pathogenic Avian Influenza A (H5N1) Virus*.
6. Dahlhausen, R. D. (2006). Implications of mycoses in clinical disorders. In G. J. Harrison & T. L. Lightfoot (Eds.), *Clinical Avian Medicine* (Vol. II, pp. 691–704). Spix Publishing.
7. Echols, S. M. (2004). The behavior of diet. In *Proceedings of the Annual Conference, Association of Avian Veterinarians* (pp. 267–270). AAV.
8. Echols, S. M. (2006). Evaluating and treating the kidneys. In G. J. Harrison & T. L. Lightfoot (Eds.), *Clinical Avian Medicine* (Vol. II, pp. 451–491). Spix Publishing.
9. Fiskett, R. F., & Reavill, D. R. (2004). Disease conditions and clinical signs of pet birds. In *Proceedings of the Annual Conference, Association of Avian Veterinarians* (pp. 245–263). AAV.
10. Gelis, S. (2006). Evaluating and treating the gastrointestinal system. In G. J. Harrison & T. L. Lightfoot (Eds.), *Clinical Avian Medicine* (Vol. I, pp. 411–440). Spix Publishing.
11. Gerlack, H. (1994). Chlamydia. In B. W. Ritchie, G. J. Harrison, & L. R. Harrison (Eds.), *Avian Medicine: Principles and Application* (pp. 984–996). Wingers Publishing, Inc.
12. Harrison, G. L., Lightfoot, T. L., & Flinchum, G. B. (2006). Emergency and critical care. In G. J. Harrison & T. L. Lightfoot (Eds.), *Clinical Avian Medicine* (Vol. II, pp. 213–240). Spix Publishing.
13. Hochleithner, M., Hochleithner, C., & Harrison, L. D. (2006). Evaluating and treating the liver. In G. J. Harrison & T. L. Lightfoot (Eds.), *Clinical Avian Medicine* (Vol. II, pp. 441–449). Spix Publishing.

14. Martin, H., & Ritchie, B. W. (1994). Orthopedic surgical techniques. In B. W. Ritchie, G. J. Harrison, & L. R. Harrison (Eds.), *Avian Medicine: Principles and Application* (pp. 1137–1169). Wingers Publishing, Inc.
15. Meehan, C. L., Millam, J. R., & Mench, J. A. (2003). Foraging opportunity and increased physical complexity both prevent and reduce psychogenic feather picking by young Amazon parrots. *Applied Animal Behavior Science, 80*, 71–85.
16. Phalen, D. N. (1997). Viruses. In R. B. Altman, S. L. Chubb, G. M. Dorrestein, & K. Quesenberry (Eds.), *Avian Medicine and Surgery* (pp. 281–322). W.B. Saunders Company.
17. Phalen, D. N. (2006). Implications of viruses in clinical disorders. In G. J. Harrison & T. L. Lightfoot (Eds.), *Clinical Avian Medicine* (Vol. II, pp. 721–746). Spix Publishing.
18. Pollock, C. G. (2006). Implications of mycobacteria in clinical disorders. In G. J. Harrison & T. L. Lightfoot (Eds.), *Clinical Avian Medicine* (Vol. II, pp. 681–690). Spix Publishing.
19. Rinder, M., Ackermann, A., Kempf, H., et al. (2009). Broad tissue and cell tropism of avian bornavirus in parrots with proventricular disease. *Journal of Virology, 83*, 5401–5407.
20. Tully, T. N. (2006). Update on *Chlamydophila psittaci*: A short comment. In G. J. Harrison & T. L. Lightfoot (Eds.), *Clinical Avian Medicine* (Vol. II, pp. 679–680). Spix Publishing.
21. United States Department of Agriculture & Animal and Plant Health Inspection Service. (2003). *Animal Disease Alert: Exotic Newcastle Disease*. Retrieved from: http://www.aphis.usda.gov/publications/animalhealth/content/printableversion/ada_ahend.pdf (accessed on 6 December 2024).

CHAPTER 7

INFECTIOUS DISEASES OF DOMESTIC RABBITS

Contents

Introduction 270
7.1. Overview of Rabbits 270
7.2. Diseases of Domestic Rabbits 285
7.3. Bacterial Diseases 306
7.4. Viral Diseases 317
References 337

INTRODUCTION

Infectious diseases in domestic rabbits present a significant concern for both pet owners and commercial breeders due to the susceptibility of rabbits to various pathogens, including bacteria, viruses, fungi, and parasites. These diseases can spread rapidly in populations due to close living conditions and the rabbits' delicate immune systems.

Viral infections such as Rabbit Hemorrhagic Disease (RHD) and Myxomatosis are among the most feared. RHD, caused by a calicivirus, leads to severe hemorrhaging and is often fatal. Myxomatosis, caused by a poxvirus, is characterized by swelling and skin lesions, and it typically results in death if untreated. Vaccination is critical in managing these viral threats.

Bacterial infections also pose significant risks. Pasteurellosis, caused by *Pasteurella multocida*, is a common respiratory disease that can become chronic, leading to conditions like pneumonia, abscesses, and septicemia. *Bordetella bronchiseptica*, another respiratory pathogen, can exacerbate respiratory illnesses, especially in young or stressed rabbits.

Parasitic diseases, including coccidiosis, caused by protozoa, and *E. cuniculi*, a microsporidian parasite, can lead to severe gastrointestinal and neurological symptoms. These infections are often managed through good hygiene, proper diet, and sometimes medication. Fungal infections, though less common, can occur, particularly dermatophytosis, which affects the skin and fur.

Prevention and control of these diseases require a multifaceted approach, including proper housing, regular veterinary care, vaccination, and strict biosecurity measures. Understanding the specific pathogens and their modes of transmission is essential for effective management and minimizing the risk of outbreaks in domestic rabbit populations. Regular monitoring and prompt treatment are crucial to ensuring the health and well-being of these animals.

7.1. OVERVIEW OF RABBITS

Small mammals in the Leporidae family include rabbits. There is only one genus of domestic rabbits, which is the European, or Old World, rabbit (*Oryctolagus cuniculus*). Cottontail rabbits (Sylvilagus) and "true" hares, or jackrabbits (Lepus), are examples of wild rabbits and hares. Since the 1800s, people have kept rabbits as pets in Western countries. They require a great deal of care and attention as pets. There are numerous breeds of rabbits available; typical breed differences include fur length, color, and size. A female rabbit is referred to as a doe, a baby is called a kit, and a male rabbit is called a buck. Rabbits are hairless and blind at birth. They are typically born and raised in underground burrows in the wild (Figure 7.1).

Figure 7.1. *Rabbit breeds.*

Source: By Kippo44-Own work, CC BY-SA 4.0, https://commons.wikimedia.org/w/index.php?curid=137892608.

7.1.1. Description and Physical Characteristics of Rabbits

Rabbits are small, hairy mammals with powerful, large hind legs, short, fluffy tails, and long ears. They have two sets of sharp incisors, or front teeth: a top pair and a bottom pair. In back of the upper incisors, they also have two peg teeth. Their teeth are designed to gnaw and continue to grow for the duration of their lives.

Rabbits hop around using their strong hind legs for propulsion. On their hind feet, they have four long, webbed toes that prevent them from spreading apart when they jump. Each of their front paws has five toes. Certain rabbit species can travel at up to 35 to 45 miles per hour, or 56 to 72 kilometers per hour. It appears that young rabbits walk rather than hop. Depending on the breed, rabbits can range in size and color from 2 to 16 pounds (one to seven kilograms). The lifespan of a pet rabbit that has received proper care and has been spayed or neutered at a young age is eight to 12 years.

7.1.2. Special Considerations for Rabbits

A rabbit's back may break if it is not picked up or held correctly. Consequently, older teens and adults may find rabbits to be a better pet than small children. With one hand under the rabbits' back legs to support their backs and the other under their chest, you should hold them firmly but gently against your body. A rabbit should never be lifted or held by its ears.

Rabbits can live outside or indoors. But rabbits are heat-sensitive animals. Hutches or vehicles with inadequate ventilation, combined with hot and muggy weather, can cause heat stroke and even death. Extended exposure to temperatures higher than 80°F (27□) could make your rabbit ill. Temperatures below 50°F (10□) are difficult for domesticated rabbits to handle. Rabbits housed outdoors in regions with extreme temperatures require a sizable shelter that can be kept cool in the summer and heated in the winter. When the rabbit is housed indoors and isn't under supervision, a sizable cage needs to be supplied. When allowing a rabbit to run loose inside, extra caution needs to be used to make sure the animal cannot get to or chew on carpeting, electrical wires, or other inappropriate objects.

Chewing is a physiological and psychological need for rabbits. Giving your bunnies an assortment of objects to gnaw on will enable them to continue wearing down their rapidly growing teeth and prevent boredom. Pine cones, hay, cardboard, straw, untreated wicker baskets, and non-poisonous wood pieces are all excellent chewing materials.

Rabbits groom themselves on a regular basis. Similar to cats, they ingest loose hair, which eventually passes through their digestive systems and emerges as excrement. Regrettably, having too much hair can result in hairballs, which can impede digestion or cause other health issues. To help avoid this, owners of rabbits should brush their pets on a regular basis using a soft brush. Long-furred rabbits need daily brushing, while short-furred rabbits need brushing at least twice a week. Furthermore, rabbits shed every three months. It's possible for one to shed lightly and the other heavily. To lessen the chance of hairballs, extra care must be taken when brushing the rabbit and removing extra hair from its hutch or cage during these times. You might observe some patches of baldness during these times of hair loss. After the period of shedding ends, these should start to fill in again. Depending on the kind of rabbit, shearing can take a day to several weeks.

Rabbit allergies occur in certain people. Furthermore, because they always need hay, rabbits might not be the ideal pet for someone who is allergic to hay.

Legal provisions prohibit the sale, exchange, or gifting of baby rabbits younger than two months in several states. Before acquiring a rabbit, make sure to check your county's and state's laws regarding any restrictions on rabbit ownership.

7.1.3. Selecting a Rabbit

Pet stores, humane societies, breeders, and rescue groups are among the places to get pet rabbits. If you're considering getting a rabbit as a pet, think about the size, color, and fur length you want as well as who will take care of it and where it will live. Make sure the rabbit is in good health and has a clean environment when choosing one. Inquire about the rabbit's nutrition, care, and handling history. Since rabbits are most active in the morning and evening, you might be able to get a better sense of the rabbit's temperament by going during those times of day.

A healthy rabbit should have glossy, well-groomed fur free of visible wounds or bare spots. Wet or matted fur, particularly under the tail or around the chin, may indicate health issues. There shouldn't be any discharge coming from the nose or eyes, and there shouldn't be any crusting in the ears. When they are active, healthy rabbits are typically aware and inquisitive. They shouldn't be stiff or lame, nor should they be reluctant to move.

7.1.4. Providing a Home for a Rabbit

Prior to bringing your rabbit home, you must set up a good diet, provide appropriate housing, and consider your pet's exercise requirements.

7.1.4.1. Housing

A common place to keep rabbits is in a hutch in the garage, basement, or back yard. Since diseases caused by neglect are common in rabbits kept in hutches at the back of the yard, the hutch needs to be easily accessible for the rabbit to receive the necessary care. Enough ventilation and safeguards against dogs and other predators should be present. Due to their vulnerability to frostbite and overheating, outdoor-housed rabbits should be brought inside during periods of extreme temperature.

When kept indoors, rabbits can integrate more easily into the household. They can be made accustomed to spending some time in a cage and trained to use a litter box. Rabbits, however, have a tendency to chew on objects and may gnaw electrical wiring, carpeting, curtains, or furniture. The rabbit runs the risk of injury and fire when it chews on electrical wiring. When left alone, rabbits ought to be kept in secure areas.

Because rabbits gnaw, cages need to be made of durable materials. It should be simple to sanitize and clean cages. To support the weight of the rabbit, 16 gauge wire is advised for the cage flooring; all wire cages with a minimum of 12 gauge wire are preferred. Wire flooring keeps the cage cleaner by allowing droppings to fall away from the rabbits' feet. To lessen pressure on the bottoms of the rabbits' feet and the possibility of sores forming there, cage floors should also have a small, smooth, level area where the rabbits can stand. This area should be made of heavy plastic or another easily disinfected material. The rabbit's size determines the hutch's dimensions. A minimum cage size of 30 by 36 inches (76 by 91 centimeters) with a height of 16 to 18 inches (41 to 46 centimeters) is needed for giant breeds weighing more than 12 pounds (5.4 kilograms). Medium-to-large breeds weighing 7–12 pounds (3–2–5–4 kg) need 24 by 30 inches (61–76 cm) of space. Cages that measure 18 by 24 inches (46 by 61 centimeters) can hold smaller breeds. These specifications are the bare minimum for one rabbit; if two or more rabbits are kept together, the cage needs to be larger.

A hay rack that is elevated off the ground, a watering system, and a feed hopper or bowl for pellets should be included in the cage. Feed hoppers made of sheet metal with holes or a screen in the bottom for the removal of small broken feed particles are commonly used for multiple rabbits sharing a cage or for a rabbit kept outside. As long as it is cleaned every day to avoid droppings getting into it, a heavy, untippable bowl filled with pelleted food works well for an individual indoor rabbit. Hay is kept off the ground and wasted less when hay racks are fastened to the cage wall.

Compared to other animals of comparable size, rabbits drink more, so they should always have access to fresh water. Unless the watering valve is made of stainless steel or has a stainless centerpiece, rabbits will frequently chew on it and eventually destroy it. Sipper tubes on water bottles are useful.

A rabbit needs something to do, so the environment in the cage needs to be improved. A cage without any furniture or toys is insufficient. Enhancing a rabbit's habitat with cardboard boxes and tubes for hiding and chewing, as well as toys made of untreated wood or wicker, is a great idea. Every day run time outside the cage is ideal for rabbits. When they're not in their cages, they should always be watched over and given access to exercise spaces that have been enclosed and secured against rabbits.

The cage needs to have a nest box if the rabbit is intended for breeding. It is recommended that nest boxes be designed in a way that makes them simple to insert into the cage and then remove for cleaning and disinfecting in between litters. Disease incidence can be decreased by disinfecting the nest box twice: once after cleaning and once more right before putting it in the cage. The size of the box should be such that it keeps the kits (newborn rabbits) warm while still preventing crowding. For medium-sized rabbits, a standard nest box dimensions are 16 by 10 by eight inches, or 41 by 25 by 20 centimeters. In both warm and cold climates, wooden, metal, or plastic nest boxes filled with hardwood shavings or straw make good nesting materials. Less success has been had when using shredded chapter, hay, leaves, and other materials. When the doe (female) hops in and out of the nest box, rough edges, like splintered wood, should be avoided as they can cause mastitis, a bacterial infection of the breast tissue.

Pens should have a non-slip floor, and to improve absorbency, they can have beds made of shredded chapter or straw covered in hay or straw. Sawdust and shavings are not the best materials because they can contain oils that irritate the respiratory systems of rabbits and are dusty. It is recommended that pen sides be at least four feet (1.22 meters) high to keep rabbits from jumping out.

Compatibility is a key consideration when organizing group housing. Assessing a person's personality for aggression and docility is important. The best rabbits are those that have grown up together, though adult rabbits can get so hostile that major fights can break out. Neutering could make compatibility better. It's a good idea to let rabbits live in separate pens next to one another before housing them together. Enough room should be provided for multiple rabbits in cages or pens. To allow for territory establishment, floor space recommendations range from 3:5 to 10 square feet (0:33 to 0:93 square meters) per rabbit. As a general guideline, some suggest 3.5 hop lengths per rabbit. All the same, group-housed rabbits need to be regularly checked on and given access to hiding and escape areas.

Cleaning schedules are determined by the kind of cage and quantity of rabbits. Usually, a corner of the cage or hutch is the elimination site of choice for rabbits. Cleaning and sanitizing must be done on a regular basis because inadequate sanitation causes illness and fatalities. As a general rule, the cage of a single indoor rabbit should be thoroughly cleaned (washed with hot, soapy water and all bedding replaced) at least once a week in addition to being spot cleaned every day to remove feces and uneaten food. After each use, nest boxes need to be cleaned. An affordable sanitizing solution, such as diluted household chlorine bleach (1 ounce of bleach in one quart of water or 30 milliliters of bleach in one liter of water), or other less corrosive disinfectants, should be used on a regular basis to thoroughly clean cages, feeders, and watering equipment. It is best to clean everything thoroughly before housing fresh bunnies. Regular removal of loose hair will reduce the risk of hairballs. Washing is one of the best ways to get hair out of cages. Every two weeks, pens or cages with wire floors need to be brushed or rinsed. To remove the rabbit urine scale from solid floor pans, an acid wash might be necessary. It's crucial to regularly remove droppings from the cage. When droppings accumulate, hazardous ammonia concentrations rise, which can irritate rabbits' respiratory tracts and hasten the onset of infection.

7.1.4.2. Litter Box Training

It is possible to teach a rabbit to use a litter box. When passing waste, rabbits naturally like to use the same spot typically a particular corner. Make a note of your rabbit's favorite spot and install a sturdy, suitably sized litter box there. The rabbit ought to have no trouble entering and exiting the box. A large box is ideal for rabbits weighing 5 to 10 pounds, while a medium-sized box is suitable for small breeds, depending on the amount of space available in the cage. Extra-large boxes might be necessary for larger breeds. When the box is first placed in the cage, a few rabbit droppings placed inside will help the rabbit associate the smell with the new location.

Litter comes in a variety of forms. Make certain that the litter you use is safe for your bunny. In addition to spending a lot of time in the litter box, your rabbit will frequently eat the litter. Rabbits can benefit from litters made of hay, chapter pulp, or recycled news chapter pellets. Hay needs to be replaced every day if it is used. Because they can obstruct the gastrointestinal tract and are indigestible if consumed, wood shavings, ground corncob, clay, and clumping products should not be used as litter. To absorb pee, place a few layers of news chapter underneath the hay.

To keep your rabbit healthy and to lessen odors, it's critical to clean the litter box frequently. Cleaning and disinfecting the litter box should be done at least once every week.

7.1.4.3. Diet

As herbivores (eaters of plants), rabbits have unique digestive systems. They are picky eaters, preferring young plant shoots and nutrient-rich leaves to older, higher fiber plant material. They can only choose the most nutrient dense plant parts to satisfy their high metabolic rate and nutritional needs. The majority of rabbit droppings are hard, tiny pellets. Additionally, rabbits release "soft droppings" known as cecotropes, which they consume. The microbial protein, vitamins, and other nutrients that rabbits require are provided by these soft droppings.

Rabbits have poor fiber digestion. For healthy digestion and to lower the risk of intestinal disease, they require a high-fiber diet (roughly 15% crude fiber). Additionally, fiber has the ability to absorb bacterial toxins and aid in their excretion through hard stools. Young rabbits fed a low-fiber diet are more likely to suffer from enterotoxemia, a toxin producing bacterial infection of the digestive tract that can result in fatal diarrhea. For your pet's health, the amount of fiber in their diet must be properly balanced with other nutrients. It is essential to consume foods high in vitamins A, D, and E. B vitamins and vitamin K are sufficiently produced by gut bacteria, so dietary supplements of these vitamins are typically not required. Alfalfa meal-based diets typically supply enough vitamin A. A diet deficient in vitamin A may result in abortion or fetal health issues. Infertility, muscular dystrophy (a progressive wasting and weakening of muscles), and fetal and neonatal deaths have all been linked to vitamin E deficiency. Stress and illness may raise the amount of vitamins needed each day. Premium commercial pelleted diets for rabbits should have all the right nutrients in them, and to prevent vitamins from being lost, they should be kept in airtight containers in a cool, dry environment.

The staples of your pet's diet protein, fiber, fat, and energy stay constant throughout its life, but the amounts required change based on the breed, lifestyle, condition, and life stage of the rabbit (growth, pregnancy, nursing, or maintenance). Ratios ought to satisfy the National Research Council's nutrient requirements. Generally speaking, rabbits should be given an infinite supply of premium grass hay, a small amount of commercial pellets made from Timothy hay (about ¼ cup per five pounds, or 59 milliliters per 2.3 kilograms, of body weight per day), and a serving of fresh vegetables (about one cup per two pounds, of body weight per day). Fruit should only be offered as a treat on rare occasions (Table 7.1).

Table 7.1. *Summary of Some Production and Health Benefits of Supplementing Rabbit Diets with Nutraceutical Products*

Nutraceutical Products	Studied Traits	Benefits
Pumpkin (*Cucurbita* sp.) seed essential oil	Growth, antioxidant status, and immune response	Supplementing growing rabbit diets with pumpkin seed oil up to 2 ml/kg is recommended to improve growth, antioxidative status, and immunological performance, particularly under heat-stress conditions.
Rosemary essential oils	Growth, antioxidant status, and immune response	Rosemary essential oils can be used, separately or in combination, in rabbit diets to improve antioxidant status and stability, along with improved kidney function, without significant effects on rabbit growth and immunological performances.

Pale purple coneflower (*Echinacea pallida*)	Reproductive performance and hematological parameters	The reproductive and hematological parameters of the does have not been affected by the pale purple coneflower (3 g/kg of diet).
Goji berries (*Lycium barbarum*)	Body weight, reproductive performance, immune response, and meat quality	Goji berries have been shown to improve rabbit reproductive and productive performance, meat quality, and immunity.
Fenugreek (*Trigonella foenum-gracum*)	Productive performance and immune response	Fenugreek administration improves milk yield and quality of rabbit does, as well as bunnies' body weight and immunological performance.
Agro-industrial by-products	Growth performance and carcass quality	The substitution of fenugreek, garden rocket, and mustard seed meals for up to 50% of soybean meal in rabbit diets had positive economical results with no negative effects on growth and carcass.
Fenugreek seeds and probiotics	Growth performance and carcass quality	Fenugreek seeds and probiotics (15 g/kg of diet and 450 mg/kg of diet, respectively) supplementation improved rabbit growth performance and nutrient digestibility without affecting carcass quality.
Bee pollen and propolis	Reproductive performance and immune response	Bee pollen and propolis, administered separately or in combination to rabbit does at 150 or 300 mg, improve reproductive performance, milk production, and immune response.
Milk whey	Growth performance and meat quality	The addition of 2.25% whey powder improved the gut health of growing rabbits and increased their growth performance and thigh muscle quality.

Insect lipids	Growth performance and gut health	Black soldier flies and yellow mealworm fats can be used as a lipid source to replace soybean oil in rabbit diets without affecting growth performance, digestive process, and gut health.

Source: El-Sabrout, Khalifah, Ciani. Current Applications and Trends in Rabbit Nutraceuticals. Agriculture. 2023; 13(7):1424. https://doi.org/10.3390/agriculture13071424.

Feed pellets for rabbits offer affordable, high-quality nutrition. Water that is fresh and clean should always be available. Due to the high calcium content of alfalfa hay and alfalfa-based pellets, adult rabbits should only be fed timothy or other lower calcium grass hays (like oat, orchard grass, or meadow grass) unless they are pregnant or nursing. Regular commercial diets that contain alfalfa meal over an extended period of time can cause kidney damage and calcium carbonate deposits in the urinary tract. For non-nursing rabbits, lowering the calcium level to 0.4–0.5% of the diet helps lessen these issues. As they get closer to one year old, young, growing rabbits should be fed lower calcium hays and pellets made of timothy. They should also be fed alfalfa hay. Avoiding certain plants that are toxic to rabbits is advised (Table 7.2).

Table 7.2. *Answers Provided to "Please Choose How Much You Agree or Disagree with the Following Statements: Rabbits Can Form Strong Bonds with Other Rabbits; Rabbits Can Form Strong Bonds with Humans; Rabbits are Happy to Live Alone" n = 1516*

	Strongly Disagree	**Disagree**	**Neither Agree nor Disagree**	**Agree**	**Strongly Agree**
"Rabbits can form strong bonds with other rabbits"	3.4% (52)	0.5% (7)	2.4% (36)	9.4% (142)	84.4 (1279)
"Rabbits can form strong bonds with humans"	2.6% (40)	0.3% (4)	2.4% (37)	27.0% (409)	67.7% (1026)
"Rabbits are happy to live alone"	26.6% (404)	28.6% (435)	29.1% (441)	12.8% (194)	2.8% (42)

Source: McMahon SA, Wigham E. 'All Ears': A Questionnaire of 1516 Owner Perceptions of the Mental Abilities of Pet Rabbits, Subsequent Resource Provision, and the Effect on Welfare. Animals. 2020; 10(10):1730. https://doi.org/10.3390/ani10101730.

7.1.4.4. Exercise

Your rabbit needs to exercise to stay healthy. Emotional and physical well-being can be preserved with a spacious cage that provides room to walk around. Give your rabbit plenty of exercise and prevent boredom by letting it spend some time each day exploring outside of its cage (under close supervision) and giving it a variety of toys and objects to chew on.

7.1.4.5. Temperament

Pet rabbits can be calm, amiable, and energetic when given the right care. Though they are gregarious, not all rabbits are fond of being touched. Examine the compatibility of the rabbits carefully before housing them together. The best times to see rabbits are at dusk and dawn. Small rabbits typically struggle to be picked up and do not enjoy being handled. Larger to medium-sized rabbits could be more docile and manageable. If a small food treat is offered along with a calm and gentle approach, most rabbits can be trained to tolerate being handled. When training more timid rabbits, it may take several weeks or months. To protect the back of a rabbit, support its hind end and legs when picking it up or holding it. It's crucial to handle and confine your rabbit with care. Strong hind limbs on rabbits have the potential to kick out and break backs if not restrained. You should never hold a rabbit by its ears. When holding a rabbit, toenails on the back limbs can cause serious scratches on uncovered arms.

7.1.5. Routine Health Care of Rabbits

Every year, rabbits should have health examinations. Veterinary professionals can examine rabbits in a manner akin to that of dogs and cats. Veterinarians usually perform oral examinations to assess dental health. Examining the rabbit's genitalia for a penis in males or a slit-like vulva (opening to the vagina) in females will reveal the gender of the animal. This distinction can occasionally be very difficult to recognize, particularly in very young rabbits. Testicles are visible in males between the ages of 10 and 12 weeks, but when a male becomes extremely anxious, they can retract their testicles back into their abdomens, making the presence of testicles an unreliable indicator of gender. The typical range for body temperature is 39.4□ to 40□, or 103°F to 104°F. When a person's body temperature is less than 38□(100.4°F) or more than 40□ (105.6°F), there should be concern.

Female rabbits must have their ovaries and uterus surgically removed before they can have unwanted litters or develop uterine cancer, a deadly condition that strikes many of them after three years of age. Male rabbits that have had their testicles surgically removed are less likely to exhibit aggressive behavior and mark their territory with urine spraying.

7.1.5.1. Routine Care of Your Rabbit

When To Do	
a. Daily	a. Give fresh water and food, including fresh hay
Da iDaily	b. Let out of the cage for supervised exercise and
c. Daily	play
d. Daily	c. Brush and remove shed hair of rabbits with
e. Twice a week	long fur
f. Once a week	d. Clean out the litter box
g. Once a year	e. Brush and remove shed hair of rabbits with
h. When needed	short fur
i. When needed	f. Clean and disinfect cage and food and water
	systems
	g. Veterinary checkup
	h. Give fresh bedding
	i. Give new chew toys

Trimming nails is recommended every one to two months, or as needed. Declawing rabbits is not a good idea. The sole vaccination against rabbit hemorrhagic disease, a deadly and highly contagious ailment brought on by infection with the RHDV-2 virus, is advised for pets on a regular basis. This vaccination is especially important for rabbits that go outside (in areas where wild rabbits have been) or come into contact with other pet rabbits because this virus affects both domestic and wild rabbits. The vaccination schedule for rabbits is two doses spaced three weeks apart, followed by a booster shot every year.

7.1.5.2. Signs of Illness in Rabbits

- Body temperature less than 100.4°F (38□) or greater than 105°F (40.6□)
- Discharge from the nose or eyes
- Fur loss or red, swollen skin
- Dark red urine
- Loss of energy
- Loss of appetite and weight
- Drooling
- Diarrhea, or no droppings for more than 12 hours
- Not hopping or moving normally
- Trouble breathing

If any of these signs occur, you should contact your veterinarian immediately.

Discharge from the nose and eyes, hair loss, red or swollen skin, dark red urine, loss of appetite, energy, or weight, drooling, diarrhea or no droppings for more than 12 hours, inability to hop or move normally, and breathing difficulties are all indicators of illness. When a rabbit is uncomfortable, it may squirm or grit its teeth while sitting. You should take the rabbit to the vet right away if any of these symptoms appear.

7.1.5.3. Giving Medication

Medication for rabbits is approved in very small numbers. Drugs authorized for use in other species, like dogs or cats, are occasionally used to treat rabbits. Antibiotics that suppress the normal, harmless bacteria in a rabbit's digestive system should be used with caution. Inappropriate antibiotic use can lead to severe diarrhea, an imbalance in gut flora, or even death. We refer to this as antibiotic toxicity. Clindamycin, lincomycin, erythromycin, ampicillin, amoxicillin/clavulanic acid, and cephalosporins are among the antibiotics that shouldn't be given to rabbits. Fipronil, a flea treatment, should not be administered to rabbits due to the possibility of toxicity for certain individuals. Before giving your rabbit any medication, make sure to consult your veterinarian.

Prior to surgery, a lengthy fast is neither necessary nor advised. Unlike other animals, rabbits are incapable of vomiting while under general anesthesia. Before surgery, your veterinarian might give you a medication to help calm you down. After surgery, it is imperative that rabbits begin eating, and administering pain medication for one to two days following the procedure can help prevent appetite loss. Water and hay are usually provided as soon as possible following surgery. After surgery, treats like bananas or alfalfa hay may increase a rabbit's appetite.

Skin sutures are chewed out by rabbits. Veterinarians therefore employ various techniques for skin closure or absorbable sutures buried beneath the skin to seal surgical incisions.

7.1.5.4. Dental Care

Because all of their teeth grow continuously throughout their lives and because they do not have the chance to wear down their teeth as they grow, dental problems are common in rabbits. Dental issues can be decreased with the right diet and dental care.

7.1.5.4.1. Dental malocclusion

Over the course of a rabbit's life, all of its teeth incisors, premolars, and molars grow. When a rabbit chews, the wearing action of opposing teeth typically keeps tooth length in check. However, malocclusion a condition in which the teeth are arranged unevenly in the jaw can lead to issues with overgrown teeth. Malocclusion, which causes both the incisors (front teeth) and cheek teeth (back teeth, known as premolars and molars) to overgrow, is most likely the most prevalent hereditary illness in rabbits. Eating and drinking becomes difficult as a result of malocclusion. Malocclusion can manifest in rabbits as the inability to close their mouths, clearly overgrown, long upper and/or lower incisors, or uneven wear of the incisors, making one side appear shorter than the other. To minimize issues, a veterinarian can anesthetize a rabbit with malocclusion and trim its teeth. Rabbits as young as three to eight weeks old can exhibit malocclusion, which is inherited. However, by tugging on the cage wire, young rabbits can also harm their incisor teeth, causing misalignment and potentially malocclusion as their teeth erupt. Furthermore, because chewing on crumbly pellets does not allow rabbits to wear down their continuously growing teeth as chewing on long-stemmed, fibrous hay does, rabbits fed only pellets instead of hay may develop overgrown teeth.

The tongue or the insides of the cheeks may get cut by the sharp, pointy edges that occasionally form when cheek teeth overgrow. A veterinarian needs to file down these points.

7.1.5.4.2. Dental abscesses

Abscesses may result from infection of the tissue encircling a tooth. Abscesses can result from a variety of factors, including foreign objects (typically plant material) lodged between the gum and tooth, cuts or sores that open up in the mouth from brushing against the sharp edges of unevenly worn teeth, exposure of the pulp, the sensitive tissue inside teeth after tooth trimming, and other medical conditions or dietary issues. Usually, multiple teeth are impacted. To confirm the diagnosis, a comprehensive dental examination and radiographs (x-rays) or computed tomography (CT scan) images are needed. It might be necessary to extract the abscessed tooth. The likelihood of recovery is low if multiple cheek teeth need to be extracted, and abscesses frequently return even after tooth extraction. The only thing that might be possible is long-term treatment with antibiotics safe for rabbits to stop the abscess from getting worse.

7.1.6. Breeding and Reproduction of Rabbits

Medium-to large-sized breeds of rabbit reach sexual maturity at 4 to 4.5 months, giant breeds at six to nine months, and small breeds (like the Dutch and Polish dwarfs) at 3.5 to 4 months. Unlike humans, female rabbits release their eggs in response to sexual activity rather than a hormone cycle. The receptivity of rabbits for mating occurs in cycles; approximately 14 out of every 16 days, they are receptive to mating. When a doe's vagina is red and wet, she is most responsive. The vagina of non-receptive does have little to no moisture and a whitish pink color. 12 days after breeding, a pregnant doe's abdomen may contain fetuses that a veterinarian can feel. False pregnancy is a common occurrence in rabbits, wherein the animal displays symptoms of pregnancy but is not truly pregnant.

About 31 to 33 days pass during a pregnancy. Pregnancies of does with small litters (typically four or fewer kits) appear to last longer than those with larger litters. Your veterinarian may induce labor if a doe has not given birth by day 32 of her pregnancy; if not, a dead litter may be delivered sometime after day 34. Pregnant women may occasionally reabsorb the fetuses or abort them due to illness or nutritional deficiencies.

After breeding, the cage needs to have nest boxes added 28–29 days later. Urine and excrement can contaminate boxes if they are added too soon. The doe removes her fur and constructs a nest in the nest box a day or two before giving birth.

7.1.6.1. Care of Newborns

Kittens of rabbits are born nude, blind, and deaf. A few days after birth, they start to grow hair, and by day 10, their eyes and ears are open. Until roughly day seven, newborn rabbits are unable to control their body temperature. The doe has a 24 hour window after giving birth to get pregnant again. 35 to 42 days after a litter is born, the majority of rabbit breeders who raise their animals for exhibition or as pets rebreed.

Large to medium-sized female rabbits typically have eight to 10 nipples, and many of them give birth to 12 or more offspring. Kits can be fostered by removing them from the nest box during the first three days and giving them to a doe that is roughly the same age but has a smaller litter if a doe is unable to nurse all of the kits successfully. Fostered kits are typically accepted if they are covered in the doe's hair and combined with the doe's own kits. The likelihood of success is increased when the larger kits are moved to the new litter rather than the smaller kits. Kits nurse for less than three minutes at a time and only once or twice a day. Kittens are weaned at the age of four to five weeks.

Though they can be raised by hand, kits have a high mortality rate. They need to be kept quiet, dry, and warm. Kitten milk replacer can be fed to the kits twice a day in the event that a nursing doe is not available to foster them. Depending on the age and breed of the kits, feedings can range from ½ teaspoon to one tablespoon. Around six weeks is when domestic rabbit kits are weaned.

Young does can kill and consume their young for a variety of reasons, such as fear, neglect (not nursing), or extreme cold. When dogs or other predators enter a rabbitry, the young are frequently scared and killed. As a natural instinct to keep the nest clean, cannibalism of the dead young happens. The doe shouldn't be used for breeding if she kills two litters in a row despite following all management procedures.

7.2. DISEASES OF DOMESTIC RABBITS

A variety of bacterial, viral, and parasitic infections can affect domestic rabbits.

7.2.1. Parasites of Rabbits

Numerous parasites that can infect domestic rabbits are hosted by wild rabbits. It is outside the purview of this book to describe every kind and species of parasite that exists in the world. Hofing and Kraus provide an extensive, illustrative account (1994). Owen provides a thorough description of the parasites that harm domestic rabbits (1992).

7.2.2. Ectoparasites

7.2.2.1. Fleas

In Europe, a common flea that infests wild rabbits is called *Spilopsyllus cuniculi*. The U.S.A is not where it happens (Kraus et al., 1984). Clusters of fleas along the edges of the pinnae are indicative of their preference for the ears. Because they can move, fleas can move from their host to their surroundings. Pet rabbits typically do not have fleas from wild rabbits. The reproductive status of the host has an impact on the life cycle of the small flea called *Spilopsyllus cuniculi*. Female reproductive hormones are necessary for the maturation of eggs. The rabbit flea can only reproduce successfully if it comes into contact with a newborn nestling or a pregnant rabbit. Fleas are drawn to late-pregnancy areas with elevated blood corticosteroid concentrations, and they cling tightly to the doe to feed. Fleas travel from the doe to the newborn babies within a few hours of parturition to feed, mate, and lay eggs in the nest. After the eggs hatch, the larvae consume the flea dirt that the adult fleas left behind after feeding on the pregnant doe. In this manner, fleas transmit from one generation to the next and play a significant role as disease vectors, particularly for myxomatosis.

Pet rabbits typically harbor *Ctenocephalides canis*, also known as felis, which is a common flea found on cats and dogs. Rabbits living in a home with dogs and cats cause infestations. Severe pruritus is brought on by infestation, and allergic dermatitis may appear. Using a fine-toothed comb, one can discover fleas and flea dirt on the rabbit.

Flea infestations are controlled similarly to other species. On rabbits, imidacloprid, and selamectin are both safe and effective, but fipronil should not be used. Treatment for all animals in contact, including those of different species, should be given, and environmental control measures should be taken.

7.2.2.2. Lice

A sucking louse known as *Haemodipsus ventricosus*, it afflicts wild rabbits and may serve as a myxomatosis vector. It is a big louse, measuring between 1.5 and 2.5 mm in length. On rare occasions, pet rabbits have it (Owen, 1992).

7.2.2.3. Mites

The common rabbit ear mite that causes crusting and ulceration of the external ear canal is called *Psoroptes cuniculi*. The large, active mites are barely perceptible with the unaided eye. When they are present in large numbers, these surface dwellers can cause intense irritation. They can occasionally be discovered in other body parts, like the skin folds on the perineum (Figure 7.2).

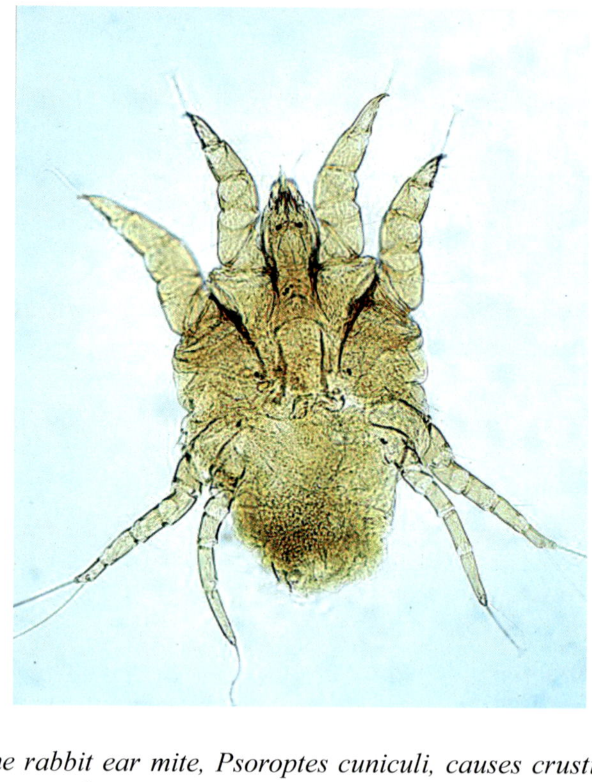

Figure 7.2. *The rabbit ear mite, Psoroptes cuniculi, causes crusting and inflammation of the external ear canal, which often extends up the pinna. Lesions are sometimes found on other parts of the body such as the perineal skin folds. Mites can just be seen with the naked eye in exudate from the lesions. Large numbers of P. cuniculi are visible on microscopic examination of the exudate, which can be softened in liquid paraffin before placing on a glass slide.*

Source: https://en.m.wikipedia.org/wiki/File:Psoroptes-cuniculi-mite.jpg.

Pet rabbits are a common host for the fur-dwelling mite *Cheyletiella parasitovorax*. The back has patches of thick, flaky, encrusted skin, particularly on the neck and above the base of the tail. By looking at skin brushings or pluckings under a microscope, the mites are easily identified. Cheyletiella infestation is frequently linked to dental disease, obesity, and spinal problems. *Cheyletiella parasitovorax* is a zoonotic bacteria that can irritate and cause erythema in humans. People who have handled infected rabbits may develop pruritic lesions on their necks or forearms. In a day, the lesions recede (Figure 7.3).

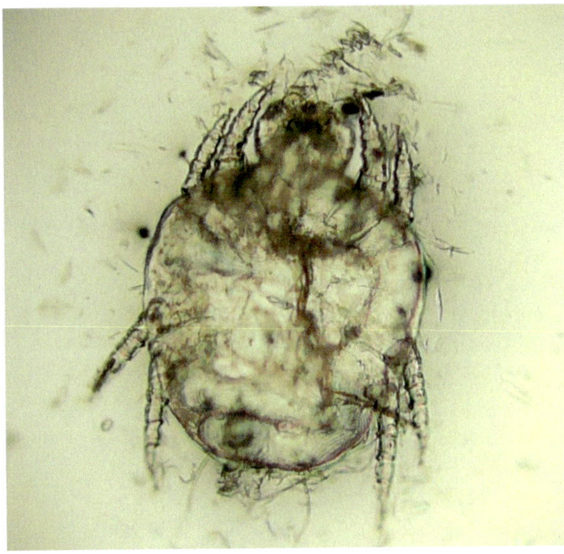

Figure 7.3. *An example of a Cheyletiella species.*

Cheyletiella parasitovorax can be found in the fur of healthy rabbits. It is not always associated with skin lesions. In large numbers, C. parasitovorax mites cause pruritus and areas of white, flaky skin. Heavy infestation is usually linked to some underlying problem with grooming, such as dental disease, obesity or spinal disorders. Mites may be seen moving among skin flakes that are combed out and placed under a bright light. Cheyletiella parasitovorax can also be detected by combing out the flakes and applying acetate strips to the exposed underlying skin. The acetate strip is placed on a microscope slide and examined on low power. In heavy infestations, a variety of nymphal stages, eggs, and adult mites are seen.

Source: By Joelmills-Own work, CC BY-SA 3.0, https://commons.wikimedia.org/w/index.php?curid=2564637.

The common fur mite of rabbits is *Leporacarus gibbus*, formerly known as *Listrophorus gibbus*. Large infestations may suggest an underlying illness, but otherwise, infestations are usually asymptomatic and of little consequence. The mite typically lives on the hair shaft, where it feeds on secretions secreted by sebaceous glands. When an infestation gives a light-colored rabbit's coat the look of pepper splattered on it, the mites are barely noticeable to the unaided eye. When the coat is damp, this effect is more noticeable (Figure 7.4).

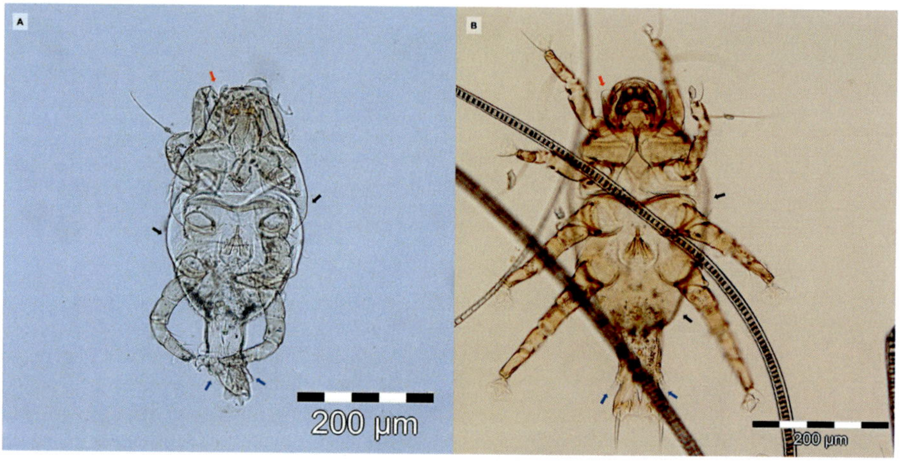

Figure 7.4. *L. gibbus male, from cat (**a**) and from rabbit (**b**), showing finely striated cuticle (black arrows), sclerotised "dark hooded" gnathostoma (red arrow), two elongated adanal processes (blue arrows).*

*Source: Dumitrache, M.O., Györke, A., D'Amico, G. et al. First case report of dermatitis associated with Leporacarus gibbus in cat. BMC Vet Res **17**, 4 (2021). https://doi.org/10.1186/s12917-020-02681-0.*

Mange in rabbits is attributed to Notoedres and Sarcoptes, two types of mites. These mites can be effectively treated with medications like selamectin (Stronghold, Pfizer), moxidectin (Advocate, Bayer), and ivermectin (various formulations). Nevertheless, it is worth mentioning that numerous mite-related problems, including those caused by Leporacarus and Cheyletiella, are frequently the result of grooming challenges. Consequently, a thorough diagnostic examination is necessary to detect any sources of discomfort or balance issues.

7.2.2.4. Warble Flies

Rabbits in the U.S.A are afflicted by warble flies called *Cuterebra horripilum* and *Cuterebra buccata*, which are not present in the U.K.

7.2.3. Endoparasites

7.2.3.1. Intestinal Worms

Around the world, wild rabbits are impacted by a variety of nematodes. Infestations in domestic rabbits, with the exception of *Passalurus ambiguus*, are uncommon and unlikely to occur, especially in pets. The large intestine and caecum contain the oxyurid *Passalurus ambiguus*. The adult worms, which are 5–10 mm in size, are not harmful to the adult animal and may even be necessary for the caecum's mechanical operation. Severe infestations in young rabbits may play a role in the enteritis complex of illnesses that arise around the time of weaning.

Affected animals' feces contain the tiny, thread like worm. There is a direct life cycle. *Passalurus ambiguus* can be affected by the majority of anthelmintics, such as fenbendazole, piperazine It is ineffective to use ivermectin (Morrisey, 1996). Limiting access to potentially infected excrement is one of the environment's control strategies, that is, routine hutch cleaning and pasture rotation, especially in areas where young rabbits are housed.

Other helminth parasites are not present in domestic pets and primarily affect wild rabbits. Among them are *Trichostrongylus retortaeformis* and *Graphidium strigosum* (Allan et al., 1999). *Obéliscoides cuniculi* is a disease that affects domestic rabbits in the United States and wild rabbits in different parts of the world (Hofing and Kraus, 1994). In clinical terms, it results in hemorrhagic diarrhea. *Obéliscoides cuniculi* has been employed as a laboratory model of the ruminant species Ostertagia and Trichostrongylus. According to Kraus et al. (1984) no trematode species have been found in rabbits.

7.2.3.2. Tapeworms

Many tapeworms that infect dogs and cats use rabbits as their intermediate host. Pet rabbits can contract the disease if they graze in gardens where dogs are kept or where foxes frequent. Since most pet owners now worm their dogs with preparations that are effective against tapeworms, the incidence of these parasites is not very high.

The tapeworm Taenia pisiformis, which primarily affects dogs and foxes, has a larval stage known as *Cysticercus pisiformis*, with rabbits serving as the intermediate host. Pastures are contaminated by tapeworm segments that are shed in feces and contain eggs. Eggs that travel via the liver to the small intestine, where the oncosphere appears, are consumed by grazing rabbits. There are several oval cysts in the mesentery. The tapeworm's inverted scolex is found inside the cysts. Distension and pain in the abdomen are brought on by severe infections. In severe situations, they may result in intestinal blockage. Necrotic foci and fibrous tracks are the end products of migration through the liver (Figure 7.5).

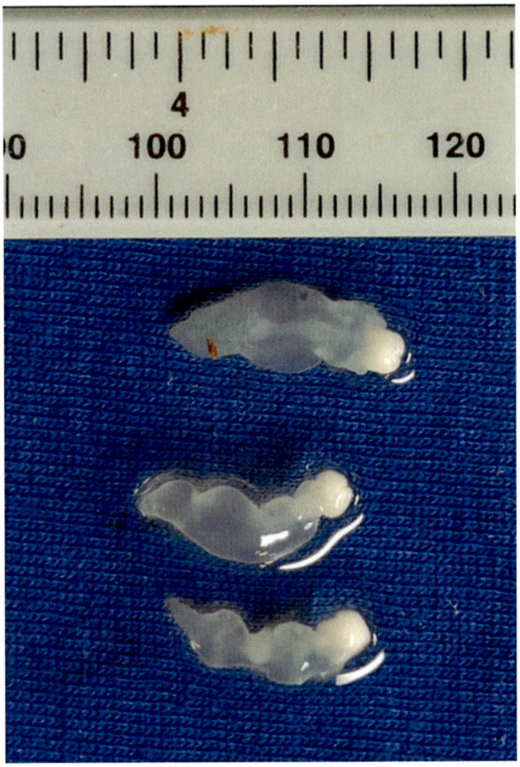

Figure 7.5. *Cysticercus pisiformis (It is the larval stage of Taenia pisiformis, which is a tapeworm that affects dogs and foxes, with rabbits acting as the intermediate host).*

Source: https://commons.wikimedia.org/wiki/File:Cysticercus_pisiformis.jpg.

The larval stage of the tapeworm *Taenia serialis*, which infects foxes and dogs, is known as *Coenurus serialis*. Primate species and even humans have the ability to serve as intermediate hosts, although naturally occurring rabbits and hares are the most common choice. The tapeworm's oncospheres move to the subcutaneous tissue, where they develop into soft swellings beneath the skin that can be felt. The cyst is filled with fluid and has inverted secondary buds with scolex in each. On rare occasions, an orbital cyst that causes retrobulbar swelling may be discovered (Wills, 2001).

Dogs and foxes are afflicted by *Echinococcus granulosus*. The majority of mammals, including humans and rabbits, are capable of serving as intermediate hosts. When compared to other tapeworm species, the adult tapeworm is small. It is 2–9 mm in size. Egg oncospheres that are swallowed travel through the mesenteric blood vessels to either the liver or the lung. Subsequently, the oncosphere grows into a massive cyst capable of generating secondary buds, each of which has an inverted scolex that can generate daughter cysts. A large cyst full of smaller cysts may form as a result of the daughter cysts' ability to create more daughter cysts. When a cyst ruptures, tiny cysts are released into the surrounding tissues, all of which have the potential to grow.

Tapeworms can also have primary hosts that are rabbits. Different regions of the world's wild rabbits have different cestode species. *Cittotaenia ctenoides* is one example, with a free-living mite serving as its intermediate host.

7.2.4. Protozoa

7.2.4.1. Coccidiosis

The pathogenicity of the at least 14 species of Emmeralia that affect rabbits varies. Serious issues with coccidiosis can arise in rabbit colonies. The two species that affect the intestine that are most pathogenic are *Ehreria irresidua* and *Ehreria magna*. *Imeria perforans*, *Imeria media*, *Imeria elongata*, *Imeria neoloporis*, *Imeria intestinalis*, *Imeria caecicola*, and *Imeria piriformis* are some additional less pathogenic species.

1. Common in colonies of rabbits.
2. Overcrowding and poor hygiene are significant factors in the development of outbreaks.
3. Oral intake of oocysts in contaminated feces leads to sporozoites infecting duodenal cells, causing damage.
4. Clinical signs include inappetence, weight loss, depression, and diarrhea.
5. In group situations medication (sulfa drugs) can be given in food or water.
6. Recovered rabbits become immune to reinfection.

7.2.4.2. Principles of Infectious Disease Control

1. Infectious and parasitic diseases are more common in groups of rabbits kept in close contact than in individual pet rabbits.
2. Diagnosis early in infection is key: a combination of clinical signs and diagnostic testing provides the most reliable information.

3. Identify source of infection/routes of transmission and take steps to reduce or eliminate further spread.
4. Consider the practicalities of treatment (groups vs individuals).
5. Consider public health considerations: is this disease a potential zoonosis?
6. Consider concurrent issues such as gut stasis, pain, and so on.
7. Address external sources of stress, for example, proximity of potential predators.
8. Address environmental issues such as poor ventilation and overcrowding.
9. Implement strategies to avoid infectious disease outbreaks in the future.

The intestinal Eimeria species has a slightly different life cycle from *Eimeria stiedae*, which causes hepatic coccidiosis. Oocysts are sensitive to dry conditions but can endure in the environment for many years. Rabbits that recover develop an immunity to infections.

7.2.4.3. Encephalitozoon cuniculi

The obligate intracellular parasite Encephalitozoon cuniculi forms spores and is a member of the phylum Microspora (Wasson and Peper, 2000). To be classified under this phylum, a spore must have a "polar filament," which protrudes during germination and is believed to aid in gaining entry into host cells. There are numerous species of Encephalitozoon, for example, any opportunist pathogens in immunocompromised human hosts, including *Entephalitozoon intestinalis*, *Entephalitozoon hellem*, *Entephalitozoon bieneusi*, and *Entephalitozoon septata*. Renal illness, keratoconjunctivitis, and diarrhea are a few of the conditions linked to human encephalitozoonosis. Regarding animals, the most significant member of the order Microsporidia is *Entephalitozoon cuniculi*. Although it can affect other species, encephalitozoon cuniculi primarily affects rabbits. The absence of mitochondria in microsporidia makes them unique; they likely obtain their nourishment from the host cells (Pakes and Gerrity, 1994). They are distinguished by a robust, highly Gram-negative capsule (Owen, 1992). It has a long, neatly wound polar filament inside of it. The spore is capped in polarity. Usually, oral ingestion of food contaminated with contaminated urine is how the host becomes infected. The spore enters the alimentary tract, makes direct contact with the mucosa, and then extrudes the polar filament to infect a host cell. Sporoplasm enters the host cell through the polar filament and travels to a vacuole where it multiplies. Along the vacuolar membrane, which is assumed to be of host origin, dividing organisms are arranged in a row (Pakes and Gerrity,

1994). However, trehalose, and chitin are also found in fungi, and because *E. cuniculi* are thought to be protozoa, the relationship to fungi may be closer than previously believed (Wasson and Peper, 2000). It is a common organism that has been found in a wide range of hosts, including humans, non-human primates, wild, and captive foxes, rabbits, shrews, mice, rats, hamsters, muskrats, guinea pigs, goats, sheep, and pigs. There are three strains: strain I affects rabbits, strain II affects rodents, and strain III affects dogs. Molecular differences could allow for the differentiation of these strains (Didier et al.,1993).

Spore intake is the method of infection. These can be excreted by infected animals in the form of urine, mucus, or feces. At 22□, spores can survive in the environment for at least four weeks (Waller, 1979). A new host can become infected by eating, breathing, or transplacentally (Baneux and Pognan, 2003). After that, spores enter host cells in the respiratory or digestive systems, specifically reticuloendothelial cells. It's unclear if phagocytosis also contributes to this or if the polar tube extrusion is the only mechanism. Among the infected cells are those of the reticuloendothelial system, which spreads the parasite throughout the body. Ultimately, the organisms transform into mature spores, which have a thick cell wall and an oval shape measuring roughly 2.5×1.5 μm (Pakes and Gerrity, 1994). The cell eventually bursts, releasing spores that infiltrate new cells as the vacuole enlarges. An inflammatory response is linked to cell rupture (Pattison et al., 1971).

Since the organism first appeared in the bloodstream, intradermal testing for *E. cuniculi* turns positive seven days after infection, and 14–28 days after infection, antibody activity can be measured (Kunstyr and Naumann, 1985).

After entering the bloodstream, the organism spreads to high-blood-flow areas at first, and 31 days after infection, the kidney, liver, and lungs can all be found to contain it (Percy and Barthold, 2001). The parasite multiplies (merogony), differentiates, and matures (sporogony) after penetrating suitable host cells. Eventually, the host cell ruptures and releases spores to complete the life cycle. It is hypothesized that the granulomatous reaction frequently linked to this illness is triggered by the rupture of host cells and the release of foreign material. This explains why a histopathological examination of the granulomatous lesions often fails to reveal any evidence of organisms. The clinical signs associated with *Entephalitozoon cuniculi* are thought to be caused by chronic granulomatous inflammation in these organs. There are three main types of clinical signs that are commonly associated with encephalitozoonosis: signs of central nervous system disease, signs of renal disease, and signs of ocular disease. Granulomatous lesions are the outcome of chronic inflammation and usually affect the brain and kidney, though they can also affect the liver. There have also been reports of myocarditis (Pakes and Gerrity, 1994). Clinical manifestations, particularly vestibular impairment and chronic renal failure, are linked to granulomatous encephalitis or nephritis. In rabbits, encephalitozoon cuniculi can also result in cataracts, pyogranulomatous uveitis, and lens rupture. The developing embryo's lens becomes infected in utero, which leads to the lens rupturing later in life (Stiles et al., 1997) (Figure 7.6).

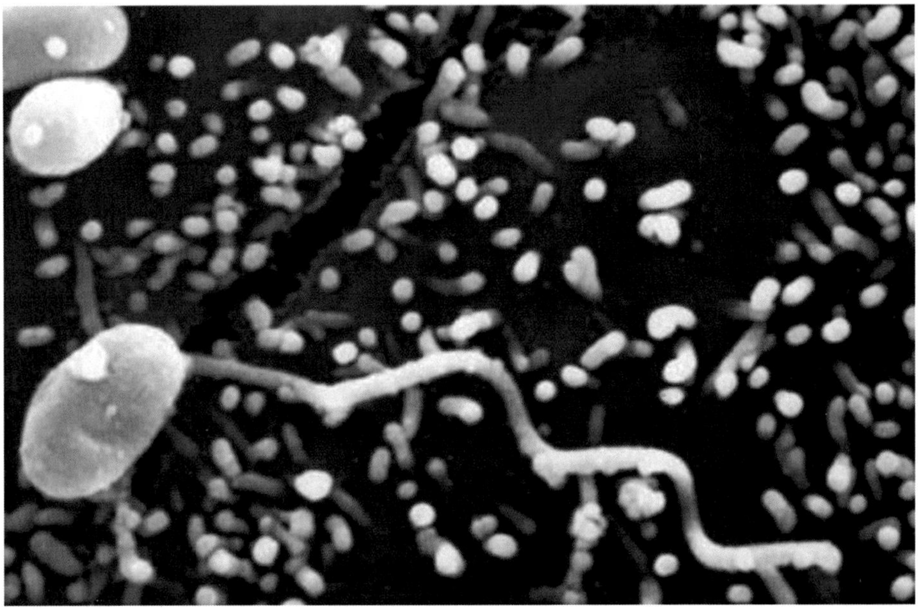

Figure 7.6. *Encephalitozoon cuniculi.*

Source: By Centers for Disease Control and Prevention-https://www.cdc.gov/dpdx/microsporidiosis/gallery.html#electron, Public Domain, https://commons.wikimedia.org/w/index.php?curid=33135596.

Spores begin to shed urine on day 42, at which point the organism has settled inside the renal tubular cells. By day 56, spore shedding reaches its peak. The antibody response reaches its peak on day 63 (Harcourt-Brown and Holloway, 2003). The organisms are only discovered to be present and causing brain lesions after 63–70 days; however, these are relatively mild in comparison to the lesions observed in the kidney at this time. This timeline makes it clear that a positive titer to *Encephalitozoon cuniculi* is conceivable before the organism most likely made its way into the brain. Any rabbit exhibiting central nervous system symptoms at this point may be mistakenly believed to be exhibiting clinical symptoms associated with this illness. This situation might become clearer if IgM and IgG titers could be distinguished. IgM titers rise in the first few days of an infection, decline, and disappear by day 38. According to Kunstyr et al. IgG increases more slowly but stays detectable for years (Sobottka et al.,

2001). 90-day post-infection urine spore shedding stops, and by day 98th, the organisms are found in the organs of predilection, the kidney, heart, and brain. Despite the fact that the heart is mentioned as a predilection site, heart disease because of *Encephalitozoon cuniculi*. It is rare to diagnose cuniculi.

7.2.4.5. Encephalitozoon cuniculi in Other Species

Numerous mammalian species are susceptible to infection by *Encephalitozoon cuniculi*, which exhibit host-specific disease variation and predilection sites. There have been reports of infections in dogs, cats, guinea pigs, hamsters, rabbits, mice, monkeys, and humans. The immunological and morphological characteristics of the various strains of *Encephalitozoon* are the same *E. cuniculi* affecting animals in laboratories.

There have also been reports of encephalitozoonosis in birds (Poonacha and Stamper, 1985). In a survey, Gannon (1980) discovered that guinea pigs housed with infected rabbits were more vulnerable than those housed separately. While cerebral granulomas were not observed in the guinea pigs, nephritis was prevalent. In 1955, a wild rabbit was reported to have contracted *Encephalitozoon cuniculi* (Pakes and Gerrity, 1994). However, subsequent serological surveys have not revealed any signs of infection in wild rabbits, despite the fact that they can contract the parasite in an experimental setting (Cox and Ross, 1980; Cox et al., 1980). There is evidence to suggest that wild rabbits' natural hygiene practices greatly reduce post-natal infection.

In the U.K., serological surveys on dogs (Hollister et al., 1989) and the African continent (Stewart et al., 1988) showed 13 and 2–23% prevalences, respectively. The latter prevalences showed the distinction between dogs with chronic renal disease (23% prevalence) and healthy control dogs (2% prevalence). The third strain (canine strain) of E. It has been demonstrated that cuniculi can cause illness in people (Weitzel et al., 2001).

Two recent studies have looked at the incidence of *E. cuniculi* in wild rodent populations. Hersteinsson et al. (1993) found serological evidence of the organism in wild mice in Iceland, and Muller-Doblies et al. (2002) isolated *E. cuniculi* from a free-ranging rat (*Rattus norvegicus*). Strain II is typically found in rodents. It has been suggested that carnivores are infected by eating infected prey and that the disease in wild rodent populations acts as a reservoir of infection for these species.

Apart from bunnies, *Alopex lagopus*, blue (Arctic) foxes, and the fur trade may suffer financial losses due to *Encephalitozoon cuniculi*. As a result, this disease has been thoroughly researched in this species. Although adult blue foxes are asymptomatic carriers, they are capable of transmitting infection transplacentally. Depending on which body system is most affected, the resulting puppies may exhibit a variety of clinical symptoms. These can include non-suppurative meningoencephalitis that is widespread, acute renal failure, and cardiac symptoms. Akerstedt (2003) showed that adult Norwegian blue foxes' humoral response lasts for a minimum of a year. Serological surveys of Arctic foxes raised in Finland have been prompted by disease in blue foxes (Akerstedt et al., 2002), untamed Arctic foxes in Greenland (Akerstedt and Kapel, 2003), and additional fox species, such as *Derbyson griseus* and *Derbyson culpaeus*, in Argentina (Martino et al., 2004). These investigations have revealed that there was no evidence of encephalitozoonosis in the populations of wild foxes in Argentina or Greenland. But the illness is endemic in foxes raised for agriculture. This is in line with the U.K.'s experience with rabbits (Cox and Ross, 1980; Blevins, unpublished data; R. Saunders, personal communication), where exposure occurs frequently in pet populations and rabbit colonies but not in the wild. It is in contrast to the U.S. and Australia (Thomas et al., 1955; 1997), in which reports of exposure in wild rabbits have been made. Regarding Australian rabbits, there is a suggestion that their seroprevalence in the wild may have shifted recently. No evidence of exposure to *Encephalitozoon cuniculi* was discovered in a 1980 study in Victoria's wild rabbit population (Cox et al., 1980), although research conducted in 1997 by Thomas et al. revealed a seroprevalence of 25% in wild rabbits in Western Australia, 1997. It is unknown what mechanism has caused the shift in prevalence in both the American and Australian wild populations. Regarding Australia, *Encephalitozoon cuniculi* was never absent from wild rabbits, and it is currently absent from those in Victoria.

7.2.4.6. Zoonotic Potential of Encephalitozoon cuniculi

Even so, *E. cuniculi* can infect a variety of hosts, but other than athymic or immunosuppressed mice, newborn dogs, and foxes, severe systemic disease is uncommon in other species. Masses of spores are discovered in the liver and other viscera of arthymic mice, who do not respond to the parasite either cellularly or humorally (Gannon, 1980). rabbits infected experimentally with *E. cuniculi* cultures inserted via a catheter into the rectum following several weeks of colonic enemas produced *E. cuniculi* infection with a predominance of hepatic lesions as opposed to the usual changes in the brain and kidney (Fuentealba et al., 1992).

Recently, the subject of *E. cuniculi* has drawn attention again because it may be able to infect people with compromised immune systems. Weber et al. (1997) reported on a case of cerebral microsporidiosis in an individual infected with the HIV virus; two pediatric cases identical to this case have also been reported (Didier, 2000). Other possible symptoms include sinusitis, pneumonia, bronchitis, and resistant diarrhea. Hollander et al. (1991) discovered serological proof of extensive human exposure to *E. cuniculi*. It's interesting to observe that the most common strain of *E. cuniculi*, strain III, which is typically linked to dogs, was discovered in humans, whereas strain I, or the rabbit strain, was more frequently discovered in Europe (Didier et al., 1995). The discovery of HIV in the 1980s and the rising number of patients receiving organ transplants and chemotherapy have made it impossible to overstate the potential zoonosis of *Encephalitozoon cuniculi* among pets.

7.2.4.7. Clinical Signs Associated With Encephalitozoon cuniculi Infection in Rabbits

In 1922, Wright, and Craighead reported the first case of *Encephalitozoon cuniculi* in rabbits displaying neurological symptoms such as hind leg paralysis. An outbreak that caused significant losses in a colony of laboratory rabbits was reported by Kimman and Akkermans (1987). Animals with the condition displayed emaciation, polydipsia, polyuria, muscle weakness, and sporadic neurological symptoms. According to other texts, encephalitozoonosis is a chronic, latent disease that affects rabbits and is noteworthy due to its impact on the outcomes of experiments. *E. coli* infection. Cuniculi in lab rabbits have complicated numerous scientific investigations. Lesions caused by the parasite have been mistakenly linked to a number of different illnesses, and subclinical infection has tainted experimental results (Wilson, 1979). The encephalitozoonosis virus can tamper with test outcomes. It has been demonstrated that the blood samples from rabbits suffering from spontaneous encephalitozoonosis exhibit notably lower catecholamine levels compared to those from healthy rabbits (Levkut et al., 1997). Rabbits kept in labs are now checked for *Encephalitozoon cuniculi* and seropositive animals. Because *E. coli* infection is common in laboratory settings, a "test and remove" policy is often implemented in conjunction with strict hygiene measures. During research trials, cuniculi may produce inaccurate results (Ansbacher et al., 1988). The reason for this is that E. The way the central nervous system responds to challenge is altered by cuniculi.

In pet rabbits, encephalitozoonosis is a common disease. There are many different manifestation signs, ranging from latent infections that don't show any symptoms of illness to acute neurological disasters. According to a serological survey conducted in Germany on 277 pet rabbits, 41% of them tested positive for antibodies (Ewringmann and Göbel, 1999). 51% (40.8%) of the rabbits that tested positive for encephalitozoonosis displayed clinical symptoms. Eight seropositive pet rabbits were found in the U.K. after a random survey of 30 of them (Carmichael, Idexx, and Harcourt-Brown, unpublished data). When the owners were questioned after the animals' seropositive status was discovered, four of them mentioned nebulous symptoms like head nodding or swaying while at rest, deafness, or diminished mental capacity. 52% of 97 clinically healthy rabbits in the U.K. were seropositive in a 2006 survey (Keeble and Shaw, 2006). Samples were collected from veterinary practices in England, Wales, and Scotland as part of a preanaesthetic or routine health screen. When filling out the survey, veterinary surgeons were asked about the health, preventive medicine regimen, vaccination history, diet, and husbandry practices of the animals. It was discovered that none of these variables was connected to the rabbits' serological condition.

Clinical signs can affect the liver, heart, or other organs, but they are typically linked to granulomatous lesions in the brain, kidney, or lens. Of the 277 rabbits surveyed in Germany, 51 (or 40% of the seropositive animals) displayed infection-related symptoms. 13 rabbits experienced uveitis, 16 had renal disease, and 23 had CNS disorders. According to Ewringmann and Göbel (1999), three animals had uveitis, renal disease, and CNS symptoms. Two rabbits had both renal and CNS disease.

7.2.4.8. Diagnosis of Encephalitozoon cuniculi

Ever since E's initial report. The diagnosis of *E. cuniculi* has always been ambiguous because it has been linked to "infectious motor paralysis" in young rabbits (Wright and Craighead, 1922). rabbits' clinical illnesses may be caused by cuniculi. According to recent research, E is seroprevalent. Ciniculi in pet rabbits in the U.K. are estimated to be 52% (Keeble and Shaw, 2006). The prevalence of the infection is high, but the illness rate is not known. Historically, although *Encephalitozoon cuniculi* is commonly found in mammals, its significance has been misunderstood. The histological and clinical manifestations of *E. coli* have been linked to the use of infected rabbits as models for poliomyelitis and syphilis being confused with *Encephalitozoon cuniculi* (Bull, 1917). Additionally, according to Wasson and Peper (2000), it has been linked to psittacosis, rabies, scrub typhus, and chemical carcinogenesis at different points in time.

Reliable serological tests are now available due to the significant effects this organism can have on research trials and the necessity of differentiating infected from non-infected rabbits. These work just as well in a laboratory setting as they do on pet rabbits. The majority of this disease's research has been done on lab rabbits. It is crucial to take into account this disease from the perspective of the pet rabbit because of the rise in the popularity of rabbits as pets and the quantity of them that are brought to veterinary surgeons for treatment. Determining whether an infection is relevant to the health status of the individual in question is often necessary when dealing with domestic rabbits. A positive titer does not indicate whether the disease is active; it only indicates exposure. In addition to being significant for the individual animal's care, this also affects the human owners of the animal, as *E. cuniculi* may represent a zoonotic threat (Deplazes et al., 1996; Weber et al., 1997; Weitzel et al., 2001).

The gold standard for diagnosing E currently exists. Using immunochemical staining and post-mortem histological examination, cuniculi allows for the precise identification of any organisms discovered (Barthold and Percy, 2001; Franzen et al., 1999). Granulomatous lesions are changes commonly linked to infection; the distribution of these lesions is determined by the interval between exposure and histological examination. According to Percy and Barthold (2001), lesions in the central nervous system are referred to as "focal non-suppurative granulomatous meningoencephalomyelitis with astrogliosis and perivascular lymphocytic infiltration," and lesions in the kidneys are referred to as "focal to segmental granulomatous interstitial nephritis." In both situations, the organism's remnants may be discovered in nearby cells but are not always present in the lesions.

To assess the sensitivity and specificity of the findings, histological examination must be contrasted with *in vivo* testing. However, since the subjects of the tests are still alive, it is frequently not possible to directly compare the results with the gold standard, especially in cases involving neurological signs. The pet rabbit's humorous reaction to *Encephalitozoon cuniculi* infection cannot be made with confidence. After an experimental infection with *E. coli*, serum antibodies start to develop in laboratory rabbits after three weeks, and the parasite excretes six weeks later. Cuneiformes (Cox et al., 1979). Infected dams pass on passive immunity to their offspring, which can have titers ranging from 1:25 to 1:800 and persist for approximately four weeks. Young rabbits seroconvert at 8–10 weeks of age in response to a natural infection after a period of seronegative (Lyngset, 1980). Therefore, the presence of antibodies does not prove that E exists rather, it only suggests exposure to the organism. Cuniculi as a pathogenic factor.

Although many domesticated rabbits have positive titers, this only suggests exposure; in fact, absolute titers do not significantly correlate with the presence of an organism in the brain, the severity of a clinical disease, or the outcome (Keeble and Shaw, 2006; Kunstyr et al., 2006). Even among rabbits that are matched for age, breed, and environment, traits can differ greatly so as to make them incomprehensible. In rabbits exhibiting chronic infection symptoms, high antibody titers have been observed in experiments (Pye and Cox, 1977). After a latent phase of 13–28 days, rabbits experimentally infected with *Encephalitozoon cuniculi* showed IgG titers of 160–2560 (Kunstyr et al., 1986). After a few weeks, some of the rabbits developed an episodic humoral response and turned seronegative. Although there was significant individual variation in the antibody response, the researchers proposed that variations in IgM and

IgG could differentiate between infections that were recent and those that were chronic. The onset of the antibody response is marked by IgM seroconversion, and concurrent detection of IgG and IgM indicates a recent infection. Jeklova et al. (2010) have proposed that the presence of an IgM titer signifies an ongoing infection and calls for medical intervention.

The fact that coexisting diseases can impact the host's immune response adds even more complexity to the situation. According to Cox (1977), rabbits that were already infected with *Encephalitozoon cuniculi*. Compared to uninfected animals, *Encephalitozoon cuniculi* showed changed immune responses to concurrent infections, with IgG levels for the new infection possibly being lower and IgM levels higher. This makes it more likely that clinical disease will develop concurrently (Harcourt-Brown and Holloway, 2003). In a similar vein, *Encephalitozoon cuniculi* may have an impact on how the central nervous system responds to external stimuli. Ansbacher et al. found that when coated platinum wires were implanted into four sites within the cerebral cortex of seropositive rabbits, the animals showed variable inflammatory responses within the same rabbit as well as between individual rabbits. Each rabbit develops an unpredictable response, making them unsuitable for a variety of research applications. It is possible that the reaction to coexisting illnesses could be just as erratic.

There are numerous serological tests available to check for *Encephalitozoon cuniculi* in rabbits, primarily because the laboratory industry needs negative rabbits for a large number of its research projects. Boot et al. (2000) examined several commercially accessible test techniques to compare their respective levels of sensitivity and specificity. A comparison was made between two carbon immunoassays, two enzyme-linked immunosorbent assays (ELISA), and two indirect immunofluorescence assays (IIF). The findings indicated that while there was no difference in the assays' ability to identify positive cases, IIF should be used for quantitative determinations rather than ELISA. It is suggested that this is because the ELISA assay is less quantitative than other methodologies, rather than a decrease in sensitivity. Percutaneous examination (Pakes et al., 1972) has shown to be both sensitive and specific in identifying rabbits that were positive for the organism, but it is not currently used for screening.

Animals with the infection can excrete *Encephalitozoon cuniculi* (Pye and Cox, 1977). The spores are visible as ovoid, Gram-positive organisms that range in size from 1.5 to 2.5 μm. Carbol fuchsin staining techniques give the organisms a pronounced purple color (Percy and Barthold, 1993). Urine testing is not a practical method for routine diagnostics in general practice, but it is theoretically a means of verifying the presence of antigen in the live animal. Urine is expelled sporadically by organisms, and collecting it can be challenging.

Regular rabbit urine frequently has sediment in it. It is now possible to identify *Encephalitozoon cuniculi* through commercial PCR testing in samples of tissue, feces, and urine.

When a pet rabbit has suggestive clinical signs and a positive titer, a clinical diagnosis of encephalitozoonosis is made after ruling out other possible causes of the clinical signs. A physical examination, complete blood count and differential, biochemistries, radiography of the axial skeleton and skull, urinalysis, and serology should, at the very least, be part of the clinical work-up. This will further guide the management of each individual case and allow the clinician to rule out or in many of the potential differential diagnoses.

The primary differentials for lead poisoning, toxoplasmosis, trauma, pasteurellosis, and neoplasia in pet rabbits are central nervous system diseases. Brain larval migration brought on by Baylisascaris is likewise distinct in the United States. Since encephalitozoonosis is associated with the formation of granulomas, clinical pathological evidence of this association may be utilized to differentiate the illness from other potential diagnoses. Only toxoplasmosis is likely to cause granulomatous lesions at the time when clinical signs start to matter among the conditions on the above list. Examining paired titers for this organism can rule out the significance of this disease, which is rarely diagnosed in U.K. rabbits. Comparably, benign tumors like embryonal nephromas, renal cysts, and malignant tumors like lymphoma and renal carcinoma are the differential diagnoses for chronic renal failure. Unlike other causes of similar clinical symptoms, granulomatous change in the target organs is the characteristic that distinguishes encephalitozoonosis. Determining the cause of many of these clinical signs presents two difficulties. The first is the unavoidable dependence on invasive or post-mortem testing to arrive at a conclusive diagnosis. Furthermore, rises in the absolute titers to interpretation challenges prevent them from being used as a diagnostic tool. Many employees don't trust *E. cuniculi* as a cause of disease on its own (it does not, in fact, satisfy Koch's postulates), believing that symptoms that are often misdiagnosed as parasite-related are actually brought on by one of the other diagnoses. A clinical cure may result from the alternative disease being successfully treated. Even so, active encephalitozoonosis cannot be ruled out if a positive diagnosis of one of the differentials for the presenting clinical signs is made.

In rabbits suffering from central nervous diseases, histological examination is not a reasonable means of making an antemortem diagnosis. When a rabbit exhibits renal symptoms, a kidney biopsy can be performed to make a diagnosis; however, lesions may go unnoticed because organisms and histological changes may not occur continuously throughout the kidney.

7.2.4.9. Encephalitozoon Cuniculi

1. *Encephalitozoon cuniculi* is a microsporidian parasite that can affect many species of animals.
2. *Encephalitozoon cuniculi* spores can be spread in the urine and feces of infected animals.
3. Spores can remain viable in the environment for 4–6 weeks.
4. Once within a host, spores are taken in the blood stream to organs of high blood flow (kidney, brain, lungs) but eventually settle in the central nervous system.
5. IgM levels rise rapidly after infection, then decrease in chronically infected individuals. IgM signifies recent or active infection.
6. IgG levels rise more slowly (they become dominant at around day 17) and can remain high for long periods. Within individual animals, IgG levels can fluctuate over time.
7. Diagnostic criteria include compatible clinical signs, exclusion of differential diagnoses and positive IgM and IgG titers.
8. Nested PCR can be used to detect microsporidial DNA in urine and fecal samples; however, as individual rabbits can shed spores in urine intermittently, this is not useful diagnostically.
9. Treatment should be aimed at controlling the spread of spores within the host and restricting access to them in the environment. Fenbendazole has been shown experimentally to clear infection in individuals currently and prevent infection in animals newly exposed to the organism.
10. Treatment of the clinical effects of encephalitozoonosis should be undertaken at the same time.

7.2.4.10. Treatment of Encephalitozoonosis

Treatment regimens for clinical encephalitozoonosis in pet rabbits have been derived from anecdotal reports or extrapolation from the management of human microsporidial disease because there have been no clinical trials looking into treatment protocols for this condition. Suber et al. (2002) provided evidence of the effectiveness of fenbendazole in the treatment of laboratory rabbits infected during experiments. Fenbendazole has been licensed for use in rabbits in the U.K. as a result of this.

A logical approach to treatment should include:

1. Reduction of inflammatory response.
2. Inhibition of spore formation.

3. Limiting access to infective spores.
4. Treating concurrent medical issues that are either directly or indirectly caused by the infection: for example, seizures, traumatic damage, renal failure, urine scalding and skin disease.

Treatment response varies greatly from person to person and is believed to be caused by variations in immune status, infectious strain, and dosage.

Potential treatments include:

1. **Benzimidazoles:** this class of drugs (including fenbendazole and albendazole) are microtubule inhibitors. In this instance they stop the host cells becoming infected by preventing the extrusion of the polar filament. Thus, they limit spore formation. Albendazole has been the treatment of choice in human microsporidial infections; however, its use in rabbits is extrapolated from this and therefore anecdotal. Adverse reactions such as bone marrow disease, pyrexia, and liver failure have been attributed to albendazole in rabbits. Albendazole is available as an oral preparation for cattle and sheep. It has been used to eliminate *Encephalitozoon* spp. from human AIDS patients, sometimes with dramatic success (De Groote et al., 1995). Elimination of the parasite is accompanied by relief of clinical symptoms in humans. Eradication of the parasite has been confirmed at autopsy (Joste et al., 1996, Sobottka et al., 1995). Albendazole has also been used to treat clinical cases of *E. cuniculi* in pet rabbits, with no reports of adverse effects and apparent improvement in clinical symptoms. The pharmacokinetic effects of the agent have been tested on rabbits (Li et al., 1995). An empirical course of 3–10 days' treatment appears to be beneficial. In-contact rabbits should be treated as a precaution, although albendazole is potentially teratogenic and is not advisable in breeding females. While it is difficult to support the use of this drug in the face of other proven less toxic compounds, its penetration into ocular tissues and its success in treating ocular manifestations of encephalitozoonosis should not be ignored. Fenbendazole has been tested clinically in experimentally infected rabbits (Suter et al., 2001) and been found to be effective in clearing the parasite from the brain, thereby reducing clinical signs in advanced cases. It is also effective in preventing the establishment of infection in exposed rabbits. Fenbendazole appears to be safer than albendazole in rabbits; however, it has been linked with bone marrow suppression, so rabbits undergoing treatment should be monitored appropriately.
2. **Glucocorticoids:** the use of glucocorticoids has been advocated

in the acute phase of encephalitozoonosis, to suppress the inflammatory response. However, rabbits are very sensitive to the immunosuppressive effects of steroids and their use in this species is always controversial. Treated individuals should be monitored closely.

3. **Fluoroquinolones:** For example, enrofloxacin, an antibiotic known to be safe for and licensed for use in rabbits, has been shown to have some antimicrosporidial action *in vitro*. The use of fluoroquinolones in the vestibular form of encephalitozoonosis is sensible, particularly as they are also effective in treating some of the main differential diagnoses such as pasteurellosis.
4. **Other Antibiotics:** oxytetracyclines and potentiated sulfonamides have also been recommended for use in treating encephalitozoonosis. While potentiated sulfonamides have been shown to be ineffective *in vitro* against *E. cuniculi*, oxytetracyclines do appear to inhibit microsporidial growth *in vitro*.
5. **Antifungal Drugs:** as *E. cuniculi* has been shown to share characteristics with fungi, and antifungal medications such as fumagillin have been evaluated as potential treatments. This drug is not well tolerated by rabbits and is not advocated as a treatment at this time.
6. **Interferon Gamma:** having been shown to be a potentially effective treatment in immunosuppressed humans, in the future interferon-gamma may become a clinically available treatment for rabbits.
7. **Chitin Synthesis Inhibitors:** For example, lufenuron, polyoxin D and nikkomycin Z may also be effective. To date, little information has been gathered about the clinical use of these compounds, either in humans or in rabbits. Some workers in the U.K. are advocating the use of lufenuron (Program Suspension Cat, Novartis Animal Health), as it has been shown to be safe in rabbits. Its efficacy is unproven.

Some cases seem to resolve on their own without any medical intervention; this is most likely because of the host's immune response. It is challenging to track the effectiveness of therapeutic agents in getting rid of the parasite in the absence of an easy diagnostic test for the antigen's presence in the live animal.

7.2.4.12. Control and Prevention of Encephalitozoonosis

Infection with *Encephalitozoon cuniculi* is common in rabbits in the United Kingdom; latent infections can occur in asymptomatic individuals, and a definitive diagnosis of the disease in a living animal is challenging. Control and prevention of this disease are made more challenging by all of these factors. It is ideal to test rabbits for encephalitozoon exposure before developing housing and management plans. Even with this approach, recently infected rabbits will be overlooked and consequently misclassified. Most of this risk would be eliminated by doing two tests four weeks apart. It is recommended that negative animals be kept apart from positive animals and not come into contact with them. Other steps could lower the risk of infection if this is not feasible. In exposed rabbits, short prophylactic courses of fenbendazole administered regularly have been demonstrated to prevent infection (with appropriate monitoring). The environment of the rabbit, including the food and water bowls and litter trays, should be cleaned and disinfected on a regular basis. Other strategies to lessen the risk of urine contamination include raising the food and water bowls off the ground, using hay nets instead of covering the floor in the hay, and controlling rodents and mice.

Serological testing is advised before introducing a new animal into a group or individual with a known status (positive or negative). Before being included in a negative group, any new members with negative titers should undergo two tests and be determined to be negative. It is challenging to support the inclusion of a negative person in a well-known positive group. Before introducing any rabbit that may have been exposed, a prophylactic 28 day course of fenbendazole has been recommended in the event that serology is not an option (Keeble, 2011). This won't stop future infections, but it will treat current infections and stop them from spreading at the moment.

Even so, *E. cuniculi* spores are extremely sensitive to disinfectants and can survive in the environment for up to four weeks. Effective disinfectants include quaternary ammonium (Trigene, F10), 1% bleach, and 70% ethanol. As usual, thorough cleaning needs to be done before disinfection.

7.2.5. Toxoplasma gondii

Rabbits can contract *Toxoplasma gondii*, just like any other mammal, though the infection is typically subclinical. The duodenum is where ingested sporulated oocysts hatch. Sporozoites infiltrate adjacent cells and travel throughout the body through the lymphatic and circulatory systems. The organisms can be found in different tissues as cysts once the host immune responses are established, and they can stay there for years (Owen, 1992). Rabbits can become infected through cat feces-contaminated feed, and symptoms have been reported in rabbits that have grazed in a cat-frequented area. The acute phase is when clinical signs in young rabbits are most prevalent. Although CNS symptoms like posterior paralysis or seizures can also occur, sudden anorexia, pyrexia, and death are the typical signs (Leland et al., 1992). There have been reports of outbreaks in the commercial rabbit industry (Harcourt, 1967; Okerman, 1988). Australia's sewerage farms have produced wild rabbits with high antibody titers (Cox et al., 1981). Ashley et al. (2011) investigated the seroprevalence of toxoplasmosis exposure in a number of Egyptian rabbit farms. Eating rabbit meat posed a risk to public health because 11% of rabbits had been exposed. A positive serology only suggests that an infection was encountered. Staffson et al. (1997) examined the variation in toxoplasmosis susceptibility between two lagomorphs: domestic rabbits (*Oryctolagus cuniculus*) and mountain hares (*Lepus timidus*). Compared to the mountain hare, domestic rabbits were found to be resistant to toxoplasmosis. Toxoplasmosis causes acute fatality in mountain hares and is typified by liver, mesenteric lymph nodes, and small intestine necrosis and tissue damage. Lesions are mild in domestic rabbits and comprise focal accumulations of mononuclear cells, primarily in the liver and heart. A pathologist's job is to diagnose. As early as 7–8 days after infection, antibodies are found (Gustaffson et al., 1997).

Toxoplasma is only transferable from rabbits to humans who handle or consume undercooked rabbit meat, despite the possibility of zoonotic toxoplasmosis. Rabbit feces do not transfer infections.

7.2.6. Other Protozoan Parasites of Rabbits

Although not identified as a primary cause of illness, several Cryptosporidium species have been reported in rabbits. Rabbits have also been reported to have *Giardia duodenalis*, though this infection does not seem to be harmful. Giardiasis has been linked to a single case of catarrhal enteritis in rabbits. No proof of transmission to humans has been found (Pakes and Gerrity, 1994). *Sarcocystis cuniculi* is a disease that affects rabbits, though reports of cases in European rabbits (*O. cuniculus*). It occurs more frequently in cottontails (*Sylvilagus floridanus*). Cysts are created by Sarcocystis in the heart and skeletal muscles. Cats are thought to be the infection's primary source.

7.3. BACTERIAL DISEASES

7.3.1. Pasteurellosis

Pasteurella multocida is a Gram-negative, ovoid, coccoid, or short rod that is extremely small, non-motile, and bipolarly stained. It is facultatively anaerobic as well as aerobic. After a 24 hour incubation period, the organism forms smooth, circular, convex colonies on blood agar. The colonies typically have a diameter of 2–3 mm and a slight iridescence, though there may be some variances. The colonies resemble mucoid tissues.

Pasteurella multocida exists in several antigenic strains. *Pasteurella multocida* is linked to several animal species. Many different animals may be harmed by the organism. Additionally, it can exist as a commensal organism. Both the human respiratory tract and the tonsils of healthy dogs have been found to contain *Pasteurella multocida*. With regard to rabbits. It is possible for *Pasteurella multocida* to live in the nasal cavity without getting sick. When pet rabbits are kept in small or individual groups, *Pasteurella multocida* rarely causes the main illness, but it is frequently identified as a secondary pathogen in any purulent or suppurative state. Pasteurellosis is an infectious disease that can be fatal in colonies of rabbits raised for breeding, meat, and fur production, or laboratory research. The disease develops when predisposing factors allow the bacteria to grow unchecked and outnumber the respiratory tract's physiological and immune defenses. Clones of pathogenic bacteria proliferate during these episodes and can subsequently spread to nearby animals with ease.

The outer membrane of *Pasteurella multocida* demonstrates a connection between the kind of protein and the animal host. Polysaccharides on bacterial capsular surfaces prevent phagocytosis. The bactericidal activity of serum and resistance to complement are conferred by bacterial lipopolysaccharides (Deeb, 1993). Various somatic and capsular serotypes of *Pasteurella multocida* are harmful to domestic animals and poultry, but very few of them are harmful to rabbits. The process of serotyping involves determining the serotypes and capsular antigen. Serotypes 12:A, 3:A, and 3:D are the most frequently found types in rabbits (Percy and Barthold, 1993). While lower respiratory tract diseases are more commonly linked to 3:A and occasionally 3:D, sniffles is most commonly associated with 12:A.

Jaglic et al. (2008) investigated serotype F's pathogenicity in people with immunocompetent and immunosuppressed conditions. While immunosuppressed people experienced severe diffuse hemorrhagic pneumonia, immunocompetent rabbits displayed fibrinopurulent pneumonia. In both groups, there was an eventual fatal septicemic syndrome. This suggests that serotype F is a pathogen with the potential to be very dangerous and could result in significant losses in commercial rabbitries. The severity of an infection differs amongst serotypes. The endotoxin produced by *Pasteurella multocida* differs depending on the serotype. Although its significance in septicemic cases is unknown, the endotoxin's role in clinical disease is unclear. Additionally, the bacteria create adhesins that help them adhere to epithelial tissue. P might benefit from the bacteria's elaborate filamentous appendages. Mucous membranes are colonized by *Pasteurella multocida* (Deeb, 1993). Different serotypes have different adhesive qualities and may play a significant role in the disease's etiology (Manning et al., 1989). When exposed to P, mucosal antibodies (IgA) are produced, which stop bacteria from growing. multocida. Elevated levels of humoral (IgG) antibodies are linked to persistent infection and have been employed to detect infected rabbits in lab colonies (Deeb, 1993).

7.3.2. Epidemiology

When there are enough *Pasteurella multocida* exchanged amongst rabbits, the upper respiratory tract becomes subclinically infected. While they are rare in the sinuses, bacteria proliferate in the mucous film that covers mucous membranes (Whittaker, 1989). When the balance between bacterial growth and mucociliary clearance is upset, clinical illness results. The disease tends to affect colony rabbits more than adult pet rabbits due to factors like genetic predisposition, bacterial serotype, poor husbandry, overcrowding, stress, pregnancy, parturition, lactation, and nutritional deficiencies.

Newborn rabbits are infected with *Pasteurella multocida* shortly after birth when they are infected and have an infection in their nasal cavity. Numerous factors can predispose young rabbits to infection, such as the age at which they are weaned, the occurrence of vaginal infections, and the colony-wide prevalence of infections. Pasteurellosis appears to have a genetic predisposition. For instance, according to Manning et al. chinchilla rabbits seem to be more vulnerable than Blue Beverans 1989. Up until roughly five months of age, the incidence of the disease rises with age. Following colonization of the upper respiratory tract, infection spreads to the tympanic bulla and the rest of the respiratory tract. It can result in otitis media, pneumonia, tracheitis, dacryocystitis, rhinitis, or conjunctivitis. There are rabbits that have P but do not exhibit any symptoms. multocida within the tentacles. These people are infectious carriers of animals they come into contact with. Other animals harbor *Pasteurella multocida* but test negative for nasal culture within the bulla tympanic.

Disease can spread between rabbits both through direct contact and through the air. Within eight days to three weeks, uninfected rabbits who come into close contact with infected rabbits develop pasteurellosis (Manning et al., 1989). A few feet' physical separation between rabbits will stop the spread of infection (Lelkes and Corbett, 1983). It has been shown that fomite spreads, and tainted water sources have been proposed as possible sources of infection (Whittaker, 1989).

7.3.3. Clinical Signs of Pasteurellosis

An acute, subacute, or chronic infection with *Pasteurella multocida* is possible. Pasteurellosis is linked to over a dozen clinical syndromes. The following syndromes, in decreasing order of severity, were discovered during pasteurellosis surveillance of rabbits in a laboratory animal facility: rhinitis, conjunctivitis, abscesses, and otitis media (DiGiacomo et al., 1983).

7.3.3.1. Rhinitis ("snuffles")

Upper respiratory tract infections are known by the colloquial term "snuffles," which is defined as a serous discharge from the nose followed by a purulent discharge. Sniffles and coughs are common symptoms in affected rabbits, and they may also make an audible snuffle sound. Snuffles is typically connected to P. infection caused by multocida, though other pathogens like *Staphylococcus aureus* may also be implicated. Similar symptoms can be caused by nasal foreign bodies and dental disease in pet rabbits. The forelegs wipe away the thick, sticky white discharge from the nose, causing a yellow stain and matting of the fur. The disease is made worse by poor husbandry, crowding, inadequate ventilation, dust, and ammonia buildup. Research on rhinitis in lab rabbits has revealed that some rabbits may experience rhinitis for as long as two weeks prior to nasal swabs are used to isolate *Pasteurella multocida*. Clinical symptoms fluctuate, but they frequently linger after treatment.

7.3.3.2. Pneumonia

Pneumonia in rabbits is caused by *Pasteurella multocida*. The illness may be severe and lethal very quickly. Rabbits can also get chronic or subacute infections without exhibiting any symptoms. Incidental pneumonic lesions are frequently discovered during post-mortem examinations of rabbits that appear to be in good health. There may be large abscesses in the thoracic cavity.

7.3.3.3. Genital Infection

In a comparatively high proportion of carrier animals, the vagina can yield *Pasteurella multocida*, and the organ can serve as a reservoir for infection (Percy and Barthold, 1993). During mating, *Pasteurella multocida* can proliferate. Bucks might have genital tract infections. Pasteurellosis can cause genital infections, pyometra, and orchitis.

7.3.3.4. Wound Infections and Abscesses

Pasteurella multocida is frequently isolated from bite wounds that are infected and abscessed. Many rabbits have the organism in their nasal cavity, which can contaminate tissues when they lick and groom. Additionally, it can spread hemogenously. Osteomyelitis following orthopedic surgery can be caused by *Pasteurella multocida*, which may be isolated from post-surgical wound breakdowns (Leibenberg and Badger, 1984). In rabbits suffering from dental disease, *Pasteurella multocida* can be isolated from facial abscesses resulting from periapical infection or tissue damage from elongated crowns.

7.3.3.5. Dacryocystitis

Dacryocystitis may be brought on by *Pasteurella multocida* (Petersen-Jones and Carrington, 1988). The organism can be isolated from purulent infections of the nasolacrimal duct, which can arise from secondary infections in ducts blocked by elongated tooth roots, particularly those of the maxillary incisors, or from the spread of infection from the nasal cavity.

7.3.3.6. Otitis Media

Pasteurella multocida can travel through the eustachian tube from the nasal cavity to the tympanic bulla. Inspissated pus in the deeper ear structures is a common radiographic and post-mortem finding. Rolling and nystagmus are two neurological symptoms of vestibular disease, which is brought on by an infection that travels down the vestibulocochlear nerve.

7.3.3.7. Detection of Pasteurellosis

To stop the disease from spreading, pasteurellosis in rabbit colonies must be confirmed. Although clinical symptoms point to infection, subclinical carriers must be found and the organism must be isolated through diagnostic testing. For bacteriology, a deep nasal swab is necessary. It can be challenging to get good swabs from a conscious animal; sedation or anesthesia is necessary. One cannot always rely on bacterial culture. False-negative results may arise from an infection that is located deep within the nasal passages or in the paranasal sinuses. Antibiotics have already been administered to a few rabbits. According to Sanchez et al. (2000) *Pasteurella multocida* does not fare well in transport media. At room temperature, it can only survive for a short time. At 34–35□, which is colder than most standard cultures, *Pasteurella multocida* grows at its best.

In the U.S.A, tests using polymerase chain reaction (PCR) and serology are available. An increasing titer indicates infection exposure. Antibodies alone, however, do not indicate that an infection is active. Sanchez et al. (2000) studied rabbits exhibiting clinical symptoms suggestive of pasteurellosis using a combination of bacterial culture, serology, and PCR testing. They discovered that serology and PCR worked better together than nasal swab culture. The researchers came to the conclusion that additional organisms, including *Pseudomonas, Bordetella*, and *Staphylococcus* species, result in clinical symptoms resembling pasteurellosis symptoms.

7.3.3.8. Control of Pasteurellosis in Rabbit Colonies

A serious issue in commercial, laboratory, or breeding rabbit colonies is pasteurellosis. The illness also causes issues for households with multiple rabbits or for sanctuaries and rescue facilities that house a large number of rabbits in a constrained area. Stress, coexisting medical conditions, crowded living conditions, and poor air quality can all cause latent infections to resurface. Good husbandry plays a key role in the control of the disease, just like it does in any intense situation involving an infectious disease. Since afflicted animals can spread the infection to other stock, they should be quickly isolated, treated, or even put to death. Disease transmission is decreased when rabbit populations are small and contact between batches of rabbits is kept to a minimum.

Clinical disease in infected colonies can be reduced to a minimum by keeping newly weaned rabbits apart from adults, managing the environment carefully, and lowering stress levels. It must be a dry, clean, well-ventilated space free of drafts. Although they can tolerate the cold, rabbits become anxious in hot weather. Disease risk is increased by closed stuffy sheds, particularly if ammonia buildup is causing poor air quality. Temperature fluctuations should be avoided; 16–20□ is the ideal range, and 50–70% humidity is the ideal range (Whittaker, 1989). With about 20 air changes per hour and ideally a filter system, the quality of the air should be good.

7.3.3.9. Prevention of Pasteurellosis

Pasteurellosis control measures have been implemented in rabbit colonies over the years, with differing degrees of success. Today, a large number of laboratory colonies are vigilantly barrier-housed to prevent the introduction of infection and are free of disease. Rabbits are kept in isolation for two to four weeks and their nasal passages are frequently cultured to select stock that is free of Pasteurella. Rabbits showing symptoms of rhinitis or positive cultures are euthanized. Rabbits that survive are bred from, and the colony is deemed disease-free after three years. Caesarean derivation combined with hand-rearing is another technique for creating disease-free rabbits (Manning et al., 1989) or fertilizing eggs and transferring them to does that are free of Pasteurella. Antibiotic use and early weaning can raise the proportion of healthy people. Periodic serological testing for antibodies to Pasteurella is required to guarantee an uninfected status. A PCR test that can be used to identify infections was recently developed (Sanchez et al., 2000).

Antibiotics have also been given to pregnant does either in their drink or feed as a preventative measure to try and avoid pasteurellosis. There have been attempts to create disease-free strains of rabbits, but there seems to be a genetic resistance to Pasteurella.

Immunizations against P. In other species, like sheep, multocida are successfully employed, and efforts have been made to create a vaccine that effectively protects rabbits against pasteurellosis. It has been discovered that using both live and dead vaccines can effectively lower clinical disease and mortality brought on by a homologous strain of the bacterium. The majority of rabbit pathogenic strains are capsule type A or D. cross-immunity is higher between strains of the same serotype, and they carry somatic antigens 3 and 12. But even with the encouraging outcomes in lab rabbits (DiGiacomo et al., 1987), there is insufficient protection against heterologous strains' clinical disease and nasal colonization, and field trials employing an intranasal vaccine against A:12 have produced unsatisfactory results (DiGiacomo and Deeb, 1989). Sukow et al. (2008) reported the results of a vaccine containing serotype D:3,12,15, which was successfully field tested. It's interesting to note that the subcutaneous immunization outperformed the intranasal one. Some rabbit breeders think that vaccination is possible using an autogenous vaccine or a vaccine designed to prevent pasteurellosis in other animals like sheep or cattle. It is challenging to assess claims of efficacy for these vaccinations.

7.3.3.10. Treatment of Pasteurellosis

The clinical signs of pasteurellosis, as well as the disease's type and financial or emotional significance, all influence the course of treatment. Other chapters address the treatment of vestibular disease, respiratory tract infections, abscesses, and dacryocystitis.

A logical approach to treatment should include:

1. Obtaining a specific diagnosis: culture.
2. Basing treatment on specificity results.
3. Reduce stressors such as overcrowding.
4. Improve environmental factors such as poor ventilation and hygiene.
5. Pasteurellosis.
6. Pasteurellosis can be spread by direct contact, aerosol, and fomites. It is endemic in many rabbit colonies.
7. Clinical disease occurs when there is a disruption of the balance between mucociliary clearance and bacterial proliferation.
8. Pasteurellosis can cause a variety of clinical signs, but these are not pathognomonic, so culture is recommended.
9. Treatment is best planned using sensitivity results on appropriate samples. PCR in combination with serology is the most specific diagnostic test for pasteurellosis.
10. Contributory factors such as stress and overcrowding must be addressed.

7.3.4. Staphylococcus Aureus

Suppurative inflammation is brought on by *Staphylococcus aureus*. The organism is often isolated from rabbit infected sites. Healthy rabbits that have *P. multocida* may carry *S. aureus* within the nasal passages. In healthy rabbits, it can also be separated from the skin and conjunctiva. It is possible to isolate *Staphylococcus aureus* from patients suffering from skin infections, abscesses, rhinitis, conjunctivitis, mastitis, and ulcerative pododermatitis. In tissues that have been harmed by trauma or another risk factor, it frequently appears as a secondary invader. Bacterial virulence and host resistance determine the severity of the disease (Delong and Manning, 1994). Staphylococcosis can result in significant losses in rabbit colonies.

7.3.5. Bordetella Bronchiseptica

Animal species from which *Bordetella bronchiseptica* has been isolated include rats, dogs, cats, guinea pigs, rabbits, and pigs. Regarding rabbits, B. Despite causing localized suppurative bronchopneumonia in laboratory rabbits treated with cortisone before the organism was inoculated into their nasal cavity, bronchiseptica appears to be relatively non-pathogenic. In guinea pigs, *Bordetella bronchiseptica* can result in a serious upper respiratory tract infection. Research indicates that infected rabbits and guinea pigs may be able to spread the infection to one another. Bronchiseptica from various species have been categorized based on their bacterial sensitivity (Boot et al., 1995). Despite the fact that there are few real cases of cross-infection, many texts advise against housing the two species together.

7.3.6. Tyzzer's Disease

A large pleomorphic, Gram-negative, obligate intracellular bacterium that forms spores and is flagellate (moving) is the cause of Tyzzer's disease (Delong and Manning, 1994). Although the bacterium can grow in tissue culture, it cannot be grown *in vitro*. The organism has been reclassified as *Clostridium piliforme* rather than *Bacillus piliformis* in recent years due to the close resemblance of the bacterial genome to that of the Clostridium species (Besch-Williford, 1997). Numerous animals, including rodents, cats, and monkeys, are susceptible to Tyzzer's disease (Delong and Manning, 1994). The illness affects the liver, intestine, and caecum. In acute cases, it causes sudden death and acute diarrhea; in chronic cases, it causes intestinal fibrosis, stenosis, and liver necrosis. There may also be effects on the myocardium. The illness can strike at any age and is frequently exacerbated by stress, but it typically strikes weanling rabbits between the ages of 6 and 12 weeks. It is a significant difference in diarrhea outbreaks in rabbit colonies. Recent developments in tissue culture have aided in the creation of diagnostic tools, and in certain nations, serological testing is now feasible. Although a PCR has been developed to detect ribosomal DNA sequences, serology, and PCR do not agree very well. Although cross-reaction with non-pathogenic bacteria can lead to false-positive results, the presence of antibody in animals that appear healthy suggests latent intestinal tract infection (Pritt et al., 2010). Overt disease can be brought on by stress or immunosuppression (Delong and Manning, 1994). Spores that remain viable in the environment for a while after an infected animal has been removed are consumed and cause transmission. Clinical disease is predisposed by stress, low dietary fiber, overcrowding, and transportation. Antibiotic therapy and supportive care are typically unsatisfactory.

7.3.7. Salmonellosis

Wild rodents can contaminate food and water with salmonella organisms. Clinical symptoms can include diarrhea, emaciation, and even death in cases of asymptomatic carriers. There has never been a report of a successful remedy.

7.3.8. Escherichia coli

In rabbits, *Escherichia coli* is typically not present in the gut flora. But *E. coli* is a major cause of enteritis and mortality in rabbit colonies. It can cause enteritis, particularly in suckling rabbits. It has been suggested that intestinal coccidiosis, which increases the growth of the organism. The pathogenicity of different strains of *E.coli* varies and numerous strains have been identified from enteritis outbreaks. In the U.K., a strain known as "attaching and effacing" has been found, with mortality rates ranging from 25 to 75% This organism clings tightly to the epithelial cells in the caecum (Dannatt et al., 2000).

7.3.9. Clostridial Enterotoxaemia

Anaerobic Gram-positive bacilli called clostridium are capable of generating potent enterotoxins that can cause serious intestinal illness. Usually, clostridial enterotoxemia results in death. A tiny number of *Clostridium* species are typical members of the rabbits' gut flora. In rabbits, enterotoxaemia can be brought on by *Clostridium spiriforme*, *Clostridium difficile*, and *Clostridium perfringens*. Affected rabbits are most frequently weanlings.

Iota toxin is produced by *Clostridium spiriforme*. A glucose substrate is necessary for the synthesis of iota toxin. It is thought that a high starch diet increases the risk of enterotoxaemia by overloading the colon with carbohydrates. When leftover starch enters the caecum, it can decompose and release glucose, which serves as a substrate for the production of iota toxins. Juvenile rabbits are more likely than adults to experience this circumstance. Starch is not effectively absorbed in the small intestine of immature rabbits, but it is in the intestines of adult animals before it reaches the caecum. Adult enterotoxaemia is typically associated with other factors like stress or antibiotic therapy, which upset the microbiota in the caecum and allow *Clostridium* spp. to proliferate.

7.3.10. Other Causes of Bacterial Enteritis

Enteric diseases in rabbits have been linked to both Vibrio and Campylobacter. There have been reports of a syndrome in Japan called "*Histiocytic enteritis*." Enteric disease outbreaks in rabbit colonies have yielded the isolation of adenoviruses, parvoviruses, rotaviruses, coronaviruses, and herpes-like viruses. It is improbable that an adult pet rabbit will contract these infections. Comprehensive accounts of these infections are provided by DiGiacomo and Mare (1994) and Percy and Barthold (1993).

7.3.11. Treponematosis

The pathogen *Treponema paraluiscuniculi* is unique to rabbits. It is a spirochaete that causes inflammatory, crusty lesions on the face and genitalia. It is spread through sex. When a young rabbit passes through the birth canal, it may become infected. The illness is sometimes referred to as "rabbit syphilis" or venereal spirochaetosis. In certain breeding colonies, treponematosis is endemic and can also sporadically be found in pet rabbits.

7.3.12. Listeriosis

Infections with *Listeria monocytogenes* are rare in rabbits. It is distinguished by abrupt death and abortion. In breeding colonies, outbreaks can be brought on by contaminated feed. In advanced pregnancy, *Listeria monocytogenes* prefers the gravid uterus. Abortion, stillbirths, and the doe's death can all result from infection. Visceral congestion, scattered pale miliary foci on the liver, and straw-colored fluid in the peritoneal cavity are examples of post-mortem symptoms. There is ecchymosis and fibrinoid exudate on the uterine serosal surface.

7.3.13. Paratuberculosis (Johne's Disease)

Mycobacterium avium subspecies *paratuberculosis*, which causes paratuberculosis (a disease known as paratuberculosis), primarily affects ruminants. Its symptoms include diarrhea, emaciation, and loss of body condition. Most animals contract the infection when they are still newborns by consuming tainted milk or water. Following a protracted subclinical phase that can extend for several years, clinical infection becomes evident. Although ruminants are the animals from which disease reports are most common, monogastric animals have been experimentally infected without exhibiting any symptoms of the disease. When a newborn rabbit contracts an oral infection, it can cause granulomatous enteritis and sporadic diarrhea akin to what is seen in cattle.

In Scotland, there is a strong correlation between the high prevalence of infection in cattle and the high incidence of paratuberculosis in wild rabbits. 67% of wild rabbits tested positive for *M. pneumococcal* disease (Greig et al., 1997). A history of Johne's disease on the farms where the wild rabbits were caught was linked to the infection in the rabbits, according to epidemiological studies (Greig et al., 1999). Additionally, foxes, and stoats taken from impacted farms were found to have *M. paratuberculosis* (Beard et al., 1999). The overall health of the wild rabbits afflicted with paratuberculosis was good, but some of them had thickened intestinal mucosa with sporadic granulomas. The lesions contained a large number of intracellular acid-fast bacilli.

7.3.14. Pseudotuberculosis

Yersinia pseudotuberculosis, the causative agent of pseudotuberculosis, is frequently found in rodents, particularly guinea pigs. Although it has been reported in captive animals, the disease primarily affects wild rabbits. The afflicted rabbits have a dull coat, a wasting disease, and sporadic diarrhea. Palpation of the abdomen may reveal nodular liver swelling (Wood, 1978). It is possible to isolate *Yersinia pseudotuberculosis* from feces or cecal contents. Large areas of caseous necrosis in the liver, spleen, and mesenteric lymph nodes are among the lesions associated with pseudotuberculosis. There may be Peyer's patch necrosis in the cecum and small intestine. Other organs like the liver and spleen may also be affected by the illness (Delong and Manning, 1994). Since yerbiniosis is linked to rodents, mice, and rats must be kept under control (Okerman, 1988).

7.3.15. Tularaemia

Francisella tularensis is the cause of tularemia, an acute septicemic illness. It is more common in hares and cottontails, but less common in domestic rabbits. The organism exhibits zoonotic potential and can impact numerous species of vertebrates. The majority of human cases connected to rabbits have occurred after exposure to the cottontail (*S. floridanus*). Delong and Manning (1994) state that no human cases of tularemia contracted from *O. cuniculus* have been documented. There are two biovars of *F. tularensis*: type B is found in Europe and Asia, while type A is primarily found in North America. Type A causes an acute febrile illness that is more severe in younger people and is more virulent for both humans and rabbits. Tularaemia can be spread by ticks and possibly other biting insects. Antibiosis (tetracyclines, fluoroquinolones) and supportive care are part of the treatment. There is a vaccine available for humans that has been proven to be successful in rabbit experiments (Pasetti et al., 2008).

7.3.16. Lyme Disease

Acute, frequently recurrent polyarthritis in humans and dogs brought on by the spirochaete *Borrelia burgdorferi* is known as Lyme disease. It is a disease spread by ticks. Antibodies to *B. burgdorferi* have been demonstrated in cottontail rabbits in regions with a high concentration of Ixodes that feed on rabbits (Telford and Speilman, 1989).

Numerous other bacterial infections in rabbits exist. Stress, overcrowding, wounds, reproduction, subpar husbandry, and concurrent diseases are linked to them. Pseudomonas, Fusiformis, and Corynebacterium are a few examples.

7.4. VIRAL DISEASES

7.4.1. Myxomatosis

The deadly disease known as myxomatosis affects European rabbits. Subcutaneous swells that, when sectioned, release a mucoid secretion are its defining feature. Lesions appear on the face, particularly on the eyelids, and around body orifices. Pet rabbits may get the illness through insect vectors or direct contact with wild rabbits that have the infection. Arthropods, particularly the European rabbit flea, *Spolopsyllus cuniculi*, are the primary vectors of the disease. Myxomatosis outbreaks in wild rabbits wax and wane based on the immune status of the native rabbit population and the virulence of the strain. When rabbit flea populations are at their highest (generally when there are does or young in the nest), outbreaks also tend to increase.

7.4.2. History of Myxomatosis

Among the first viruses to be identified was the myxoma virus. In 1896, it had an impact on a group of lab rabbits in Uruguay (Fenner and Fantani, 1999). Aragao identified virus particles in stained smears in 1927 and drew attention to the virus's striking similarities to smallpox and fowlpox. Later research identified the myxoma virus as a pox virus (Fenner and Ross, 1994). Researchers in Brazil discovered that fleas and mosquitoes mechanically spread the virus.

In Europe today, myxomatosis is a disease that affects wild rabbits exclusively. It was brought illegally into France in 1952, and it wasn't until 1953 that it was acknowledged in England. Before this, as a purposeful pest control experiment, infected rabbits had been released into the Outer Hebrides of Heisker Islands in 1952. Despite the significant mortality caused by myxomatosis, two years later, in 1954, the rabbit population was as large as ever (Fenner and Fantani, 1999). Despite efforts to eradicate myxomatosis in the U.K., the disease became endemic in the late 1950s after spreading quickly through the wild rabbit population in the summer of 1953. Compared to other countries, the U.K. had a different perspective on myxomatosis. Many people kept rabbits as pets, and when sick or blind rabbits were seen stumbling along roads, common areas, and other public spaces, it caused outcry. Consequently, using or permitting the use of an infected rabbit to spread the disease to an uninfected population was made illegal in 1954. It was challenging to enforce this law.

7.4.2.1. Epidemiology of Myxomatosis

In its natural host, either *Sylvilagus bachmani* (brush rabbit), which is native to California, or *Sylvilagus brasiliensis* (Tapeti, forest rabbit, found in Mexico or Argentina), the myxoma virus causes a minor infection. The myxoma virus infects European rabbits, leading to a dangerous and potentially fatal condition. Hares are susceptible to myxomatosis, although infection is uncommon and typically mild.

Wild rabbits are susceptible to various strains of myxomatosis, including, for example, the more pathogenic Lausanne strain and the common laboratory strain (Moses). In contrast to the protuberant lesions caused by the Lausanne strains, the standard laboratory strain results in comparatively flat skin lesions (Fenner and Ross, 1994). Some variations result in massive pulmonary oedema but are linked to fewer, smaller skin lesions.

In the wild, fleas, and mosquitoes are the primary insect vectors of myxomatosis, though any insect that pierces the skin can spread the disease. Additionally, direct contact or inhalation between rabbits can transmit the disease. When unvaccinated rabbits are placed in hutches that have been contaminated with fluid from lesions from infected rabbits, the virus survives and spreads. *Cheyletiella parasitovorax* has the potential to serve as a disease-transmission vector (Fenner and Fantani, 1999).

The pattern of disease outbreaks and the epidemiology of myxomatosis are influenced by the life cycle of the insect vector. In many regions of the world, mosquitoes are the primary vectors. Myxomatosis spreads quickly in nations where mosquitoes are the primary carrier of the disease, and it is commonly found in pet rabbits kept in hutches. The seasonal incidence is high. Isolated pockets of infection typically characterize disease outbreaks in the U.K., and pet rabbits are only infrequently affected by the illness. The distinction in the life cycle of insect vectors is responsible for the variation in epidemiology. The European rabbit flea, *S. cuniculi* is present in the U.K. as opposed to Aedes and Anopheles species of mosquitoes, is the primary insect vector. Fleas can remain contagious throughout the winter and serve as a reservoir of infection for the following year even when their host is not present. Because of their life cycle, which is synchronized with the reproductive status of the doe and causes heavy flea infestations in susceptible neonates, fleas are an effective means of transmission.

The virulence of the myxoma virus varies amongst its strains. Highly virulent strains of the disease kill rabbits so quickly that it is not spread as easily as it is with less virulent strains. Mortality rates are also impacted by environmental temperature since the disease is more deadly in colder climates. Some people are genetically resistant to myxomatosis.

7.4.2.2. Clinical Signs of Myxomatosis

Myxomatosis pathogenesis is similar to other infections caused by the pox virus (Fenner and Ross, 1994). The site of inoculation and the local lymph node are the sites of sequential virus replication. Cell-associated viraemia and widespread infection of the body occur after it. The illness initially manifests as a skin lesion that usually appears 4–5 days after the virus is injected and grows to a diameter of roughly 3 cm 9–10 days after infection. Since the virus replicates throughout the lymphoid system, the rabbit is viremic. On the ninth day, the eyelids thicken and finally close completely, accompanied by a semi purulent ocular discharge. All over the body, secondary lesions appear; these usually appear on the nares, lips, eyelids, base of the ears, external genitalia, and anus. Pneumonic symptoms are one of the characteristics of outbreaks in intensively farmed rabbits that can be caused by aerosol infection. According to Fenner and Fantani (1999), this syndrome is distinguished by a longer incubation period of 1–3 weeks, as well as lacrimation and mucopurulent nasal discharge. Litter abandonment and sterility are associated with myxomatosis.

Inanition, a secondary bacterial infection, or, in the case of wild rabbits, predation are the usual causes of myxomatosis death. Infected testicles can cause a buck to become infertile for upto a year in rabbits that recover (Fenner and Fantani, 1999). Unless they have some passive immunity, very young rabbits are especially vulnerable to infection and pass away more quickly than adult animals.

How long rabbits live after infection and whether they survive myxomatosis are determined by a number of factors. According to Fenner and Ross (1994), infected rabbits develop an immune response that can be seen *in vitro* tests approximately seven days after infection and peaks in 28 days later. Antibodies give complete immunity for several months and endure for extended periods of time. In newborn rabbits, immunity is transferred by the mother and lasts for

4–5 weeks. Because of their genetic resistance to infection, some rabbits are less likely to die during outbreaks in the wild. Countries and populations of rabbits have different levels of genetic resistance to infection. Compared to Australian rabbits, British rabbits took longer to become resistant (Fenner and Ross, 1994). Also discussed is a phenomenon referred to as "paternal resistance." It has been found that when bucks mate within seven months of infection, the mated doe's offspring are occasionally born with partial resistance. There have been rumors that semen contains an immunogenic substance (Fenner and Ross, 1994).

7.4.2.3. Relationship of Myxomatosis with Shope Fibroma Virus

The Leporipoxvirus genus, which causes fibromas in its natural hosts, is home to the viruses that cause myxomatosis. The native forest or brush rabbits of North and South America are the natural hosts of the myxoma virus, not the European rabbit. The rabbit fibroma virus, also known as the Shope fibroma virus, is a significant member of the Leporipoxvirus genus and naturally affects cottontail rabbits. The benign fibroma in European rabbits is caused by the Shope fibroma virus.

In the eastern United States, cottontails are endemic to the Shope fibroma virus. It results in fibromas that are localized and may last for several months. Generalized fibromatosis can happen in newborns or immunocompromised people (Fenner and Fantani, 1999). The Shope fibroma virus is transmitted by insects, just like myxomatosis. Mosquito transmission is more common than with European rabbits. Adults may develop generalized fibromatosis in environments where cottontails and mosquitoes are prevalent because repeated mosquito bites result in fibromas at each location (Fenner and Fantani, 1999).

Within three weeks of injection, Shope fibroma virus-caused fibromas in European rabbits regress. The superficial layers of fibromas induced by the Shope fibroma virus in its natural host, *S. floridanus* contain a large amount of virus, in contrast to the European rabbit's fibromas. Shope fibroma cannot be established as an enzootic disease in European rabbits, but cross-immunity between Shope fibroma virus and myxomatosis occurs and European rabbits that have recovered from infection with Shope fibroma virus are immune to myxomatosis.

7.4.2.4. Immunization

As with other pox viruses, live vaccinations are necessary to impart resistance to myxomatosis because dead vaccines are unlikely to be effective. Myxoma virus vaccinations using live attenuated strains have been attempted, however virulence issues and potential immunosuppression have been reported (Fenner and Fantani, 1999). A live vaccine containing Shope fibroma virus was developed as a result of the identification of the virus and its cross-immunity with myxomatosis. Up until recently, Shope fibroma virus grown in cell-line tissue culture was the only vaccine available in the U.K. (Nobivac Myxo, Intervet), which was a live, attenuated virus vaccine that was freeze-dried. A new vaccine called Nubivac Myxo-RHD was introduced in 2012 (Spibey et al., 2012). It was based on an attenuated strain of myxomatosis with a capsid protein added to prevent RHD. This new vaccine provides immunity for a full year.

A few points on myxomatosis:

1. Myxomatosis is characterized by subcutaneous swellings that exude a mucoid secretion when sectioned. Lesions occur around body orifices and on the face, especially the eyelids.
2. The disease is mainly spread by arthropods, especially the European rabbit flea, *Spilopsyllus cuniculi.* Mosquitoes are vectors in many parts of the world.
3. In wild rabbits, outbreaks of myxomatosis wax and wane according to the virulence of the strain and the immune status of the native rabbit population.
4. Myxomatosis can occur in hares but is rare and usually mild.
5. Environmental temperature has an effect on mortality rates, with the disease being more lethal at low temperatures.
6. Myxomatosis starts with a skin lesion at the site of inoculation. The rabbit becomes viremic, with virus replication taking place throughout the lymphoid system. Secondary lesions develop throughout the body, typically on the nares, lips, eyelids, and base of the ears and on the external genitalia and anus.
7. Aerosol infection can result in pneumonic signs, which may be a feature of outbreaks in intensive farmed rabbits.
8. Myxomatosis is accompanied by sterility and abandonment of litters.
9. It is possible, on rare occasions, for rabbits to recover from myxomatosis. Ambient temperature affects the course of the disease, with high environmental temperature increasing recovery rate (85°F).
10. Antibiotics, a warm environment, good nursing and non-steroidal analgesics can be used to treat myxomatosis. Corticosteroids are contraindicated due to their immunosuppressive effects. Opioid analgesics are ineffective in ameliorating signs of pain.

7.4.2.5. Recovery from Myxomatosis

Rabbits that suffer from myxomatosis can recover. In addition to the virus strain's virulence, a few environmental elements influence the rabbit's resistance to myxomatosis, that is, temperature of the surroundings and concurrent infection. The course of the illness is influenced by the ambient temperature; a high temperature (85°F) accelerates recovery. While the prognosis for myxomatosis is not good, some pet rabbits have survived with the help of antibiotics, a warm environment, and proper nursing. There is always a chance of developing secondary issues like pasteurellosis or gastrointestinal stasis. While non-steroidal analgesics are helpful, corticosteroids should not be used because of their immunosuppressive properties. It seems that opioid analgesics are ineffective at reducing pain signals. There was no difference in survival time in a study looking at how buprenorphine affected the course of myxomatosis in lab rabbits. According to Robinson et al (1999), treated rabbits had lower rectal temperatures just before they passed away and refused food and water one day earlier than untreated rabbits. Even after receiving a myxomatosis vaccination, rabbits may still exhibit clinical symptoms and develop an unusual kind of the disease known as lumpy bunny syndrome. The disease can affect a rabbit in good health with just one lump to one in very bad health with numerous skin lesions. For many of these people, the prognosis is not as dire because they can recover with the right supportive care.

7.4.3. Viral Hemorrhagic Disease (VHD)

Rabbits can contract viral hemorrhagic disease (VHD), a highly contagious disease with a high death rate. A host-specific calicivirus is the cause of it. Only the European rabbit is afflicted by VHD.

RHD is one name for the illness, and rabbit hemorrhagic disease virus (RHDV) is the name of the virus. The phrase "rabbit calicivirus disease" (RCD) is occasionally employed. The People's Republic of China, which at the time was the biggest exporter of rabbit meat worldwide, was where VHD first started in 1984. A colony of newly imported angora rabbits into Germany became infected (Fenner and Fantani, 1999). All of the rabbits—aside from the sucklings—died within a week, and the disease reached Italy and Europe in less than nine months after spreading over 50,000 km^2. By 1988, VHD had been documented in commercial rabbits across numerous nations, most likely due to exposure to rabbit flesh. The illness began to spread throughout the wild rabbit population in Europe. VHD made its way to Scandinavia in 1990. In less than a week, wild rabbits on the crowded island of Gotland almost went extinct (Gavier-Widén, 1996). Numerous rabbits perished in their burrows, and hundreds were discovered dead in the fields. Pet rabbits fed commercial food and kept indoors began to die, suggesting that humans can spread the virus. At the same time, European brown hare syndrome (EBHS), another illness, was spreading throughout the continent. EBHS is brought on by a distinct strain of calicivirus.

A non-pathogenic virus that caused seroconversion and shielded rabbits from VHD was discovered in 1996 while breeding rabbits in Italy. After isolation, the virus was recognized as a calicivirus. Evidence suggests that this virus predated the development of VHD (Capucci et al., 1997).

7.4.3.1. Pathogenesis of VHD

A calicivirus that replicates in the cytoplasm of hepatocytes is the cause of viral hepatitis (VHD). Three to four days after infection, rabbits used in experiments pass away.

In essence, VHD is a necrotizing hepatitis that frequently results in splenic necrosis. Hemorrhages are caused by fibrinous thrombi that form within small blood vessels in most organs, particularly the kidneys, heart, and lungs, as a result of disseminated intravascular coagulation. Liver failure or widespread intravascular coagulopathy are the causes of death (Figure 7.7).

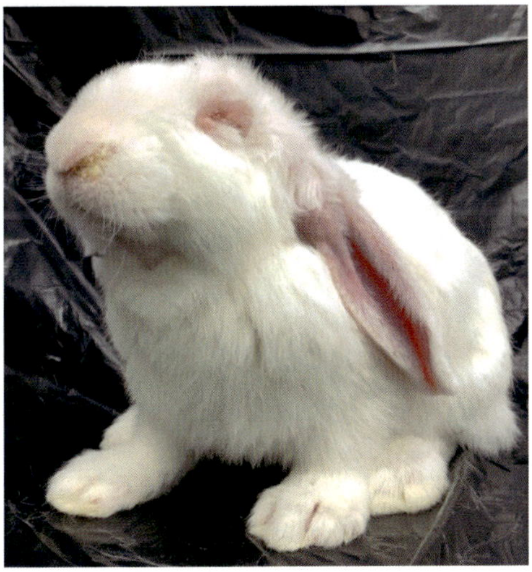

Figure 7.7. *Myxomatosis and rabbit hemorrhagic disease (RHD).*

Source: Rosell JM, de la Fuente LF, Parra F, Dalton KP, Badiola Sáiz JI, Pérez de Rozas A, Badiola Díez JJ, Fernández de Luco D, Casal J, Majó N et al. Myxomatosis and Rabbit Hemorrhagic Disease: A 30-Year Study of the Occurrence on Commercial Farms in Spain. Animals. *2019; 9(10):780. https://doi.org/10.3390/ani9100780.*

7.4.3.2. Epidemiology of VHD

The virus that causes EBHS and the calicivirus that causes VHD are similar antigenically. Heterologous virus attempts to cross-infect rabbits and hares have not been successful in causing illness. Only European rabbits are afflicted by VHD; cottontails and other small mammals like chinchillas, guinea pigs, rats, and mice are not affected.

The VHD calicivirus has a long half-life outside of its host. On a cloth, viable viruses have been found to remain active for up to 105 days (Fenner and Fantani, 1999). The virus can survive for 22–35 days at 22□ but only for 3–7 days at 37□, so environmental temperature plays a critical role in its ability to survive. The VHD virus can be transmitted orally, nasally, or parenterally. Infected rabbits' urine and feces can also harbor the virus. Foods tainted with contaminants may spread illness.

In susceptible populations, the mortality rate from VHD can reach 90–100% in rabbits older than two months. When the disease becomes endemic, the morbidity and mortality rate decreases because infected young rabbits live and develop immunity. The illness seems to erupt in wild rabbits every other year. The virus in viremic blood is mechanically transferred from one animal to another by insects. The VHD virus can persist in skin and carcasses for a few weeks. The disease is known to be transmitted by fleas, blowflies, and mosquitoes (Fenner and Fantani, 1999). According to PCR methods, a virus can survive in the bodies of blowflies and bushflies for up to nine days and seven days, respectively. Fly "spots," or excrement, can contaminate pasture and spread disease. Because they are light and easily propelled by the wind, flies have the ability to spread disease over great distances. Further evidence that domestic and wild carnivores can be significant in the epidemiology of VHD has been provided by the ability to extract virulent material from feces following experimental oral inoculation. Even after freezing and thawing, the virus remains highly stable in carcasses (Lumeij, 1997). In Great Britain, VHD was a disease that needed to be reported until October 1996. Because of its resistance and ease of transmission, it is now endemic and presents a serious risk to pet rabbits. There have been reported deaths all over the country whose cause of death was determined to be VHD at post-mortem.

7.4.3.3. Clinical Signs of Viral Hemorrhagic Disease

The incubation period for VHD is only three to four days. Animals that have been eating and acting normally for a few hours can be found dead due to a peracute form of the disease. Acute cases typically have a higher respiratory rate, are quiet, pyrexic, and die in 12 hours or less. A characteristic of the illness is a sharp drop in blood pressure, which makes it challenging to locate a vein for intravenous fluids or blood sample collection. Rabbits that are dying appear pale, shocked, and collapsed. There may be hematuria, hemorrhagic vaginal discharges, or bloody or foamy nasal discharge. Brain vascular infarctions can happen, and occasionally convulsions or other neurological symptoms appear shortly before death. There have been descriptions of cyanosis and agonal vocalizing (Donnelly, 1995). A dead rabbit in opisthotonus with a hemorrhagic nasal discharge is depicted in the "classic" image. Occasionally, a rabbit will recover from the acute phase, but a few days later, it will develop jaundice and die. If exposed to the disease, young rabbits younger than four weeks of age are immune to it for the rest of their lives and are not affected. Before their physiological resistance to the virus vanishes between 6 and 10 weeks of age, unexposed rabbits become more and more vulnerable. The increase in hepatic transaminase production that happens after five weeks of age has been linked to the physiological age immunity of young rabbits (Donnelly, 1995). Adult rabbits have a high mortality rate. The afflicted rabbits do not have a treatment.

7.4.3.4. Diagnosis of VHD

Any sudden death is suspected of being caused by VHD, particularly if multiple rabbits in the household have passed away. A healthy rabbit with undamaged food in the stomach and hard fecal pellets in the distal colon may be seen in the post-mortem image, indicating a sudden demise. Even though the severe histopathological alterations may not be visible on the surface, the liver is always affected. The liver has a clear lobular pattern and is enlarged, pale, and friable. Moreover, the spleen is larger. While they can occur in any organ, hemorrhages most frequently occur in the lungs. Often, a foamy exudate fills the trachea. Hematological conditions that cause multiple organ failure as a result of general circulatory dysfunction include fibrin thrombi, lymphopenia, a decrease in platelets, and a failure of other blood-clotting factors. One aspect of the pathophysiology of VHD that is distinctive is disseminated intravascular coagulation (Chasey, 1997). Acute hepatic necrosis is confirmed by histopathology. Numerous additional alterations, like acute nephropathy or alveolar hemorrhage, could also exist. Any organ can experience congestion and bleeding as a result of terminal intravascular coagulation. Although the typical histopathological alterations in the liver are typically diagnostic, a variety of additional tests, such as electron microscopy and hemagglutination tests, can confirm the diagnosis. By examining the liver under an electron microscope, a large number of distinctive caliciviruses can be found (Chasey et al., 1995). The lab requires a liver that is fresh. Additionally, ELISA tests are offered.

7.4.3.5. Vaccination

Owing to the severe consequences of VHD in China, a vaccine was expeditiously created utilizing inactivated virus extracted from the liver and spleen of affected rabbits. When given inactivated vaccines, the immune system responds well. Since attenuated strains of the VHD virus are difficult to grow in tissue culture, none have been created. Virus antigen obtained from rabbits in experiments is rendered inactive using formalin or β-propiolactone to create vaccines that are viable and sold commercially. It is advisable to vaccinate every pet rabbit. To immunize against both myxomatosis and VHD at the same time, genetically modified vaccines that incorporate the VHD coat protein gene into the attenuated myxoma virus are being developed (Barcena et al., 2000; Spibey and colleagues, 2012). The U.K and Europe can now get this vaccine (Nobivac Myxo-RHD, MSD Animal Health).

1. Viral hemorrhagic disease.
2. VHD is a highly infectious lethal disease of rabbits with a high mortality rate. It is caused by a host-specific calicivirus. VHD only affects the European rabbit (*Oryctolagus cuniculi*).
3. The disease may be called 'rabbit hemorrhagic disease' (RHD) and the virus known as 'rabbit hemorrhagic disease virus' (RHDV). Sometimes the term 'rabbit calicivirus disease' (RCD) is used.
4. The causative calicivirus of VHD has a predilection for hepatocytes. VHD is essentially a necrotizing hepatitis. Death is usually due to disseminated intravascular coagulopathy.
5. VHD calicivirus can survive for long periods outside the host. Viable virus has been detected for as long as 105 days on cloth. Infection does not require direct contact.
6. VHD virus is spread by oral, nasal, and parenteral transmission and is present in urine and feces from infected rabbits. Contaminated foods can be a source of infection.
7. When VHD is introduced into a susceptible population, the mortality rate is high and can be 90–100% in rabbits over two months of age.
8. VHD has a short incubation period of 3–4 days. The disease can be peracute, with animals being found dead within a few hours of eating and behaving normally.
9. The 'classic' picture is a dead rabbit in opisthotonus with a hemorrhagic nasal discharge.
10. Hematuria, hemorrhagic vaginal discharges or foamy exudate from the nostrils may be seen.

11. Vascular infarcts can occur within the brain and occasionally convulsions or other neurological signs are seen just before death.
12. There is no specific treatment for affected rabbits.
13. Young rabbits, less than four weeks of age, remain unaffected and develop lifelong immunity if they are exposed to the disease. Unexposed rabbits become increasingly susceptible until 6–10 weeks of age when physiological resistance to the virus disappears.
14. The typical histopathological changes in the liver are usually diagnostic.
15. An effective vaccine against VHD is available in the U.K.

7.4.4. Papillomatosis

Two papillomaviruses that can infect rabbits have been described. In cottontail rabbits, Shope papillomavirus causes a benign disease; however, in European rabbits, it can result in malignant neoplasms that mimic squamous cell carcinomas. The illness affects domestic rabbits in some U.S. commercial units as well as the wild cottontail population in the eastern United States.

Oral papillomatosis is caused by the Shope papillomavirus, which differs from other papillomaviruses immunologically. On the ventral side of the tongue and in other areas of the oral mucosa, wart-like growths are the hallmark of oral papillomatosis. The virus is spread through oral secretions that have wart-derived sloughed cells in them. Over the course of six to nine months, the papillomas grow slowly, with young rabbits being the most susceptible. When the animals develop immunity, the papilloma's base becomes irritated, which leads to the tumor's sloughing, the development of ulcers, and ultimately the reepithelialization of the tumor. It is unknown if rabbit oral papillomas can experience carcinomatous transformation (Kraus et al., 1984).

7.4.5. Corona virus

The first report of coronavirus infection in rabbits dates back to 1968. The afflicted rabbits had a high mortality rate, pulmonary edema, and pleural effusion, and were pyrexic. The illness has been linked to iridocyclitis. There has been a comparison drawn to infectious peritonitis in cats. Weanling rabbit enteric disease outbreaks have also been linked to coronavirus.

Due to the lack of *in vitro* propagation, it is unknown if the virus is a naturally occurring rabbit pathogen or a virus from a different species that has evolved to fit rabbits in contaminated treponemal stocks (DiGiacomo and Mare, 1994). When rabbits were injected with suspensions of rabbit testes containing *Treponema pallidum* (human syphilis), the disease was initially identified in the

1960s. The agent has subsequently been found in rabbit tissue contaminated with *T. pallidum* everywhere in the world. Coronavirus infection has only been reported in laboratory rabbits and is used experimentally to create a rabbit model of cardiomyopathy. Antibodies against the virus exhibit cross-reactivity with coronaviruses from mammals, including humans (DiGiacomo and Mare, 1994).

7.4.5. Mycotic Infections

7.4.5.1. Dermatophytosis

Ringworm, or dermatophytosis, can occasionally affect rabbits. The two species that are most frequently described are *M. canis* and *Trichopyton mentagrophytes* (Percy and Barthold, 1993). Although they can affect other parts of the body like the paws, lesions are typically located on the muzzle and base of the ears. There may be asymptomatic carriers. The majority of affected rabbits are young ones (Vangeel et al., 2000). From rabbits, *Dermatophilus congolensis* has been isolated.

7.4.5.2. Aspergillosis

Laboratory rabbits have been reported to have pulmonary aspergillotic granulomas (Percy and Barthold, 1993). In rare cases, rabbits with chronic rhinitis may also have nasal aspergillosis.

We end this chapter with hope: the promise of nutraceuticals in rabbit stress has been presented below (Table 7.3).

Table 7.3. *Nutraceuticals' Potential to Alleviate Heat Stress Impacts in Rabbit Production*

Additives	Level	Heat Stress Conditions	Animal	Main Impacts
Vitamin C	0.5 g/kg diet from 5 to 14 wk of age	32.44□ and 84.67% relative humidity	Giant Flander male growing rabbits	· Enhanced feed intake and feed conversion ratio. · Improved serum concentration of SOD and TAC. · Reduced serum concentrations of triglycerides, hydrogen peroxide, and MDA. · Decreased *Staphylococcus aureus, E. coli*, and total bacterial counts in cecum.
Vitamin C	200 mg/kg diet from 6 to 12 wk of age	28–39□ and 60% relative humidity	New Zealand White growing rabbits	· Improved body weight gain and feed conversion ratio. · Restored cortisol, leptin, IFN-γ, TNF-α, and IL-10 plasma concentrations to normal levels. · Decreased HSP70 and forkhead box P3 (FOXP3) gene expression in liver and kidney tissues.
Vitamin E	0.25 g/kg diet from 7 to 14 wk of age	36.4□ and 97% relative humidity	Californian unsexed growing rabbits	· Improved body weight, weight gain, feed intake, and feed conversion ratio. · Improved carcass yield% and total edible parts%.
Vitamin E	100 mg/kg diet for three months	32.9□ and 80.38% relative humidity	New Zealand White rabbits does	· Improved pregnancy rate and litter size. · Improved serum GSH, GSH-Px, SOD, CAT, GST, and TAC. · •Enhanced lysozyme, IgG, and IgM concentrations.

Vitamin A	12,000 IU/kg diet	30–34□	Rex rabbits	· Increased hair follicle density and fur quality. · Regulated Wnt10/β-catenin, insulin-like growth factor-1 (IGF-I), fibroblast growth factor 5 (FGF5), noggin-BMP, and sonic hedgehog (SHH).
Selenium	25 and 50 mg of nano-Se/kg diet from 7 to 13 wk of age	33□ and 90% relative humidity	Domestic rowing rabbits	· Enhanced live body weight, body weight gain, feed intake, and feed conversion ratio. Enhanced carcass%. · Improved serum GSH and SOD concentrations. · Regulated inflammatory cytokines (IFNγ and IL-4).
Selenium	0.3 mg organic Se/kg diet for 12 week	31□ and 75% relative humidity	Adult V-line male rabbits	· Decreased rectal temperature and respiration rate. · Decreased MDA concentration in seminal plasma. · Enhanced TAC concentration in seminal plasma. · Enhanced total functional sperm counts and percentages of integrated sperm membranes.
Zinc	20, 40, 60, and 80 mg nano-Zn/kg diet for 60 d	38.20–40.10□ and 45–50% relative humidity	New Zealand White male growing rabbits	· Enhanced body weight, weight gain, feed intake, and conversion ratio. · Improved carcass% and hind parts%. · Reduced carcass fat%. · Reduced meat lipid oxidation (TBARS content). · Improved water-holding capacity.

Zinc	75 mg $ZnSO_4$/kg diet or 75 mg Zn picolinate/kg diet from 32 to 42 week of age	30.7–37.6□ and 70–80% relative humidity	New Zealand White rabbit bucks	· Decreased serum concentration of AST, ALT, and glucose. · Improved antioxidative status (serum concentration of SOD and TAC). · The increased serum concentration of testosterone.
Copper	200 mg Cu-methionine/kg diet or 200 mg copper-glycine/kg diet for five wk	30.12□ and 82.40% relative humidity	V line unsexed growing rabbits	· Enhanced body weight, body weight gain, and feed conversion ratio. · Improved serum concentrations of TAC, GSH-Px, and SOD. · Decreased serum MDA concentration. · Increased serum concentrations of HDL-cholesterol.
Copper	100 mg Cu-acetate/kg diet or 50 mg nano-Cu/kg diet from 5 to 14 wk of age	17 and 22□	New Zealand White unsexed growing rabbits	· Improved final body weight, body weight gain, and feed conversion ratio. · Enhanced nutrient digestibility (nitrogen-free extract, dry, and organic matter). · Enhanced serum concentrations of complement component-3, IgM, and lysozyme activity.
Chromium	0.4–1.6 mg organic Cr/kg diet	30□ and 80% relative humidity	Growing rabbits	· Improved final body weight, body weight gain, and feed intake. · Enhanced cell-mediated immunity.
Chromium	2.5 mg Cr-yeast/kg diet for 198 d	29.3□ and 71% relative humidity	New Zealand White male rabbits	· Improved final body weight, body weight gain, and feed conversion ratio. · Enhanced serum testosterone level. · Improved advanced-sperm motility, live spermatozoa, and morphologically normal spermatozoa.

Probiotics	5 × 10^6 CFU *Clostridium butyricum*, 2 × 10^8 CFU *Enterococcus faecium*, or 2.5 × 10^6 CFU *C. butyricum* + 1 × 10^8 CFU *E. faecium*/kg diet from 35 to 91 d of age	31.78□ and 60.19% relative humidity	New Zealand White male growing rabbits	· Enhanced final body weight, body weight gain, and feed conversion ratio. · Improved serum concentrations of total protein, globulin, HDL-cholesterol, complement component 3, and lysozyme activity. · Improved cecal fermentation (cecal concentrations of total volatile fatty acids and butyric acids). · Improved duodenal histomorphometry indices (higher villus height, thicker muscular layer thickness, lower crypt depth, and lower villus width).
Probiotics	3 × 10^9 CFU *Saccharomyces cerevisiae*/kg or 3 × 10^9 CFU *Lactobacillus acidophilus*/ kg from 5 to 13 week of age	33□ and 81% relative humidity	New Zealand White growing rabbits	· Enhanced final body weight, body weight gain, and feed conversion ratio. · Supported the beneficial yeast species (*Yarrowia lipolytica*). · Reduced pathogenic bacteria (*Salmonella* spp., *C.* spp., and *Enterobacteria* spp.).
Prebiotics	0.3% mannan-oligosaccharides or 0.05% isomalto-oligosaccharide from 6 to 16 wk of age	19□	New Zealand White male growing rabbits	· Improved final body weight and feed intake. · Enhanced carcass criteria (dressing % and giblets %). · Increased duodenal villi area and length. · Increased spleen white bulb area and length. · Elevated liver GSH-Px, insulin-like growth factor 1, SOD gene expression, and muscle insulin-like growth factor one receptor expression. · Reduced spleen IL-6 expression gene.

Prebiotics	3 g Bio-Mos®/ kg diet from 4 to 12 week of age	31.50□ and 79.07% relative humidity	New Zealand White male growing rabbits	· Improved final body weight, body weight gain, and feed conversion ratio. · Increased serum levels of hemoglobin, total protein, globulin, and albumin. · Decreased respiration rate, rectal temperature, and heart rate.
Moringa oleifera	200 mg *Moringa oleifera* leaves powder/kg BW daily for four wk (from 32 to 32 wk of age)	35□ and 80% relative humidity	New Zealand White male rabbits	· Decreased rectal temperature and respiration rate. · Enhanced intestinal histomorphology. · Down-regulated the mRNA expression of TNF-α, HSP-A2, GSH-Px, and IL-1α. · Elevated IL-6 expression. · Decreased jejunal contents of lipopolysaccharide, pro-inflammatory cytokines, and myeloperoxidase. · Elevated mRNA expression of tight junction proteins. · Reduced MDA concentration in jejunal mucosa. · Modulated intestinal microbiota composition positively. · Enhanced mucosal immunity (down-regulation of gene expression of TNFRSF13C, LBP, and COX2, while up-regulation of the gene expression of protein digestion and absorption pathway, including PRSS2, LOC100349163, CPA1, CPB1, SLC9A3, SLC1A1, and SLC7A9).

Moringa oleifera	50 mg *Moringa oleifera* leaves ethanolic extract/kg BW for 12 consecutive weeks	31.11□ and 87% relative humidity	V-line rabbit bucks	· Reduced rectal temperature. · Increased serum total antioxidant capacity concentration. · Increased sperm concentration, sperm progressive motility, sperm viability, sperm normal morphology, intact acrosome sperm, and sperm with the integrated cell membrane.
Ginger	7.5 g ginger powder/kg diet from 5 to 13 week of age	33□ and 74.5% relative humidity	APRI growing rabbits	· Enhanced final body weight and feed conversion ratio. · Increased carcass weight. · Enhanced digestibility of dry matter, organic matter, crude protein, and nitrogen-free extract. · Minimized serum levels of triglycerides, total cholesterol, HDL-cholesterol, and LDL-cholesterol. · Increased plasma concentration of total antioxidant capacity. · Reduced plasma concentration of MDA.
Ginger	250 mg ginger/doe/d for eight week	35□ and 80% relative humidity	New Zealand White virgin female rabbits	· Decreased rectal temperature, skin temperature, ear temperature, and respiration rate. · Improved body weight and feed intake. · Enhanced the reproductive traits (conception rate, litter size, and litter weight at birthing and weaning). · Increased plasma concentrations of progesterone, T3, T4, total protein, albumin, and globulin. · Decreased plasma concentrations of cortisol, urea, and creatinine.

Thyme	16 g thyme/kg diet for 90 d	39□ and 30–35% relative humidity	New Zealand White male rabbits	· Enhanced body weight gain, feed intake, and feed conversion ratio. · Increased the ejaculate volume, sperm viability, and sperm motility. · Reduced serum concentrations of serum ALT, AST, urea, and creatinine.
Thyme	100 thyme essential oil mg/kg diet from six to nine months of age	33□ and 80% relative humidity	APRI rabbit does	· Improved reproductive performance (milk production, ovulation rate, total litter size, live litter size at birth, litter size at weaning, viability rate at birth, and viability rate at weaning). · Enhanced antioxidative status (increased TAC and GSH). · Decreased serum lipid peroxidation (MDA concentration). · Improved serum concentrations of immunoglobulins (IgG and IgM).
Turmeric	2.5 g turmeric nanoparticles/kg diet from 5 to 13 wk of age	32.77□ and 43.23% relative humidity	APRI growing rabbits	· Improved final body weight, body weight gain, and feed conversion ratio. · Reduced serum concentrations of total lipids, total cholesterol, triglycerides, and LDL-cholesterol, while increasing HDL-cholesterol. · Increased serum immunoglobulins. · Improved antioxidative status (serum concentrations of TAC, SOD, GSH-Px, and GSH). · Reduced liver lipid peroxidation (MDA concentration).

Maca (*Lepidium meyenii*)	400 or 600 mg maca extract/ head twice weekly	33□ and 80% relative humidity	V-line rabbit bucks	· Enhanced live body weight, feed intake, and feed conversion ratio · Reduced serum concentration of total bilirubin, creatinine, glucose, cholesterol, and triglycerides. · Enhanced cortisol and testosterone levels. · Improved antioxidative status (serum concentration of TAC and SOD). · Decreased serum lipid peroxidation (MDA concentration). · Improved semen quality parameters.
Pumpkin	2 ml pumpkin seed essential oil/kg diet from 5 to 13 wk of age	38□ and 60% relative humidity	New Zealand White male growing rabbits	· Improved body weight, weight gain, feed intake, and conversion ratio. · Increased serum levels of total protein, albumin, and globulin. · Lowered plasma total bilirubin, ALT, AST, creatinine, uric acid, glucose, cortisol, T3, and corticosterone concentrations. · Reduced anti-inflammation agent (exhibiting negative caspase-3 immunoreactivity surrounding portal tract).

Source: Ebeid TA, Aljabeili HS, Al-Homidan IH, Volek Z, Barakat H. Ramifications of Heat Stress on Rabbit Production and Role of Nutraceuticals in Alleviating Its Negative Impacts: An Updated Review. Antioxidants. 2023; 12(7):1407. https://doi.org/10.3390/antiox12071407.

REFERENCES

1. Akerstedt, J. (2003). Humoral immune response in adult blue foxes (*Alopex lagopus*) after oral infection with *Encephalitozoon cuniculi* spores. *Veterinary Parasitology, 113*, 203–210.
2. Akerstedt, J., & Kapel, C. M. (2003). Survey for *Encephalitozoon cuniculi* in arctic foxes (*Alopex lagopus*) in Greenland. *Journal of Wildlife Diseases, 39*, 228–232.
3. Akerstedt, J., Nordstoga, K., & Mathis, A. (2002). Fox encephalitozoonosis: Isolation of the agent from an outbreak in farmed blue foxes (*Alopex lagopus*) in Finland and some hitherto unreported pathologic lesions. *Journal of Veterinary Medicine B: Infectious Diseases and Veterinary Public Health, 49*, 400–405.
4. Allan, J. C., Craig, P. S., & Sherington, J. (1999). Helminth parasites of the wild rabbit *Oryctolagus cuniculus* near Malham Tarn, Yorkshire, UK. (Abstract) *Journal of Helminthology, 73*, 289–294.
5. Ansbacher, L., Nichols, M. F., & Hahn, A. W. (1988). The influence of *Encephalitozoon cuniculi* on neural tissue responses to implanted biomaterials in the rabbit. *Laboratory Animal Science, 38*, 689–695.
6. Ashmawy, K. I., Abuakkada, S. S., & Awad, A. M. (2011). Seroprevalence of antibodies to *Encephalitozoon cuniculi* and *Toxoplasma gondii* in farmed domestic rabbits in Egypt. *Zoonoses and Public Health, 58*, 357–364.
7. Baneux, P. J., & Pognan, F. (2003). In utero transmission of *Encephalitozoon cuniculi* strain type I in rabbits. *Laboratory Animals, 37*, 132–138.
8. Barcena, J., Morales, M., & Vazquez, B. (2000). Horizontal transmissible protection against myxomatosis and rabbit haemorrhagic disease by using a recombinant myxoma virus. *Journal of Virology, 74*, 1114–1123.
9. Beard, P. M., Henderson, D., & Daniels, M. J. (1999). Evidence of paratuberculosis in fox (*Vulpes vulpes*) and stoat (*Mustela erminea*). *Veterinary Record, 145*, 612–613.
10. Besch-Williford, C. (1997). Rabbit medicine and procedures for practitioners. Program and abstracts. Tyzzer's disease in rabbits (pp. 113–117). House Rabbit Society Veterinary Conference, USA.
11. Boot, R., Hansen, A. K., & Hansen, C. K. (2000). Comparison of assays for antibodies to *Encephalitozoon cuniculi* in rabbits. *Laboratory Animals, 34*, 281–289.
12. Boot, R., Thuis, H., & Wieten, G. (1995). Multifactorial analysis of antibiotic sensitivity of *Bordetella bronchiseptica* isolates from guinea pigs, rabbits, and rats. *Laboratory Animals, 29*, 45–49.

13. Bull, C. G. (1917). The pathologic effects of streptococci from cases of poliomyelitis and other sources. *Journal of Experimental Medicine, 25*, 557–580.

14. Capucci, L., Nardin, A., & Lavazza, A. (1997). Seroconversion in an industrial unit of rabbits infected with a nonpathogenic rabbit haemorrhagic disease-like virus. *Veterinary Record, 140*, 647–650.

15. Chasey, D. (1997). Rabbit haemorrhagic disease: The new scourge of *Oryctolagus cuniculus. Laboratory Animals, 31*, 33–44.

16. Chasey, D., Lucas, M. H., & Westcott, D. G. (1995). Development of a diagnostic approach to the identification of rabbit haemorrhagic disease. *Veterinary Record, 137*, 158–160.

17. Cox, J. C. (1977). Altered immune responsiveness associated with *Encephalitozoon cuniculi* infection in rabbits. *Infection and Immunity, 15*, 392–395.

18. Cox, J. C., & Ross, J. (1980). A serological survey of *Encephalitozoon cuniculi* infection in the wild rabbit in England and Scotland. *Research in Veterinary Science, 28*, 396.

19. Cox, J. C., Edmonds, J. W., & Shepherd, R. C. (1981). Toxoplasmosis and the wild rabbit *Oryctolagus cuniculus* in Victoria, Australia with suggested mechanisms for dissemination of oocysts. (Abstract) *Journal of Hygiene, 87*, 331–337.

20. Cox, J. C., Hamilton, R. C., & Attwood, H. D. (1979). An investigation of the route and progression of *Encephalitozoon cuniculi* infection in adult rabbits. *Journal of Protozoology, 26*, 260–265.

21. Cox, J. C., Pye, D., Edmonds, J. W., & Shepherd, R. (1980). An investigation of *Encephalitozoon cuniculi* in the rabbit *Oryctolagus cuniculus* in Victoria, Australia. *Journal of Hygiene (London), 84*, 295–300.

22. Dannatt, L., Gunning, R., & Higgins, R. (2000). Attaching and effacing *E. coli* in rabbits. *Veterinary Record, 147*, 524.

23. De Groote, M. A., Visvesvara, G., & Wilson, M. L. (1995). Polymerase chain reaction and culture confirmation of disseminated *Encephalitozoon cuniculi* in a patient with AIDS: Successful therapy with albendazole. *Journal of Infectious Diseases, 171*, 1375–1378.

24. Deeb, B. (1993). Update for veterinary practitioners on pasteurellosis in rabbits. *Journal of Small Exotic Animal Medicine, 2*, 112–113.

25. Delong, D., & Manning, P. J. (1994). Bacterial diseases. In P. J. Manning, D. H. Ringler, & C. E. Newcomer (Eds.), *The Biology of the Laboratory Rabbit* (2nd ed., pp. 131–170). Academic Press.

26. Deplazes, P., Mathis, A., & Baumgartner, R. (1996). Immunological and

molecular characteristics of *Encephalitozoon*-like microsporidia isolated from humans and rabbits indicate that *Encephalitozoon cuniculi* is a zoonotic parasite. (Abstract) *Clinical Infectious Diseases, 22*, 557–559.

27. Didier, E. S. (2000). Immunology of microsporidiosis. In F. Petry (Ed.), *Cryptosporidiosis and Microsporidiosis* (Vol. 6, pp. 193–208). Karger.
28. Didier, E. S., Didier, P. J., Snowden, K. F., & Shadduck, J. A. (2000). Microsporidiosis in mammals. *Microbes and Infection, 2*, 709–720.
29. Didier, E. S., Vossbrinck, C. R., & Baker, M. D. (1995). Identification and characterization of three *Encephalitozoon cuniculi* strains. *Parasitology, 111*, 411–421.
30. DiGiacomo, R. F., & Deeb, B. J. (1989). Pasteurellosis in rabbits–sequelae and vaccination. *Journal of Applied Rabbit Research, 12*, 10–13.
31. DiGiacomo, R. F., & Mare, J. (1994). Viral diseases. In P. J. Manning, D. H. Ringler, & C. E. Newcomer (Eds.), *The Biology of the Laboratory Rabbit* (2nd ed., pp. 171–197). Academic Press.
32. DiGiacomo, R. F., Deeb, B. J., & Bernard, B. L. (1987). Safety and efficacy of a streptomycin-dependent live *Pasteurella multocida* vaccine in rabbits. *Laboratory Animal Science, 37*, 187–190.
33. DiGiacomo, R. F., Garlinghouse, L. E. Jr., & Van Hoosier Jr, G. L. (1983). Natural history of infection with *Pasteurella multocida* in rabbits. *Journal of the American Veterinary Medical Association, 183*, 1172–1175.
34. Donnelly, T. M. (1995). Emerging viral diseases of rabbits and rodents: Viral haemorrhagic disease and hantavirus. *Seminars in Avian and Exotic Pet Medicine, 4*, 83–91.
35. Ewringmann, A., & Göbel, T. (1999). Untersuchungen zur Klinik und Therapie der Encephalitozoonose beim Heimtierkaninchen (Article in German, English summary). *Klientierpraxis, 44*, 357–372.
36. Fenner, F., & Fantani, B. (1999). *Biological Control of Vertebrate Pests: The History of Myxomatosis–An Experiment in Evolution.* CABI Publishing.
37. Fenner, F., & Ross, J. (1994). Myxomatosis. In H. V. Thompson & C. M. King (Eds.), *The European Rabbit: The History and Biology of a Successful Coloniser* (pp. 205–235). Oxford University Press.
38. Franzen, C., Müller, A., & Hartmann, P. (1999). Polymerase chain reaction for diagnosis and species differentiation of microsporidia. *Folia Parasitologica (Praha), 45*, 140–148.
39. Fuentealba, I. C., Mahoney, N. T., & Shadduck, J. A. (1992). Hepatic lesions in rabbits infected with *Encephalitozoon cuniculi* per rectum. *Veterinary Pathology, 29*, 536–540.

40. Gannon, J. (1980). The course of infection of *Encephalitozoon cuniculi* in immunodeficient and immunocompetent mice. *Laboratory Animals, 14*, 189–192.

41. Gavier-Widén, D. (1996). Viral hepatitis of rabbits and hares in Scandinavia. In *Zoo and Wildlife Medicine* (pp. 322–325). W.B. Saunders.

42. Greig, A., Stevenson, K., & Henderson, D. (1999). Epidemiological study of paratuberculosis in wild rabbits in Scotland. *Journal of Clinical Microbiology, 37*, 1746–1751.

43. Greig, A., Stevenson, K., & Perez, V. (1997). Paratuberculosis in wild rabbits (*Oryctolagus cuniculus*). *Veterinary Record, 140*, 141–143.

44. Gustaffson, K., Wattrang, E., & Fossum, C. (1997). *Toxoplasma gondii* infection in the mountain hare (*Lepus timidus*) and domestic rabbit (*Oryctolagus cuniculus*). *Journal of Comparative Pathology, 117*, 351–368.

45. Harcourt, R. A. (1967). Toxoplasmosis in rabbits. *Veterinary Record, 81*, 91–92.

46. Harcourt-Brown, F. M., & Holloway, H. K. R. (2003). *Encephalitozoon cuniculi* in pet rabbits. *Veterinary Record, 152*, 427–431.

47. Hersteinsson, P., Gunnarsson, E., Hjartardóttir, S., & Skirnisson, K. (1993). Prevalence of *Encephalitozoon cuniculi* antibodies in terrestrial animals in Iceland, 1986 to 1989. *Journal of Wildlife Diseases, 29*, 341–344.

48. Hofing, G. L., & Kraus, A. L. (1994). Arthropod and helminth parasites. In P. J. Newcomer, D. H. Ringler, & C. E. Newcomer (Eds.), *The Biology of the Laboratory Rabbit* (2nd ed., pp. 231–258). Academic Press.

49. Hollister, W. S., Canning, E. U., & Viney, M. (1989). Prevalence of antibodies to *Encephalitozoon cuniculi* in stray dogs as determined by an ELISA. *Veterinary Record, 124*, 332–336.

50. Hollister, W. S., Canning, E. U., & Willcox, A. (1991). Evidence for widespread occurrence of antibodies to *Encephalitozoon cuniculi* (Microspora) in humans provided by ELISA and other serological tests. *Parasitology, 102*(Pt 1), 33–43.

51. Jaglic, Z., Jeklova, E., & Leva, L. (2008). Experimental study of pathogenicity of *Pasteurella multocida* serogroup F in rabbits. *Veterinary Microbiology, 126*, 168–177.

52. Jeklova, E., Jekl, V., & Kovarcik, K. (2010). Usefulness of detection of specific IgM and IgG antibodies for diagnosis of clinical encephalitozoonosis in pet rabbits. *Veterinary Parasitology, 170*, 143–148.

53. Joste, N. E., Rich, J. D., Busam, K. J., & Schwartz, D. A. (1996). Autopsy verification of *Encephalitozoonosis intestinalis* (microsporidiosis)

eradication following albendazole therapy. (Abstract) *Archives of Pathology and Laboratory Medicine, 120*, 199–203.

54. Jungherr, E. (1955). *Encephalitozoon* encephalomyelitis in a rabbit. *Journal of the American Veterinary Medical Association, 127*, 518.
55. Keeble, E. J. (2011). Encephalitozoonosis in rabbits–what we do and don't know. *In Practice, 33*, 426–435.
56. Keeble, E. J., & Shaw, D. J. (2006). Seroprevalence of antibodies to *Encephalitozoon cuniculi* in domestic rabbits in the United Kingdom. *Veterinary Record, 158*, 539–544.
57. Kimman, T. G., & Akkermans, J. P. (1987). *Encephalitozoon cuniculi* in a rabbit-breeding colony. (Article in Dutch, English abstract) *Tijdschrift voor Diergeneeskunde, 112*, 1405–1409.
58. Kraus, A., Weisbroth, S. H., Flatt, R. E., & Brewer, N. (1984). *Laboratory Animal Medicine*. Academic Press. Biology and diseases of rabbits (pp. 207–237).
59. Kunstyr, I., & Naumann, S. (1985). Head tilt in rabbits caused by pasteurellosis and encephalitozoonosis. *Laboratory Animals, 19*, 208–213.
60. Kunstyr, I., Lev, L., & Naumann, S. (1986). Humoral antibody response in rabbits to experimental infection with *Encephalitozoon cuniculi. Veterinary Parasitology, 21*, 223–232.
61. Leibenberg, S. P., & Badger, V. M. (1984). Suppurative osteomyelitis in the foot of a rabbit. *Journal of the American Veterinary Medical Association, 185*, 1382.
62. Leland, M. M., Hubbard, C. B., & Dupey, J. P. (1992). Clinical toxoplasmosis in domestic rabbits. *Laboratory Animal Science, 42*, 318–319.
63. Lelkes, L., & Corbett, M. J. (1983). A preliminary study of the transmission of *Pasteurella multocida* in rabbits. *Journal of Applied Rabbit Research, 6*, 125–126.
64. Levkut, M., Horvath, M., & Balent, P. (1997). Catecholamines and encephalitozoonosis in rabbits. *Veterinary Parasitology, 73*, 173–176.
65. Li, T., Qiao, G. L., & Hu, G. Z. (1995). Comparative plasma and tissue pharmacokinetics and drug residue profiles of different chemotherapeutants in fowls and rabbits. *Journal of Veterinary Pharmacology and Therapeutics, 18*, 260–273.
66. Lumeij, J. T. (1997). Disease risks with translocations of rabbits and hares into, out of, and within Europe. *Journal of the British Veterinary Zoological Society, 2*, 19–25.

67. Lyngset, A. (1980). A survey of serum antibodies to *Encephalitozoon cuniculi* in breeding rabbits and their young. *Laboratory Animal Science, 30*, 558–561.

68. Manning, P. J., DiGiacomo, R. F., & DeLong, D. (1989). Pasteurellosis in laboratory animals. In C. Adlam & J. M. Rutter (Eds.), *Pasteurella and Pasteurellosis* (pp. 264–289). Academic Press.

69. Martino, P. E., Montenegro, J. L., & Preziosi, J. A. (2004). Serological survey of selected pathogens of free-ranging foxes in southern Argentina, 1998–2001. *Revue Scientifique et Technique, 23*, 801–806.

70. Morrisey, J. K. (1996). Parasites of ferrets, rabbits, and rodents. *Seminars in Avian and Exotic Pet Medicine, 5*, 106–114.

71. Müller-Doblies, U. U., Herzog, K., & Tanner, I. (2002). First isolation and characterisation of *Encephalitozoon cuniculi* from a free-ranging rat (*Rattus norvegicus*). *Veterinary Parasitology, 107*, 279–285.

72. Okerman, L. (1988). *Diseases of Domestic Rabbits*. Blackwell Scientific Publications.

73. Owen, D. G. (1992). *Parasites of Laboratory Animals*. Laboratory Animal Handbooks No 12. Royal Society of Medicine Services.

74. Pakes, S. P., & Gerrity, L. W. (1994). Protozoal diseases. In P. J. Manning, D. H. Ringler, & C. E. Newcomer (Eds.), *The Biology of the Laboratory Rabbit* (2nd ed., pp. 205–224). Academic Press.

75. Pakes, S., Shadduck, J., & Olsen, R. (1972). A diagnostic skin test for encephalitozoonosis (nosematosis) in rabbits. *Laboratory Animal Science, 22*, 870–877.

76. Pasetti, M. F., Cuberos, L., & Horn, T. L. (2008). An improved *Francisella tularensis* live vaccine strain (LVS) is well tolerated and highly immunogenic when administered to rabbits in escalating doses using various immunization routes. *Vaccine, 26*, 1773–1785.

77. Pattison, M., Clegg, F. G., & Duncan, A. L. (1971). An outbreak of encephalomyelitis in broiler rabbits caused by *Nosema cuniculi*. *Veterinary Record, 88*, 404–405.

78. Percy, D. H., & Barthold, S. W. (1993). Rabbit. In *Pathology of Laboratory Rodents and Rabbits* (pp. 179–223). Iowa State University Press.

79. Percy, D. H., & Barthold, S. W. (2001). *Pathology of Laboratory Rodents and Rabbits* (2nd ed.). Wiley.

80. Petersen-Jones, S. M., & Carrington, S. D. (1988). *Pasteurella dacryocystitis* in rabbits. *Veterinary Record, 122*, 514–515.

81. Poonacha, K. B., & Stamper, R. D. (1985). Encephalitozoonosis in a parrot. *Journal of the American Veterinary Medical Association, 186*, 700–702.

82. Pritt, S., Henderson, K. S., & Shek, W. R. (2010). Evaluation of available diagnostic methods for *Clostridium piliforme* in laboratory rabbits (*Oryctolagus cuniculus*). *Laboratory Animals, 44*, 14–19.

83. Pye, D., & Cox, J. C. (1977). Isolation of *Encephalitozoon cuniculi* from urine samples. *Laboratory Animals, 11*, 223–224.

84. Robinson, A. J., Muller, W. J., Braid, A. L., & Kerr, P. J. (1999). The effect of buprenorphine on the course of disease in laboratory rabbits infected with myxoma virus. *Laboratory Animals, 33*, 252–257.

85. Sanchez, S., Mizan, S., Ritchie, B. W., & Lee, M. D. (2000). Pasteurellosis in rabbits. *Compendium on Continuing Education, 22*, 344–350.

86. Sobottka, I., Albrecht, H., & Schafer, H. (1995). Disseminated *Encephalitozoon* (Septata) intestinal infection in a patient with AIDS: Novel diagnostic approaches and autopsy-confirmed parasitological cure following treatment with albendazole. *Journal of Clinical Microbiology, 33*, 2948–2952.

87. Sobottka, I., Iglauer, F., & Schüler, T. (2001). Acute and long-term humoral immunity following active immunization of rabbits with inactivated spores of various *Encephalitozoon* species. *Parasite Research, 87*, 1–6.

88. Spibey, N., McCabe, V. J., & Greenwood, N. M. (2012). Novel bivalent vectored vaccine for control of myxomatosis and rabbit haemorrhagic disease. *Veterinary Record, 170*, 309.

89. Stewart, C. G., Reyers, F., & Snyman, H. (1988). The relationship in dogs between primary renal disease and antibodies to *Encephalitozoon cuniculi*. *Journal of the South African Veterinary Association, 59*, 19–21.

90. Stiles, J., Didier, E., & Ritchie, B. (1997). *Encephalitozoon cuniculi* in the lens of a rabbit with phacoclastic uveitis: Confirmation and treatment. *Veterinary Comparative Ophthalmology, 7*, 233–238.

91. Suckow, M. A., Haab, R. W., Miloscio, L. J., & Guilloud, N. B. (2008). Field trial of a *Pasteurella multocida* extract vaccine in rabbits. *Journal of the American Association for Laboratory Animal Science, 47*, 18–21.

92. Suter, C., Müller-Doblies, U. U., Hatt, J. M., & Deplazes, P. (2001). Prevention and treatment of *Encephalitozoon cuniculi* in rabbits with fenbendazole. *Veterinary Record, 148*, 478–480.

93. Telford, S. R., & Spielman, A. (1989). Enzootic transmission of the agent of Lyme disease in rabbits. *American Journal of Tropical Medicine and Hygiene, 41*, 482–490.

94. Thomas, C., Finn, M., & Twigg, L. (1997). Microsporidia (*Encephalitozoon cuniculi*) in wild rabbits in Australia. *Australian Veterinary Journal, 75*, 808–810.

95. Vangeel, I., Pasmans, F., & Vanrobaeys, M. (2000). Prevalence of dermatophytes in asymptomatic guinea pigs and rabbits. *Veterinary Record, 146*, 440–441.

96. Waller, T. (1979). Sensitivity of *Encephalitozoon cuniculi* to various temperatures, disinfectants and drugs. *Laboratory Animals, 13*, 227–230.

97. Wasson, K., & Peper, R. L. (2000). Mammalian microsporidiosis. *Veterinary Pathology, 37*, 113–128.

98. Weber, R., Deplazes, P., & Flepp, M. (1997). Cerebral microsporidiosis due to *Encephalitozoon cuniculi* in a patient with human immunodeficiency virus infection. *New England Journal of Medicine, 336*, 474–478.

99. Weitzel, T., Wolff, M., & Dabanch, J. (2001). Dual microsporidial infection with *Encephalitozoon cuniculi* and *Enterocytozoon bieneusi* in an HIV-positive patient. *Infection, 29*, 237–239.

100. Whittaker, D. (1989). Pasteurellosis in the laboratory rabbit: A review. *Veterinary Annals, 29*, 285–291.

101. Wills, J. (2001). Coenurosis in a pet rabbit. *Veterinary Record, 148*, 188.

102. Wilson, J. M. (1979). The biology of *Encephalitozoon cuniculi. Medical Biology, 57*, 84–101.

103. Wood, C. (1978). The pet rabbit–veterinary problems. *Veterinary Record, 102*, 304–308.

104. Wright, J. H., & Craighead, E. M. (1922). Infectious motor paralysis in young rabbits. *Journal of Experimental Medicine, 36*, 135–140.

CHAPTER 8

INFECTIOUS DISEASES OF FERRETS

Contents

Introduction 346

8.1. Overview of Ferrets 346

8.2. Infectious Diseases of Ferrets 358

8.3. Neoplasia of Ferrets 364

8.4. Hormonal Disorders of Ferrets 369

References 373

INTRODUCTION

Infectious diseases in ferrets pose significant health challenges, particularly because ferrets are susceptible to a range of pathogens, including viruses, bacteria, and parasites. As popular pets, their close contact with humans and other animals increases the risk of disease transmission, making prevention and control critical for their health.

Viral infections are among the most serious threats to ferrets. Canine distemper virus (CDV) is the most feared, as it is almost always fatal in ferrets. CDV causes symptoms such as fever, nasal discharge, and neurological issues, and there is no effective treatment once the disease takes hold. Vaccination is the primary means of prevention. Ferrets are also susceptible to influenza viruses, particularly human influenza, which can cause respiratory symptoms similar to those seen in humans. While not usually fatal, flu can be severe in ferrets, and transmission from humans to ferrets is a common concern.

Bacterial infections can also significantly impact ferret health. *Helicobacter mustelae* is a common bacterium in ferrets, leading to gastrointestinal ulcers and gastritis. If untreated, these conditions can result in severe discomfort and even death. Ferrets are also prone to infections by *Mycobacterium* spp., which can cause tuberculosis, a serious and often fatal disease.

Parasitic infections, while less common, can occur, with ectoparasites like fleas and ear mites being the most frequent. Endoparasites, such as gastrointestinal worms, can also infect ferrets, leading to symptoms like diarrhea and weight loss.

Preventing infectious diseases in ferrets involves regular veterinary care, vaccination, and maintaining good hygiene and environmental controls. Quarantine measures for new or sick ferrets, proper nutrition, and minimizing stress are also vital. Understanding the risks and early signs of disease allows for prompt treatment, reducing the potential for severe outcomes.

8.1. OVERVIEW OF FERRETS

The domestic ferret, *Mustela putorius* furo, belongs to the family Mustelidae and order Carnivora. It has been kept in captivity for more than 2000 years. Although they were once mostly kept as pets, ferrets are now kept as pets all over the world. They are also utilized as study animals, frequently in respiratory system investigations, specifically those involving influenza virus infections, and as a model for *Helicobacter* sp. infections.

Ferrets are slender and have a body shape that is typical of mustelids. Their pelage comes in a variety of colors, but some of the most popular ones are sable, chocolate, cinnamon, and white with pink eyes. It is typical to have white patches on the toes or throat. Seasonal variations in coat color and density cause it to darken and thicken in the winter and lighten and shed in the summer.

Figure. 8.1. *A pet. This is the basic (wild type) color of the ferret, but many color variations exist.*

Source: By Alfredo Gutiérrez-Own work, CC BY-SA 4.0, https://commons.wikimedia.org/w/index.php?curid=4478158

At six months of age, ferrets typically reach their full size, and at one year old, they are deemed adults. Although ferrets typically live six to 10 years, many of them have shorter lifespans as a result of diseases or accidents.

When intact, male ferrets, also known as hobs, can weigh up to four pounds (2 kilograms), while neutered ferrets typically weigh 2.5 pounds (1.2 kg). When intact, females can weigh up to 2.5 pounds (1.2 kg), and after neutering, their average weight drops to 1.75 pounds (0.8 kg). In addition to having a larger, rounder head and more muscle, males also typically have a narrower, daintier head with a thin, pointed nose, compared to females.

8.1.1. Description and Physical Characteristics of Ferrets

Ferrets and cats are similar in appearance. A male ferret, known as a hob, has a maximum weight of four pounds (2 kg). A jill, or female, may weigh as much as 2.5 pounds, or 1.2 kilograms. Ferrets can be any color, but common hues include sable, chocolate, white with pink eyes, and cinnamon. It is typical to have white patches on the toes or throat. The color and density of a ferret's coat vary with the season; in the winter, it gets darker and thicker, and in the summer, it gets lighter and sheds. Ferrets are regarded as adults after a year of age, and they typically reach full size at five to six months. Although most individuals live between six and 10 years, many people's lives are cut short by illnesses or accidents (Figure 8.2).

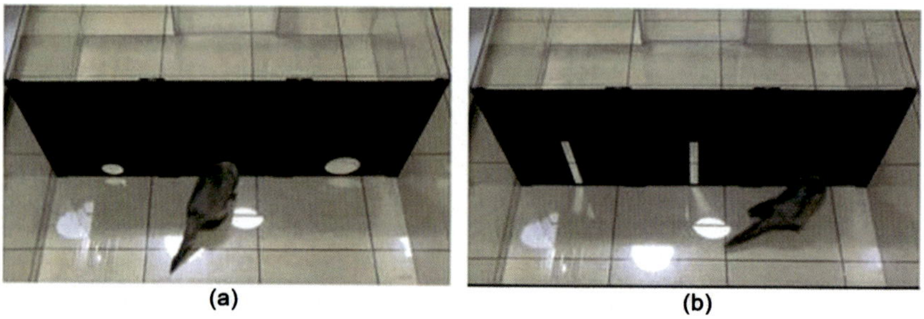

Figure 8.2. *Ferrets can detect dimensions.*

Source: Khvatov IA, Sokolov AY, Kharitonov AN. Ferrets (Mustela furo) are aware of their dimensions. Animals. *2023; 13(3):444. https://doi.org/10.3390/ani13030444.*

Early removal of the gonads from ferrets reduces their musky smell, which is typical of this family of animals. Typically, this also includes the removal of the anal scent glands. A lot of ferrets undergo these procedures before being purchased. If not, it's critical to spay a female ferret before her first heat cycle to avoid potentially fatal outcomes like the onset of adrenal gland disease, which is not the same as hyperadrenocorticism (an excess of cortisol in other species) but rather an overproduction of sex hormones like estrogen.

Due to their increased erythrocyte production, many adult ferrets have large spleens. A veterinarian should be consulted to ensure that an enlarged spleen is not causing any problems for a ferret, even if it does not exhibit any symptoms of illness.

8.1.2. Special Considerations for Ferrets

Ferrets are prone to stumbling into hazardous situations. They can fit through extremely tiny spaces and are incredibly persistent, fearless, and extremely curious. Some characteristics may cause fatalities or severe injuries. Additionally, ferrets enjoy chewing on soft or plastic items from the house, like foam, pencil erasers, rubber bands, buttons, and other items. These items may cause intestinal blockages or present a choking hazard. An intestinal blockage typically necessitates emergency surgery and can be fatal. It's imperative to "ferret-proof" your house before bringing in a ferret or ferrets. Seal all openings bigger than one inch in ductwork, walls, floors, and appliances like washers, refrigerators, stoves, and televisions. Instead of using tape, which ferrets can remove, use wire mesh. Stow away anything that could be consumed. Remember to keep toxic items like houseplants in mind. Lock cabinets, drawers, and doors, but keep in mind that a ferret may still get past childproof locks. They have been observed to untwist bottle caps and open zippers thanks to their exceptional object manipulation skills with the front paws. Lower yourself to a ferret's eye level and search for any tiny openings behind furniture and around fixtures like pipes and radiators, particularly those that taper and narrow, as these could be potential places for a ferret to become lost, suffocated, or stuck. Sleeper sofas and recliners are dangerous as well. Ferrets ought to be confined when proper supervision is not available for them (Figure 8.3).

	African Green Monkey	Ferret	Syrian Golden Hamster	Mouse
Pathogenesis	+++	++	++	++*
Treatment	+++	++	++	+**
Vaccine	+++	++	++	+**
Transmission	++	+++	++	+
Immune Response	+++	+	+	++**
Neuroinfection	++	++	++	+++*
Cost/Husbandry	+	++	++	+++

+++ highly suited, ++ moderately suited, + poorly suited, *immunocompetent, **immunocompromised

Figure 8.3. *The utility in virology.*

Source: Pigeaud DD, Geisbert TW, Woolsey C. Animal Models for Henipavirus Research. Viruses. *2023; 15(10):1980. https://doi.org/10.3390/v15101980.*

Ferrets have also been stepped on under rugs and sat on in couches due to the "tunneling" instinct. To sate the tunneling instinct, ferrets may benefit from prefabricated enclosures. Think about rugs, huge piping, blankets, and flower pots. Use of objects that could break or get chewed up is not advised.

All states in the union allow the pet ownership of ferrets, with the exception of California and Hawaii. Nonetheless, some cities forbid the ownership of ferrets, including New York City. Before buying a ferret, make sure to check with your local department of fish and game or wildlife.

Families with small children should receive extra consideration. Ferrets bite and scratch to protect themselves from harsh handling as well as during play. Harsh treatment can also cause harm to the ferret.

8.1.2.1. Biting

Ferrets bite as part of their natural behavior. They might bite to play or to attract attention. This is especially true for young male ferrets, ages three to four months, who have lived most of their lives in a pet store with other ferrets and little exposure to people. The rough play that ferrets engage in with other ferrets may also extend to human interaction. Put your ferret in its cage for a little "time-out" if it starts biting at you. If this behavior is repeated, the animal will come to understand that biting is inappropriate.

8.1.3. Providing a Home for a Ferret

Because they don't take up much room, ferrets are great in apartments and other small spaces. However, when ferrets are left alone or unattended, proper housing needs to be supplied. To stay healthy, ferrets also need a high-quality diet, proper grooming, and the right amount of exercise and care.

8.1.3.1. Housing

A ferret can be kept in a wire cage that is at least three feet wide, three feet deep, and two feet high, providing it with plenty of space to roam around. Examine the cage for any openings big enough for the ferret to escape, as well as any sharp, jagged, or rough wire. Either place a cushion on the cage's bottom or get one with a solid bottom. An unprotected ferret's paw pads could get damaged.

The majority of ferrets like to sleep in cramped, small spaces. Hammocks and fabric huts are popular sleeping arrangements for ferrets. A fresh towel or tiny blanket works great for bedding; however, if you use fabric, make sure the ferret won't eat it or unravel it. Don't use wood chips or news chapters. These things may produce dust or harbor bacteria that irritates the respiratory system. Oil released by cedar trees may be hazardous to ferrets.

There should be areas for feeding and watering inside the cage. Use sturdy containers as ferrets frequently tip over bowls of food and water. Certain kinds of food bowls and water bottles can be fastened to the cage's side.

It is possible to teach a ferret to use a litter box. They favor using the same location for both urinating and defecating, away from sleeping and eating areas. Additionally, ferrets use their hindquarters as a means of self-cleansing after urinating or defecating. This is typical ferret behavior and not an indication of a disease.

You can use cat litter and a reduced-size litter box. On the other hand, stay away from using clumping cat litter since it can lead to respiratory issues in ferrets. The ferret won't use the litter box if it is placed too close to its sleeping or eating quarters in the cage.

Avoid putting the cage in front of windows and keep it away from moisture and drafts. Due to the lack of sweat glands, ferrets are susceptible to heat and moisture. Typically, a comfortable temperature range is between 65°F and 75°F (18°C and 23°C). They may eat twice as much when exposed to low temperatures, despite their preference for cold over heat.

8.1.3.2. Diet

Fresh water and food should always be available. Ferrets are prone to dehydration and heat stroke. They also need to eat frequently due to their short digestive tracts and high metabolisms.

Commercial ferret food should be fed to ferrets; their diets should be low in carbohydrates, moderate in fat, and high in protein. Ferrets were previously fed cat or kitten food, but there are now a variety of commercial diets made especially for them that better suit their nutritional needs than cat or kitten food. Ferrets should only eat meat since they are carnivores. Bladder stones can result from a diet high in plant proteins and ash, which is present in poor-quality foods. When buying commercial food, read the ingredient label.

Dry food, or kibble, is the suggested diet. Long-term consumption of soft, wet food causes gum and tooth disease. Avoid giving milk, other dairy products, or carb-rich foods to your child. Ferrets have trouble digesting any of these things. Ferrets are particularly vulnerable to insulinomas, which are tumors in the pancreas that cause excessive insulin production and sharp drops in blood sugar. Sugary foods like honey, raisins, fruits, and other high-carb treats should not be given to ferrets as they can cause the pancreas to release more insulin and exacerbate the symptoms of insulinomas.

Treats should not account for more than 5% of a ferret's daily caloric intake. A diet high in treats could result in malnutrition. Meat, eggs, and freeze-dried organ or muscle meat that is sold as treats for dogs or cats are healthy options.

Frequent dietary changes do not sit well with ferrets. If adjustments are necessary, make sure the animal is not stressed by the sudden changes. For instance, gradually increase the quantity of the new food given by combining it with the regular diet for a few days. Supplementing with vitamins is not required unless certain circumstances exist or the ferret is elderly. You can get advice from your veterinarian if these are necessary.

8.1.3.3. Exercise

Ferrets are gregarious, active animals that need lots of play and attention. Ferrets must spend at least two to four hours each day outside of their cage to stay healthy. Keep the cage door ajar to provide the ferret with access to food, water, and the litter box; however, make sure that the animal is always under supervision or confined to a ferret-proofed area where it cannot cause trouble. Exercise also guarantees that your ferret won't object to being confined when needed.

8.1.3.4. Temperament

It's common to compare ferrets to kittens. Even as adults, they are gregarious, lively, inquisitive, and playful. They pick things up quickly and can be trained to perform tricks. They also have short attention spans and lack fear. If left outside, ferrets will stray since they lack a homing instinct and a strong sense of territory. A loud noise or sudden movement can startle a young ferret. It will screech, puff out its fur, and hunch its back. This behavior is reserved for older ferrets that are actually being attacked by another animal. If you want to ensure the animal's safety, do not try to pick it up; instead, speak quietly to it. It will subside in due course.

Ferrets may run, jump, twist in the air, and crash violently into everyday objects when they are first released from a cage. This is a typical, healthy behavior for ferrets and could indicate general well-being or even high spirits. Ferrets have an 18-hour sleep cycle.

Young ferrets may be particularly hard to wake up from their deep sleep, even when they are pinched or pounded on the chest. Many first-time pet owners fear their pet is unconscious or dead. A warm, wet pink mouth, slow, regular breathing, and a sound sleep are all signs of a healthy ferret. Additionally, it's common for ferrets to shiver and shake when they wake up, so this shouldn't be concerning.

8.1.4. Routine Health Care for Ferrets

Locating a veterinarian with ferret treatment experience could be challenging. For recommendations, check online resources like your state veterinary medical association, local phone directories, and ferret clubs. Avert a disease by doing this prior to purchasing a ferret or, in the event of a move. For pets that live indoors, a yearly checkup and vaccinations are advised; this is a great way to build a rapport with your veterinarian. Verify that the veterinarian is close by and has experience with ferrets. Find an emergency clinic that treats ferrets after hours in case your pet gets hurt or sick late at night and your primary care veterinarian isn't open 24/7.

8.1.4.1. Vaccination Schedule

Every year, ferrets should receive rabies and canine distemper vaccinations. Ferrets are susceptible to rabies, just like other mammals, but reported cases of the disease are uncommon. Another viral illness that kills ferrets is canine distemper.

In the U.S., only one rabies vaccine is authorized for use in ferrets. Ferrets should receive the vaccination for the first time at least 16 weeks of age, and then again every year after that (based on local laws).

Find out from your vet which distemper vaccine is appropriate for your ferret. Given that they may spread the disease, vaccines derived from mink or ferret tissue culture, which are included in some dog vaccines, shouldn't be administered to ferrets. Canine distemper vaccinations should be administered as a series of booster shots to adult ferrets and kits who have never received a vaccination. See your veterinarian about the recommended vaccination schedule for your ferret.

Immunization reactions in ferrets can often result in vomiting and diarrhea. Certain ferrets may die from shock as a result of vaccine reactions. Following vaccination, ferret owners should keep their pet in the veterinarian's office for 20 to 30 minutes to look for these symptoms. It is not advisable to give rabies and canine distemper vaccinations to ferrets on the same day. Reactions may happen hours or even days after immunization. After vaccinations, ferrets frequently exhibit mild lethargy for a few days.

8.1.4.2. Dental Care

Ferrets primarily use their mouths and teeth to feel their way around, which allows them to explore their surroundings. Even as young animals, ferrets have remarkably sharp teeth. They will bite to have fun. A veterinarian should examine your ferret's teeth and gums once a year. General anesthesia is necessary for dental cleaning to remove plaque, stop gum disease, and stop tooth loss. To thoroughly examine a ferret's mouth, general anesthesia might also be necessary. Plaque accumulation can be reduced by giving your ferret dry food and brushing their teeth. Fluoride-containing human toothpaste is not recommended for use on ferrets as it is not meant for ingestion and may be harmful to your pet. Veterinarians sell specially formulated pet toothpaste with a poultry flavor that is meant to be used for brushing ferrets' teeth (Andrews and Illman, 1987; Antinoff et al., 2011).

8.1.4.3. Claw Trimming

Because they are so sharp, ferret claws need to be filed down every one to two weeks. If they are not regularly trimmed, claws can grow painfully long and become more challenging to trim. Additionally, there's a chance the ferret will hurt itself or yank out the claw. Long claws may snag on toys, carpets, towels, and other objects more readily. To trim nails, use a human-sized nail clipper. When the ferret is standing, the trimmed claw end should be parallel to the ground thanks to the angle at which the claws are trimmed. Take care not to cut any of the visible dark veins in each claw. You will inflict bleeding in addition to striking the nerve. If you happen to cut the vein, stop the bleeding right away by applying styptic powder (8.4).

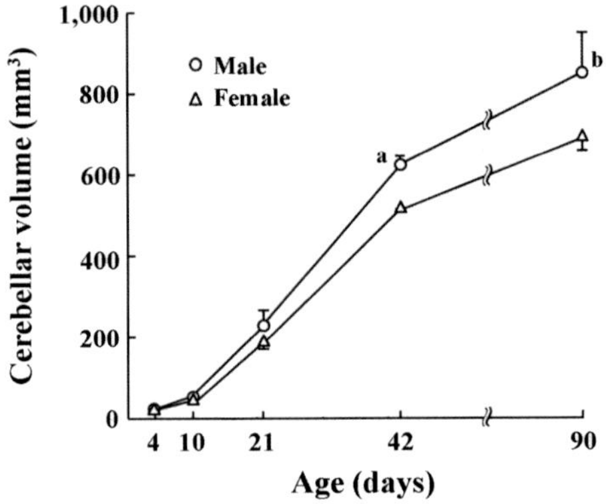

Figure 8.4. *An age-related significantly greater cerebellar cortex volume in males than in female ferrets.*

Source: Sawada K, Aoki I. Age-Dependent Sexually-Dimorphic Asymmetric Development of the Ferret Cerebellar Cortex. Symmetry. *2017; 9(3):40. https://doi.org/10.3390/sym9030040.*

Ferrets should not be declawed. Ferrets need their claws to walk and to grasp objects.

8.1.4.4. Ear Care

To get rid of the reddish wax buildup that is frequently found in their ears, ferrets' ears should be cleaned once a month. Apply a few drops of a commercial ear cleaning solution into the ear, following a veterinarian's instructions, and gently massage the base of the ear to help the solution work into the wax. After that, ferrets will shake or toss the wax. Only the outer part of the ear needs to have any leftover wax removed with a cotton swab. The cotton swab should not be inserted into the ear canal. Work gently and cautiously because the ear canal is extremely thin.

Every week, check for ear mites. Your ferret probably has ear mites if the wax is granular or gray in color or if the ear smells bad. Your veterinarian can identify whether mites are present and recommend treatment to get rid of them.

8.1.4.5. Hairballs

Ferrets are skilled self-sclerators who don't need much human assistance. Hairballs may form because they shed in the spring and fall every year. Hairballs may result in gastrointestinal obstruction, reduced appetite, or vomiting. Brush the fur with a gentle brush. One way to manage loose hair is to change your bedding once a week. Use a weekly ferret or cat laxative with a malt base if your pet is shedding a lot. But a lot of these have a lot of sugar in them. Your veterinarian can suggest suitable brands.

8.1.5. Emergencies of Ferrets

A trip to the vet is advised whenever a ferret exhibits signs of fatigue or exhibits an abrupt shift in behavior. Sick ferrets can quickly deteriorate into critically ill animals because they often conceal symptoms of discomfort or illness until they are quite advanced (Table 8.1).

Table 8.1. *Infection with Influenza Virus Types A and B Show Variable, Strain-Dependent Clinical Pictures in Ferrets*

		Clinical Parameters					
Subtype	**Strain**	**Fever** [a]	**Weight Loss** [b]	**Nasal Secretions** [c]	**Sneezing** [c]	**Neurotropism (Yes/No)** *	**Other**
H1N1	A/AA/Marton/ 1943	++	+	+++	+	NI	
	A/FortMonmouth /1/1947	+	+	+	+	NI	
	A/USSR/90/1977	+	++	+++	+	NI	
	A/Taiwan/1/1986	+	+	+++	+	NI	
	A/New Caledonia/20/1999	++	++	+++	+	NI	
	A/Solomon Islands/03/2006	+	+	++	+	NI	
	A/Brisbane/59/2007	+	+	+	+	NI	
	A/New York/18/2009	++	++	++	++	NI	
H3N2	A/Port Chalmers/1/73	+	+	+	+	NI	Depression
	A/Sydney/05/1997	+	NI	NI	NI	Yes	
	A/Panama/2007/1999	+	NI	+++	+	Yes	
	A/Brisbane/10/2007	++	++	+++	++	NI	
	A/Perth/16/2009	+++	++	++	++	NI	
	A/Wisconsin/15/2009	+	+	+	+	NI	
	A/Victoria/210/2009	++	++	+	+	NI	
		Clinical Parameters					

Subtype	Strain	Fever [a]	Weight Loss [b]	Nasal Secretions [c]	Sneezing [c]	Neurotropism (Yes/No) *	Other
H1N1	1918 H1N1 virus	NI	++	++	+++	Yes	
	A/California/07/2009	++	++	++	+	NI	
	A/Mexico/4108/2009	++	++	++	++	NI	
	A/Utah/20/2009	++	+	++	+	NI	
	A/South Carolina/2/2010	+++	++	+++	+	NI	
		Clinical Parameters					
Subtype	**Strain**	**Fever [a]**	**Weight Loss [b]**	**Nasal Secretions [c]**	**Sneezing [c]**	**Neurotropism (Yes/No) ***	**Other**
H5N1	A/Hong Kong/483/1997	+++	+++	++	+	Yes	Diarrhea in some animals
	A/Hong Kong/486/1997	+++	++	+	++	Yes	Mucopurulent nasal discharge, diarrhea in some animals
	A/Vietnam/1203/04	+++	+++	++	++	Yes	Diarrhea, mortality in some animals
H6NX	A/teal/HK/W312/1997 A/quail/HK/1721-30/1999A/mallard/Alberta/89/1985 A/duck/HK/182/1977	+	+	+	+	NI	
H7N2	A/Tky/VA/4529/1902	+	++	++	++	Yes	
	A/NY/107/2003	+	++	++	++	Yes	
H7N7	A/Netherlands/230/2003	++	++	+	+	Yes	
	A/Netherlands/219/2003	++	+++	+++	+++	Yes	Diarrhea in some animals
H7N9	A/Shanghai/2/2013	+	+	+++	+++	Yes	
	A/Anhui/1/2013	++	++	NI	NI	Yes	Lethargy, inappetence, breathing difficulties
		Clinical Parameters					
Lineage	**Strain**	**Fever [a]**	**Weight Loss [b]**	**Nasal Secretions [c]**	**Sneezing [c]**	**Neurotropism (Yes/No) ***	**Other**
Yamagata	B/Florida/04/2006	++	+	+++	+	NI	
	B/Hubei-wujiagang/158/2009	++	+	+	+	NI	
	B/Wisconsin/01/2010	+	+	++	+	NI	

Victoria	B/Brisbane/60/2008	++	++	+++	+	NI	Lower respiratory tract infection
	B/Bolivia/1526/2010	++	+	++	+	NI	

[a] Fever: mild (+) = increase < 2% for a single day; severe (+++) = 40□ or 103% baseline for > one day; [b] Weight loss: mild (+) = < 5% lost; severe (+++) = > 15% lost; [c] Respiratory symptoms: mild (+) = observed in < 30% animals; severe (+++) = observed in > 60% animals; * NI = not investigated.

Source: Rioux M, Francis ME, Swan CL, Ge A, Kroeker A, Kelvin AA. The Intersection of Age and Influenza Severity: Utility of Ferrets for Dissecting the Age-Dependent Immune Responses and Relevance to Age-Specific Vaccine Development. Viruses. *2021; 13(4):678. https://doi.org/10.3390/v13040678.*

When the disease gets worse, ferrets with an intestinal blockage will stop eating and pooping and may choke, cough, or start vomiting. Ferrets can easily become dehydrated and weak, so any vomiting or diarrhea that persists for longer than 24 hours should be examined by a veterinarian. Usually, it is not a concern if you occasionally have soft or irregular stool. For various reasons, both young and old ferrets may urinate bloody droppings. No matter how old your ferret is, you should take immediate action to get your pet veterinary care if you notice any bloody or dark, tarry droppings.

Due to their poor vision and innate curiosity, ferrets are prone to falls and other forms of trauma that can cause internal injuries or broken bones. Limping or having trouble coordinating the back legs, as well as limbs that is bent or disjointed, could be signs of a broken bone or other issues. A veterinarian should always be consulted if a ferret exhibits any of the numerous issues that can cause weakness in their back legs. Insulinomas, or excess insulin-producing tumors in the pancreas, are particularly common in ferrets. These tumors lower blood sugar levels, causing weight loss, weakness, hind leg dragging, and occasionally seizures. You should get your ferret checked out by a veterinarian as soon as possible if it exhibits any of these symptoms.

8.1.6. Breeding and Reproduction of Ferrets

Ferrets mature sexually between the ages of four and eight months. The first spring following birth marks the onset of sexual maturity. Before they are six weeks old, the great majority of ferrets are neutered or spayed. The main reason for this is that females enter heat when they reach sexual maturity and will stay in heat until they mate. When they come into season twice a year, male ferrets that have not been neutered smell strongly of musky. They also urinate on their territory to mark it.

Speak with your veterinarian about spaying or neutering your ferret if it is not done at the time of purchase. At the proper age, every pet ferret should be neutered or spayed (Andrews and Illman, 1987; Antinoff et al., 2011).

8.2. INFECTIOUS DISEASES OF FERRETS

8.2.1. Bacterial Diseases in Ferrets

Similar to other carnivores, ferrets are commonly infected with bacteria. Nonetheless, certain bacterial infections are unique to ferrets. The two most significant ones are *Lawsonia intracellularis* enteritis and *Helicobacter mustelae* gastritis. Furthermore, it has been discovered that a variety of mycobacteria can cause atypical disease in ferrets (Figure 8.5).

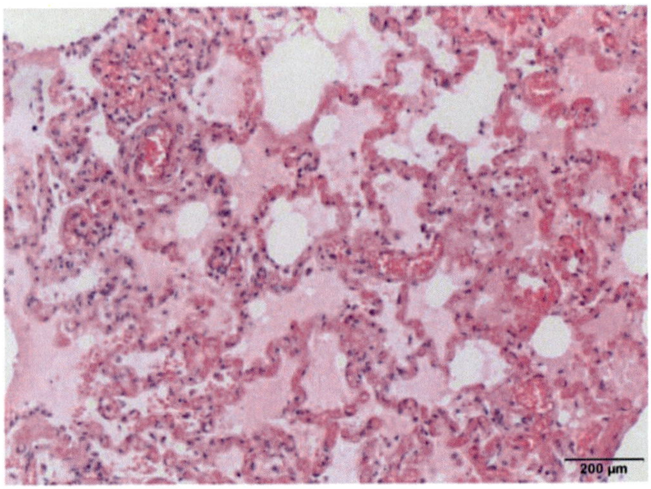

Figure 8.5. *Microscopic appearance of lung sample from ferret lung congestion and edema are present; hematoxylin–eosin stain, 100× magnification.*

Source: Golke A, Jańczak D, Szaluś-Jordanow O, Dzieciątkowski T, Sapierzyński R, Moroz-Fik A, Mickiewicz M, Frymus T. Natural Infection with Highly Pathogenic Avian Influenza A/H5N1 Virus in Pet Ferrets. Viruses. *2024; 16(6):931. https://doi.org/10.3390/v16060931.*

After they are weaned, ferrets' stomach and duodenum contain *Helicobacter mustelae*. It is an opportunistic pathogen that can cause persistent, chronic gastritis and the development of ulcers, much like gastric ulcer disease in humans. Chronic cases of gastric lymphoma are possible. Infected ferrets may not exhibit any clinical symptoms at all, but they may also exhibit diarrhea, vomiting, bruxism, inappetence, melena, and hypersalivation. There may also be weight loss, lethargy, and dehydration. When the cranial abdomen is palpated, ferrets may exhibit pain when ulcers are present. While Helicobacter can be discovered during histologic analysis of surgical or endoscopic biopsy samples, its existence shouldn't be regarded as diagnostic unless it's linked to ulcers.

Treatment usually consists of a combination of medications, such as amoxicillin (20 mg/kg, PO every 12 hours) or clarithromycin (50 mg/kg, PO every 24 hours); metronidazole (20–25 mg/kg, PO every 12 hours); omeprazole (1–4 mg/kg/day, PO); and sucralfate (25–125 mg/kg, PO every 8–12 hours). Treatment is usually started when clinical signs become apparent before the presence of *H. mustelae* is confirmed. Treatment lasts for 21 days on average. Due to the opportunistic nature of this pathogen, it's critical to manage stress and take trichobezoar formation into account.

Proliferative bowel disease is caused by the strict intracellular bacterium *Lawsonia intracellularis*, and it is most frequently observed in ferrets up to 16 weeks of age that are housed in large groups under unfavorable conditions. The bacteria cause the mucosa to swell up in the colon and ileum. Rectal prolapse, weight loss, and diarrhea are examples of clinical symptoms. It has been reported that treatment with chloramphenicol (25 mg/kg, PO, twice daily) for 14–21 days is successful; however, improving diet and housing appears to have little effect in preventing the illness.

8.2.2. Viral Diseases of Ferrets

The most common viral illnesses in ferrets are coronavirus infections, influenza A and B (caused by anorthomyxovirus infection), canine distemper (caused by infection with the morbillivirus of the family Paramyxoviridae), and Aleutian disease (caused by a parvovirus).

Although the canine distemper virus can infect ferrets, most ferrets kept as pets are currently vaccinated, so clinical signs of the disease are rare. Aerosol droplets, skin, urine, feces, conjunctival or nasal exudates, or a combination of these can all be the means of transmission. Clinical signs, which typically appear 7–10 days after infection, include fever, lymphopenia, anorexia, mucous membrane erythema, and serous to mucopurulent nasal and ocular discharge. Dermatologic symptoms also manifest as hyperkeratotic and crustaceous dermatitis of the face (chin, eyelids), abdomen, inguinal, and perineal regions, and footpad hyperkeratosis. Signs related to the respiratory and gastrointestinal systems emerge quickly and worsen as a result of secondary bacterial infections. Although they have been documented, neurologic symptoms of acute encephalomyelitis are not always evident in afflicted animals. The history, clinical symptoms, and findings of the relevant samples' histopathologic examination are used to make the diagnosis. The mortality rate is almost 100%, and the average time to die after infection is 12–14 days. Vaccination is therefore strongly advised (Figure 8.6).

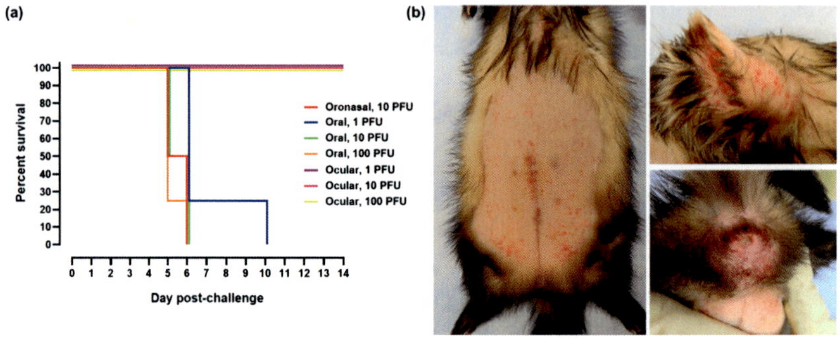

Figure 8.6. *Survival and petechial rash following EBOV mucosal challenge. (a) Kaplan–Meier survival curve following oronasal, oral, and ocular virus challenge. (**b**) Representative images of petechial rash noted in ferrets administered EBOV via oronasal and oral routes. Rash was prominent on the abdomen (left image), head/ ears (top right), and perianal surface (bottom right).*

Source: Brasel T, Comer JE, Massey S, Smith J, Hyde M, Kocsis A, Gainey M, Niemuth N, Triplett C et al. Mucosal Challenge Ferret Models of Ebola Virus Disease. Pathogens. 2021; 10(3):292. https://doi.org/10.3390/pathogens10030292.

Ferrets that contract the influenza virus experience fever, lethargy, anorexia, nasal discharge, sneezing, and depression. Early on, an infection can easily be misdiagnosed as canine distemper (and vice versa). For secondary infections, amantadine (6 mg/kg, nasally, twice daily) has been proposed as an experimental antiviral treatment; otherwise, supportive care with antimicrobials is the norm. Usually, recovery takes 7 to 14 days. Strict hygienic precautions must be followed when veterinary personnel or ferrets are suspected of having influenza infections since ferrets can contract the disease from humans and can also infect humans (Andrews and Illman, 1987; Antinoff et al., 2011).

In ferrets, disease is caused by at least two coronaviruses. *Catarrhal enteritis* is a disease caused by the ferret enteric corona virus. The highly contagious ferret enteric corona virus is frequently introduced into a group of ferrets by an asymptomatic juvenile. Clinical symptoms appear two to 14 days after the new ferret is introduced or following exposure to fomites. Following infection, the virus blunts the intestinal villi, which results in maldigestion and malabsorption. This can cause anorexia, vomiting, mucoid or green diarrhea, melena, dehydration, fatigue, and weight loss, among other clinical symptoms. The emergence of clinical symptoms has become less common in recent years. It may take months for an older ferret to fully recover from disease, if it is detected at all. If a secondary bacterial infection is suspected, supportive treatment consists of broad-spectrum antibiotics, GI protectants, fluids, and nutritional support. New ferrets should be quarantined, bedding, and toys should be thoroughly cleaned, and hands should be washed and changed after handling healthy ferrets to prevent infection (Table 8.2).

Table 8.2. *Predominant Histopathological Findings Following EBOV Mucosal Challenge in Ferrets*

Organ	**Finding**	**Oronasal**	**Oral**			**Ocular**		
		10 PFU [2]	**1 PFU** [2]	**10 PFU** [2]	**100 PFU** [2]	**1 PFU** [2]	**10 PFU** [2]	**100 PFU** [2]
Liver	Hepatocellular degeneration	4/4	4/4	4/4	3/4	0/4	1/4	0/4
	Inflammation	4/4	4/4	3/4	3/4	0/4	1/4	0/4
	Hepatocellular necrosis	3/4	3/4	4/4	3/4	0/4	0/4	0/4
	Vacuolation	3/4	2/4	4/4	2/4	0/4	0/4	0/4
Lung	Perivascular edema	4/4	3/4	3/4	3/4	0/4	1/4	0/4
	Alveolar inflammation	3/4	3/4	1/4	3/4	0/4	3/4	2/4
	Perivascular inflammation	4/4	3/4	3/4	3/4	0/4	1/4	0/4
	Perivascular necrosis	1/4	0/4	0/4	1/4	0/4	0/4	0/4
Spleen	Decreased lymphocytes	4/4	0/4	1/4	1/4	0/4	0/4	0/4
	Inflammation	4/4	4/4	4/4	3/4	0/4	1/4	0/4
	Necrosis	4/4	3/4	3/4	2/4	0/4	0/4	0/4
Kidney	Tubule vacuolation	3/4	3/4	4/4	1/4	0/4	1/4	0/4
Bladder	Mucosal hemorrhage	1/4	0/4	0/4	0/4	0/4	0/4	0/4
Stomach	Mucosal hemorrhage	1/4	0/4	0/4	0/4	0/4	0/4	0/4

Lower GI [3]	Mucosal hemorrhage	0/4	0/4	1/4	0/4	0/4	0/4	0/4
Brain	Hippocampus necrosis	1/4	0/4	1/4	0/4	0/4	0/4	0/4

[1]Results are presented as the number of animals in which the specified histopathological finding was noted out of the total number of animals in the challenge dose/route group.

[2]Target challenge dose; PFU, plaque forming units. [3] GI, gastrointestinal tract; includes rectum and/or colon.
Source: Brasel T, Comer JE, Massey S, Smith J, Hyde M, Kocsis A, Gainey M, Niemuth N, Triplett C et al. Mucosal Challenge Ferret Models of Ebola Virus Disease. Pathogens. *2021; 10(3):292. https://doi.org/10.3390/pathogens10030292.*

Similar to the dry form of feline infectious peritonitis, the second related corona virus, ferret systemic corona virus, causes a systemic pyogranulomatous inflammatory disease. As a result, the illness is now commonly known as ferret infectious peritonitis. This illness progresses over several weeks to months and is most common in young ferrets (average age of 11 months). Anorexia, weight loss, diarrhea, enlarged intra-abdominal lymph nodes, and, less frequently, peripheral lymph nodes are among the clinical symptoms. As the illness worsens, hypergammaglobulinemia, anemia, and CNS symptoms become apparent. About two months is the average survival time. If treatment is received, it is mainly supportive; anecdotally, prednisolone use has been shown to somewhat extend survival times.

The parvovirus that causes Aleutian disease was first discovered in mink, but at least two different strains of the virus have been found in ferrets. The virus leads to immune complex deposition in organs, causing a range of nonspecific clinical symptoms, including ataxia, hepatomegaly, splenomegaly, and progressive weight loss. In laboratory testing, the most consistent result is severe hypergammaglobulinemia. Clinical symptoms and the existence of hyperglobulinemia serve as the basis for an assumed diagnosis. In the United States, serologic testing (conducted by the University of Georgia's Infectious Diseases Laboratory) and PCR assay are the most widely used methods for diagnosing viral infections. The most widely used test to find parvovirus antibodies in other nations is counter immunoelectrophoresis (CIEP). Because seemingly normal ferrets in shelters may have positive titers, definitive diagnosis is challenging. Furthermore, the virus has been detected in the blood, feces, and urine of animals that exhibit symptoms as well as those that do not.

Treatment with immunosuppressants and anti-inflammatories, such as cyclophosphamide and prednisolone, can be explored and may be beneficial clinically. For ferrets, there is no vaccine against this illness (Andrews and Illman, 1987; Antinoff et al., 2011).

8.2.3. Fungal Diseases of Ferrets

Although clinical symptoms of disease are uncommon, ferrets are susceptible to *M. canis* and *Trichomonas mentagrophytes* infection. Transmission occurs through direct touch or through fomites, and it's frequently linked to cat exposure and overcrowding. Kits and young ferrets are more likely to contract an infection, which is frequently seasonal and self-limiting. The diagnosis and course of treatment are the same as those outlined for ringworm in cats and dogs. *Microsporum nanum* has been linked to both fungal pododermatitis and pyogranulomatous dermatitis. *Cryptococcal meningitis* and blastomycosis, which result in granulomatous meningoencephalitis, are two other fungal diseases that affect ferrets. Although it is rare, *Blastomyces dermatitidis* and *Cercodiaides immitis* can cause fungal pneumonia in ferrets in endemic areas. Ferrets have been diagnosed with cryptococcosis caused by *Cryptococcus bacillisporus* and *Cryptococcus neoformans var grubii*. Pneumonia, pleuritis, rhinitis, and enlargement of the lymph nodes locally are among the symptoms.

8.2.4. Parasitic Diseases of Ferrets

Otodectes cynotis is the causative agent of ear mite infection, the most prevalent ectoparasitic disease in ferrets. The same ear mite can spread between species and is present in both cats and dogs. As with dogs and cats, otoscopy is used for diagnosis. Numerous medications are approved for use in ferrets, and the course of treatment is comparable to that for dogs and cats. Ferrets frequently carry fleas, which they can spread to other household pets. The same methods of treatment for dogs and cats are used for diagnosis and treatment planning. Fipronil and other long-acting topical medications have a longer half-life in ferrets due to their higher sebum concentration in their coats. *Sarcoptes scabiei* is the cause of mange in ferrets. It can present as a generalized dermatitis or as a pedal form, which only affects the toes and foot pads and is specific to ferrets.

Ferrets are susceptible to heartworm disease, which is brought on by *Dirofilaria immitis*, particularly if they are allowed outside in endemic areas. A single worm is capable of causing disease. Ascites, coughing, dyspnea, and lethargy are examples of clinical signs. The typical number of worms infected in ferrets is 1–20, which makes diagnosis challenging. Because the parasites frequently obstruct blood flow and result in right-side heart failure, echocardiography is necessary. Finding the worms in the pulmonary arteries, vena cavae, and right ventricle may also be aided by echocardiography. Since peripheral microfilaremia is rare in ferrets, antigen testing is a better option. Although it is possible to begin long-term treatment with adulticides and antithrombotic medications, side effects are possible. In endemic areas, selamectin (18 mg/kg, topically) is currently the advised course of treatment.

In young ferrets, coccidiosis—a disease caused by Eumyces or Isospora species—can manifest as lethargy and diarrhea. The course of diagnosis and treatment is the same as for dogs. Rectal prolapse is typically resolved following treatment for the underlying condition, although it can also occur with coccidiosis. Another protozoal illness that can affect ferrets is giardiasis. Weight loss and diarrhea are the most typical symptoms when they are present. Metronidazole (20 mg/kg, PO, twice a day) is the treatment (Andrews and Illman, 1987; Antinoff et al., 2011).

8.3. NEOPLASIA OF FERRETS

In ferrets, neoplasia is frequently diagnosed. While tumors can arise in any organ system, the endocrine and hemolymphatic systems account for about half of all ferret tumor diagnoses. The table naturally occurring neoplasms (tumors) reported in ferrets provides a summary of neoplastic disease (tumors) diagnosed in these animals. The section "Endocrine Disorders of Ferrets" discusses endocrine neoplasia.

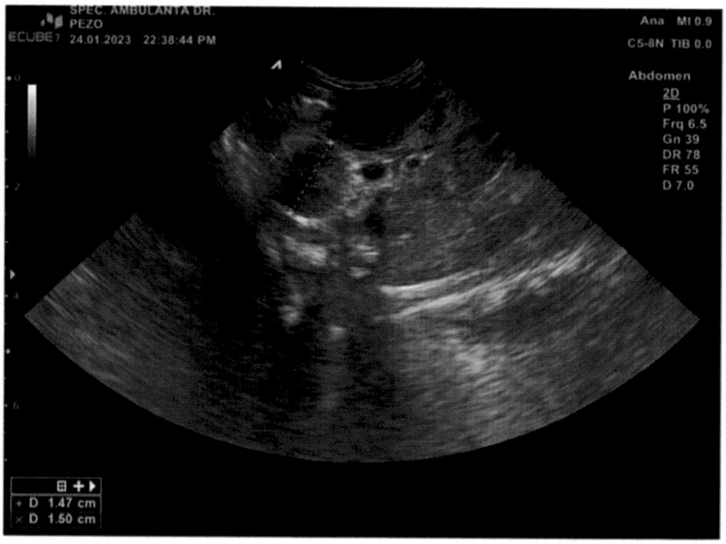

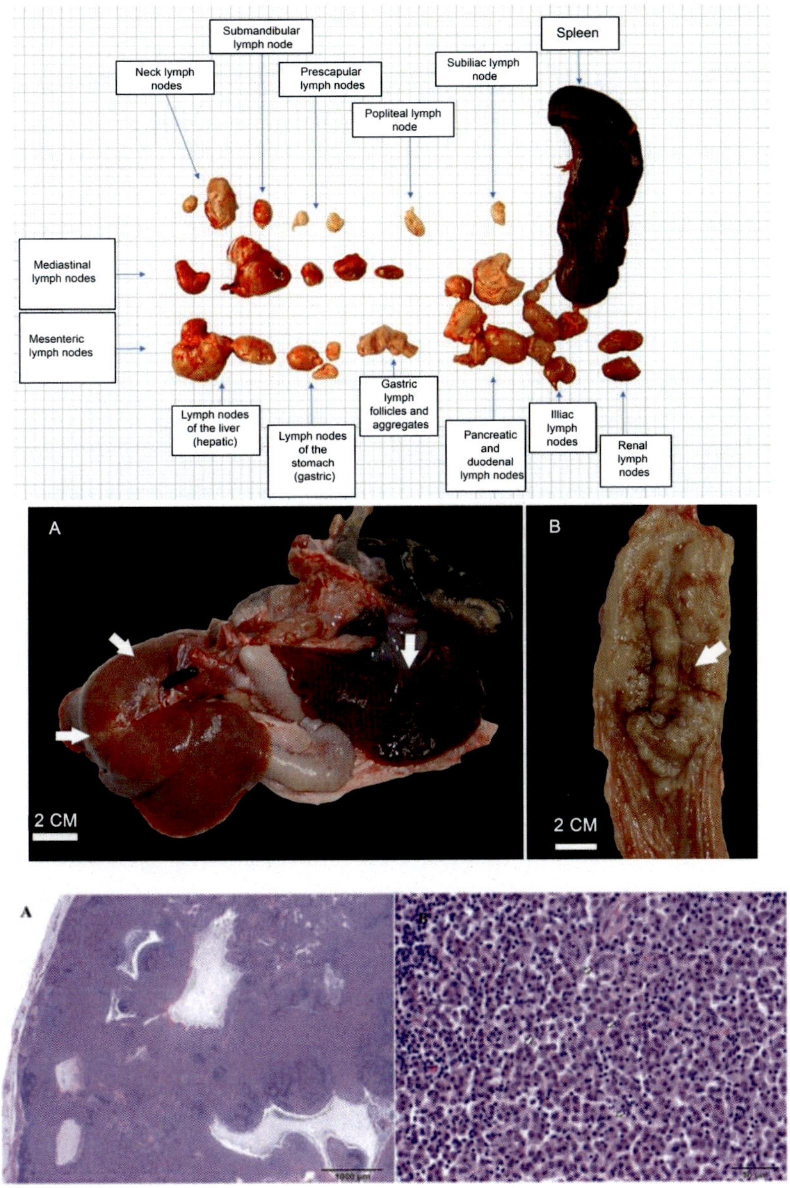

Figure 8.7. *M. xenopi mycobacteriosis in a pet ferret.*

Source: Mihaljević Ž, Reil I, Habuš J, Štritof Z, Naletilić Š, Jurkić Krsteska G, Kovač T, Zdelar-Tuk M, Duvnjak S, Špičić S. Mycobacteriosis in a Pet Ferret (Mustela putorius furo) Caused by Mycobacterium xenopi: A Case Report on Neglected Risk of Zoonotic Transmission. Pathogens. *2024; 13(4):328. https://doi.org/10.3390/pathogens13040328.*

Ferrets' most common integumentary (skin) tumors are sebaceous epitheliomas and cutaneous mast cell tumors. On the body, these tumors can develop anyplace. Both varieties manifest as elevated, asymmetrical, and frequently crusted masses. Although they rarely show symptoms, the tumors may bleed when touched. The tumors are thought to be benign, despite the fact that they may look malignant. Excision is the method of treatment. However, after the original tumors are surgically removed, new tumors may develop anywhere on the skin (Figure 8.7).

As early as nine months of age, a ferret's lymphoma can be diagnosed, making it the third most common tumor in this species. Gender preference is unknown. Related or cohabiting ferrets have been observed to exhibit clusters of lymphoma, and a transmission study has demonstrated that certain forms of lymphoma are transmissible, potentially due to a viral agent.

Any hemolymphatic organ, such as the bone marrow, spleen, liver, or lymph nodes, can be the site of a lymphoma. Ferrets can also have diseases of the spine and central nervous system (CNS), as well as skin cancer, also known as mycosis fungoides. In contrast to adult ferrets, young ferrets with lymphoma typically experience a rapid progression of the disease.

Ferrets with lymphoma frequently present with nonspecific clinical symptoms, such as appetite loss, weight loss, and enlargement of peripheral lymph nodes (Andrews and Illman, 1987; Antinoff et al., 2011).

A CBC, radiographs, ultrasonography, computed tomography, fine-needle aspirations, and cytology of suspected tissues are a few examples of diagnostic procedures. Ultrasonography is the method of choice for diagnosis because it makes it possible to identify the impacted tissues and acts as a useful tool for precise tissue/cytologic sample collection.

Depending on the owner's preferences and the lymphoma's stage, several treatment protocols can be used for ferrets with the disease. If there is only one affected anatomic structure, surgery can be done. It is also possible to employ chemotherapy alone or in conjunction with radiation therapy. According to reports, the average overall mean survival time for ferrets with lymphoma is six months, but it can range from two weeks to 19 months. Given that chemotherapy may be linked to severe neutropenia, treatment with glucocorticoids or chemotherapy is dubious because these interventions do not substantially alter survival times.

Ferrets have been reported to have chondrosarcomas and choredomas. An intraosseous neoplasm derived from the notochord of mesoderms, a chordoma usually presents as a firm mass on the tail. Paresis and ataxia may also result from these tumors if they develop in the cervical region. Anywhere along the spine, ribs, or sternum can develop a chondrosarcoma, which typically causes spinal cord compression and related clinical symptoms. Chordomas on the tail are relatively easy to remove due to their location and slow rate of development; remove one or two extra coccygea to ensure a clean margin.

Although splenomegaly is frequently observed in adult ferrets, its practical significance is still debatable. It is easy to palpate the spleen, and in severe cases, it may protrude into the pelvic inlet. On CBC, anomalies are typically not discovered. Extramedullary hematopoiesis is the most frequent histologic diagnosis for these enlarged spleens, though hemangiosarcoma and lymphoma can also develop. Therefore, nodules should be aspirated and ultrasonography should be used to investigate an irregular or firm spleen. Plenectomy is only indicated in cases where there is severe discomfort or if there is a tumor (such as a lymphoma) on the spleen. Prior bone marrow biopsies are advised in cases where the spleen is removed to guarantee adequate hemopoietic activity (Andrews and Illman, 1987; Antinoff et al., 2011).

8.3.1. Endocrine Disorders of Ferrets

In ferrets, endocrine disorders are among the most prevalent conditions observed; insulinomas and hyperadrenocorticism make up the bulk of endocrinopathies. Hyperthyroidism, hypoparathyroidism, diabetes mellitus, and persistent estrus are among the other endocrine conditions that can affect ferrets. Although they have not been reported, growth hormone excess or deficiency, diabetes insipidus, hyperthyroidism, hyperparathyroidism, and spontaneous Addison disease should not be ruled out of the differential. When a ferret exhibits clinical signs similar to those of endocrine diseases in dogs, cats, or both, additional diagnostic testing for these conditions is necessary.

Ferrets older than two or three years old frequently have insulinomas. These functional beta cell tumors of the pancreas typically measure 0–2 mm, and because of their excessive and uncontrollably secreted insulin, they cause hypoglycemia. Clinical signs can range from weakness in the hind limbs, lethargy, and stargazing to complete collapse, generalized seizures, and coma. Additionally, there may be symptoms of nausea, ptyalism, and mouth pawing. Additionally, ferret owners may notice a glassy appearance in their pets' eyes. One noteworthy characteristic of ferrets with insulinomas is that their clinical symptoms frequently go away after they eat.

In ferrets, insulinomas hardly ever spread. Clinical symptoms combined with a low blood glucose level are used to make the diagnosis. < 60 mg/dL [< 3.3 mmol/L]; reference range: 90–125 mg/dL [5.0–7.0 mmol/L]) after the ferret has had food withheld for at least four hours. Plasma insulin concentrations will usually be increased (> 773 pmol/L or > 108 mIU/L), although it might fall inside the reference range. Ferrets that were healthy and those that were hypoglycemic did not significantly differ in their fructosamine concentrations. Most of the time, other blood parameters are normal.

There are options for both medical and surgical treatments, and choosing between them depends on a number of factors. Any ferret with an insulinoma needs to be fed freely, and its owners are responsible for making sure it eats all day long. To remove the source of excess insulin production, surgical removal of the insulinoma(s) appears to be the best option. Depending on the quantity and location of the pancreatic nodules, either a nodulectomy, a partial pancreatectomy, or both may be performed. The most widely used medications to treat insulinomas are diazoxide (5–30 mg/kg, PO, twice a day; titrated to effect) and prednisone (0–2 mg/kg, PO, twice a day). Prednisone is often mentioned as the first line of treatment because it promotes gluconeogenesis. Long-term glucocorticoid therapy may cause weight gain and decreased hair growth in ferrets, though adverse effects are rare. This is indicative of iatrogenic hyperadrenocorticism. Diazoxide stimulates hepatic gluconeogenesis and glycogenolysis, inhibits cellular uptake of glucose, and inhibits the release of insulin from beta cells. Both medications can be taken singly or in combination for maximum effect. There is no significant difference in the disease-free interval between medical and surgical treatment.

In ferrets, hyperadrenocorticism also called adrenocortical disease or adrenal gland disease is the most prevalent endocrine tumor. It is thought likely, based on earlier research, that the persistent stimulation of the adrenal cortex following neutering, due to the absence of negative feedback, causes increased concentrations of gonadotropins, which in turn cause adrenocortical growth. Neutering timing is no longer considered to be important.

The most noticeable symptoms of hyperadrenocorticism in ferrets are pruritus, symmetrical alopecia, a swollen vulva in female ferrets who have had their sex neutered, and male ferrets who have returned to sexual behavior following neutering. Male ferrets may exhibit prostatic enlargement and urinary obstruction from periprostatic or periurethral cysts, while female ferrets may occasionally exhibit mammary gland enlargement.

A physical examination and history are required before a presumptive diagnosis is made. It is possible to palpate the enlarged adrenal glands craniomedial to either kidney (it is easier to palpate the left adrenal than the right). The most helpful method for figuring out whether an ovarian remnant the most significant differential diagnosis in females or one or both adrenal glands are affected is ultrasonography. Another useful method for determining the size of the adrenal glands is to use intravenous contrast medium in conjunction with computed tomography. Measuring the levels of the adrenocortical hormones, androstenedione, estradiol, and 17-hydroxyprogesterone (a hormone panel that the University of Tennessee can measure) is the only way to identify adrenal cortex hyperfunction. Elevation of one or more of these hormones, however, is insufficient to distinguish between an active ovarian remnant and hyperadrenocorticism.

The management of hyperadrenocorticism in ferrets entails the administration of a long-acting GnRH analog or surgery. Surgery to remove the left adrenal gland is relatively simple and involves only ligating the vena phrenicoabdominalis. The right adrenal gland's dorsolateral attachment to the caudal vena cava and its close proximity to the liver make resection of the gland more challenging.

Ferrets with hyperadrenocorticism may be treated with the depot GnRH-agonist deslorelin-containing implant. By continuously releasing deslorelin into the bloodstream, deslorelin suppresses the release of gonadotropins by overriding the pulsatile release of GnRH that is necessary for the release of gonadotropins. Furthermore, there is a down-regulation of GnRH receptors in the pituitary gland. Over time, the adrenal gland may begin to produce steroids on its own, which could lead to a decline in the implant's effectiveness and a return of the clinical symptoms.

The average disease-free period following implant treatment is 16 +/– 5 months (range: 3–30 months); this period is 13 +/– 6 months (range: 0–38 months) for surgical intervention. Even though medical treatment has no effect on the tumor itself. These results have led to the preference for medical treatment over surgical treatment.

Moreover, preliminary data indicate that the incidence of adrenal disease in ferrets may be reduced by using the depot deslorelin-containing implant rather than surgical neutering (Andrews and Illman, 1987; Antinoff et al., 2011).

8.4. HORMONAL DISORDERS OF FERRETS

Insulinomas and hyperadrenocorticism are two common hormonal (endocrine) disorders in ferrets.

8.4.1. Insulinomas

An insulinoma is a pancreatic tumor. When a ferret is older than three years old, these tumors are quite common. More insulin is produced by insulinomas than the body requires. The reverse of diabetes, known as hypoglycemia, is brought on by high insulin levels and low blood sugar. Ferrets with insulinoma have signs such as:

1. weakness.
2. lethargy.

3. slight or partial paralysis of the rear legs.
4. increased salivation.
5. teeth clenching or grinding.

Seizures may happen in severe situations. Evidence of hypoglycemia and corresponding normal or elevated insulin levels are used to make the diagnosis. An insulin fasting test is used for this. The ferret will need to go without food for approximately four hours prior to the appointment; during that time, a blood sample will be taken. The results of other tests are typically normal. These tumors are only sporadically visible on ultrasonography.

There is no known cure, but there are possible medical and surgical treatments. During surgery, the mass itself or the area of the pancreas containing the mass are removed. It is possible that some insulinomas remain after surgery because they start off as several small tumors dispersed throughout the pancreas. After surgery, some ferrets experience a period of normal blood glucose (sugar) levels; however, the majority of ferrets need ongoing medical care. Reducing the severity of signs, making management easier, and somewhat lengthening survival time are some advantages of surgery. Medical therapy does not directly reduce the tumor; rather, it counteracts its effects for the rest of one's life. Blood glucose tests are required for ferrets with insulinomas every few months and following adjustments to medication dosages.

Try giving your ferret a small amount of honey or corn syrup if it exhibits hypoglycemia symptoms, such as weakness or collapse. Never put anything in the mouth of a seizure-suffering ferret for safety's sake. Ferrets exhibiting hypoglycemia symptoms ought to be taken to the vet right away.

8.4.2. Adrenal Gland Disease

The overproduction of sex hormones (progesterone, testosterone, and estrogen) by the adrenal gland is the cause of adrenal gland disease in ferrets. This is not the same as hyperadrenocorticism, also known as Cushing disease, which affects other species and is characterized by an excess of cortisol hormone instead of sex hormones. Ferrets as young as 1.5 years old exhibit it. Spaying your ferret will help ward off this illness. The most typical symptom is hair loss that starts on the rump and tail and moves up the body to the head. Additionally, female ferrets may have enlarged nipples and a swollen vulva. Due to an enlarged prostate gland, male ferrets may become aggressive and experience urination difficulties. Serious blood estrogen elevation in either sex can be followed by bone marrow suppression. Ferrets exhibiting these symptoms ought to be seen by a veterinarian right away.

A preliminary diagnosis will be made by a veterinarian based on the ferret's physical examination and medical history. It is possible to feel the enlarged adrenal glands in front of the kidneys. The results of routine blood tests are typically normal. Although your veterinarian needs to measure the sex hormones to make an accurate diagnosis, ultrasonography may reveal enlarged glands.

Medical and surgical options are available for the treatment of this condition. Although surgical removal of the adrenal gland(s) is more likely to cure the disease than medical management, approximately 40% of affected ferrets will relapse. Partially removing the adrenal glands is an option if both sides are impacted.

Based on a review of gland tissue, one of three stages of the disease progression—hyperplasia, adenoma, or adenocarcinoma may be identified (cancer). All three grades function similarly, and it is unlikely that cancer cells will spread outside of the glands. Ferrets with both adrenal glands removed entirely or partially may experience other issues as a result of not having enough of the other hormones that the adrenal glands produce and the body needs. Medication is an option for treating these ailments.

Medical management can lessen the symptoms, but it does not shrink the adrenal gland tumor that is the source of the issue. For the remainder of the ferret's life, medical care is required to control disease signs. If ferrets are receiving long-term medical care, they should be closely observed. Melatonin does not stop the growth of the adrenal tumor, despite the fact that it has been used to treat hair loss in ferrets with diseases of the adrenal glands. Veterinarians are now using other medications to suppress the effects of the sex hormones that ferrets' cancerous adrenal glands overproduce. One such medication is injectable leuprolide acetate, a hormone that is used to treat certain medical conditions in humans. Leuprolide acetate, however, does not stop the tumor from growing, so to keep the clinical signs of the tumor under control, the ferret needs to receive this medication every few weeks for the rest of its life (Andrews and Illman, 1987; Antinoff et al., 2011).

8.4.3. Noninfectious Diseases of Ferrets

In young ferrets, gastric foreign bodies are frequently diagnosed. Ferrets frequently swallow soft rubber or plastic objects, which can obstruct their digestive tracts. Gastric trichobezoars can also develop if hair is consumed during molting. Anorexia, bruxism, hypersalivation, cranial abdominal pain, diarrhea, and/or melena are examples of clinical signs. Abdominal ultrasonography or plain radiography can be used to make the diagnosis. Surgery or endoscopic removal are used as treatments. Treatment for gastritis should begin as soon as the foreign body is removed.

Ferrets with cardiovascular diseases are frequently diagnosed with them. The most prevalent cardiac condition in ferrets, dilated cardiomyopathy, is typically identified in animals older than four years old. Lethargy, weakness, increased respiratory effort, coughing, exercise intolerance, and/or ascites are examples of nonspecific clinical signs. The diagnosis needs to be verified with echocardiography. To check for pleural effusion or lung edema, radiographs may be useful. Similar to other companion species, furosemide, pimobendan, and enalapril may be used in the course of treatment. For information on dosage, refer to a formulary.

Ferrets often exhibit cardiac arrhythmias, though not all of them are deemed pathological. For example, an insulinoma may coexist with a second-degree AV block, which resolves when the insulinoma's clinical symptoms are addressed. Clinical signs of severe bradycardia (e.g., third-degree AV block) can include weakness, lethargy, intolerance to exercise, and syncope. The altered rhythm and frequency can be felt when the pulse is palpated and the heart is auscultated. To determine the type of arrhythmia, an ECG is required. The course of treatment is comparable to that given to dogs and cats and may involve pacemaker implantation.

In ferrets, renal disease is commonly observed. There are no known specific etiologic factors. Pale mucous membranes, vomiting, decreased appetite, and polyuria/polydipsia may be observed. Examining the blood and urine yields the diagnosis. For the latter, concentrations of urea, calcium, and phosphate are particularly significant, while creatinine is frequently found to be nondiagnostic. Phosphate concentrations will usually be higher than calcium concentrations when renal disease is present. Moreover, nonregenerative anemia might be discovered. Abdominal ultrasonography may reveal renal cysts, though these are usually benign unless they are numerous. Since there is no known cure for renal disease, the best course of treatment for a ferret with this condition is usually supportive care, which includes feeding it a high-quality kidney diet for cats, managing proteinuria with ACE inhibitors (like enalapril), antiemetics (like metoclopramide), antacids (like omeprazole), and/or gastric protectants (like sucralfate).

Although ferrets are not typically affected by urolithiasis, there appears to be a rise in cases in recent years. The most common type of urolith used to be struvite, but as plant proteins have been less common in diets, the composition of uroliths has shifted toward cystine uroliths. Since the cause of these uroliths is currently unknown, preventive measures are also unknown. The standard course of treatment involves surgically removing the urolith using a method akin to that used on dogs and cats (Andrews and Illman, 1987; Antinoff et al., 2011).

REFERENCES

1. Aasted, B., Alexandersen, S., & Hansen, M. (1988). Treatment of neonatally Aleutian disease virus (ADV) infected mink kits with gammaglobulin containing antibodies to ADV reduces the death rate of mink kits. *Acta Veterinaria Scandinavica, 29*(3, 4), 323–330.
2. Akerman, C. J., Tolhurst, D. J., Morgan, J. E., Baker, G. E., & Thompson, I. D. (2003). The relay of visual information to the lateral geniculate nucleus and the visual cortex in albino ferrets. *Journal of Comparative Neurology, 461*, 217–235.
3. Allen, R. P., Evans, E. V., & Sibbald, I. R. (1964). Energy: protein relationships in the diets of growing mink. *Canadian Journal of Physiology and Pharmacology, 47*, 733.
4. Anderson, E. (1989). The phylogeny of mustelids and the systematics of ferrets. In U. S. Seal, E. T. Thorne, M. A. Bogan, & S. H. Anderson (Eds.), *Conservation Biology and the Black-Footed Ferret* (pp. 123–145). Yale University Press.
5. Andrews, P. L. R., & Illman, O. (1987). The ferret. In T. Poole (Ed.), *UFAW Handbook on the Care and Management of Laboratory Animals* (6th ed., pp. 124–128). Longmans.
6. Andrews, P. L., Bower, A. J., & Illman, O. (1979). Some aspects of the physiology and anatomy of the cardiovascular system of the ferret, *Mustela putorius furo*. *Laboratory Animals, 13*, 215–220.
7. Andrews, P., Illman, O., & Mellersh, A. (1979). Some observations of anatomical abnormalities and disease states in a population of 350 ferrets (*Mustela furo*). *Zeitschrift für Versuchstierkunde, 21*, 346–353.
8. Antinoff, N. (1997). Musculoskeletal and neurological diseases. In E. V. Hillyer & K. E. Quesenberry (Eds.), *Ferrets, Rabbits, and Rodents: Clinical Medicine and Surgery* (pp. 126–130). Saunders.
9. Antinoff, N., & Williams, B. H. (2011). Neoplasia. In K. Quesenberry & J. Carpenter (Eds.), *Ferrets, Rabbits, and Rodents* (3rd ed., pp. 118–119). Saunders.
10. Apfelbach, R. (1986). Imprinting on prey odours in ferrets (*Mustela putorius f. furo L.*) and its neural correlates. *Behavioral Processes, 12*(4), 363–381.
11. Barron, H. W., & Rosenthal, K. L. (2011). Respiratory diseases. In K. Quesenberry & J. Carpenter (Eds.), *Ferrets, Rabbits, and Rodents* (3rd ed., pp. 80–81). Saunders.
12. Batchelder, M. A., Bell, J. A., Erdman, S. E., Marini, R. P., Murphy, J. C., & Fox, J. G. (1999). Pregnancy toxemia in the European ferret (*Mustela putorius furo*). *Laboratory Animal Science, 49*, 372–379.

13. Baum, M. J. (1998). Use of the ferret in reproductive neuroendocrinology. In J. G. Fox (Ed.), *Biology and Diseases of the Ferret* (pp. 521–536). Williams & Wilkins.
14. Baum, M. J., Lynch, H. J., Gallagher, C. A., & Deng, M. H. (1986). Plasma and pineal melatonin levels in female ferrets housed under long or short photoperiods. *Biology of Reproduction, 34*(1), 96–100.
15. Beck, F., Schion, H., Mould, G., Swidzinska, P., Curry, S., & Grauwiler, J. (1976). Comparison of the teratogenic effects of mustine hydrochloride in rats and ferrets. The value of the ferret as an experimental animal in teratology. *Teratology, 13*, 151.
16. Bell, J. (1997). Periparturient and neonatal diseases. In E. V. Hillyer & K. E. Quesenberry (Eds.), *Ferrets, Rabbits, and Rodents: Clinical Medicine and Surgery* (pp. 58–60). Saunders.
17. Bell, J. A. (1996). Ensuring proper nutrition in ferrets. *Veterinary Medicine, 91*, 1098.
18. Bell, J. A. (1997). Helicobacter mustelae gastritis, proliferative bowel disease, and eosinophilic gastroenteritis. In E. V. Hillyer & K. E. Quesenberry (Eds.), *Ferrets, Rabbits, and Rodents: Clinical Medicine and Surgery* (pp. 37–43). Saunders.
19. Bell, J. A. (1999). Ferret nutrition. *Veterinary Clinics of North America: Exotic Animal Practice, 2*, 169–192.
20. Bell, J. A., & Manning, D. D. (1990). A domestic ferret model of immunity to *Campylobacter jejuni*-induced enteric disease. *Infection and Immunity, 58*, 1848–1852.
21. Bell, J. A., & Manning, D. D. (1990). Reproductive failure in mink and ferrets after intravenous or oral inoculation of *Campylobacter jejuni*. *Canadian Journal of Veterinary Research, 54*, 432–437.
22. Bell, J. A., & Manning, D. D. (1991). Evaluation of *Campylobacter jejuni* colonization of the domestic ferret intestine as a model of proliferative colitis. *American Journal of Veterinary Research, 52*, 826–832.
23. Belser, J. A., Katz, J. M., & Tumpey, T. M. (2011). The ferret as a model organism to study influenza A virus infection. *Disease Models & Mechanisms, 4*, 575–579.
24. Bernard, S. L., Gorham, J. R., & Ryland, L. M. (1984). Biology and diseases of ferrets. In J. G. Fox, B. J. Cohen, & F. M. Loew (Eds.), *Laboratory Animal Medicine* (pp. 394). Academic Press.
25. Bernard, S. L., Leathers, C. W., Brobst, D. F., & Gorham, J. R. (1983). Estrogen-induced bone marrow depression in ferrets. *American Journal of Veterinary Research, 44*, 657–661.

26. Besch-Williford, C. L. (1987). Biology and medicine of the ferret. *Veterinary Clinics of North America: Small Animal Practice, 17*, 1155–1183.

27. Blanco, M. C., Fox, J. G., Rosenthal, K., Hillyer, E. V., Quesenberry, K. E., & Murphy, J. C. (1994). Megaesophagus in nine ferrets. *Journal of the American Veterinary Medical Association, 205*, 444–447.

28. Blancou, J., Aubert, J. F. A., & Artois, M. (1982). Rage expérimentale du furet (*Mustela putorius furo*). *Revue de Médecine Vétérinaire, 133*, 553.

29. Blankenship-Paris, T. L., Chang, J., & Bagnell, C. R. (1993). Enteric coccidiosis in a ferret. *Laboratory Animal Science, 43*, 361–363.

30. Bleavins, M. R., & Aulerich, R. J. (1981). Feed consumption and food passage time in mink (*Mustela vison*) and European ferrets (*Mustela putorius furo*). *Laboratory Animal Science, 31*, 268–269.

31. Bradbury, C., Saunders, A. B., Heatley, J. J., Gregory, C. R., Wilcox, A. L., & Russell, K. E. (2010). Transvenous heartworm extraction in a ferret with caval syndrome. *Journal of the American Animal Hospital Association, 46*(1), 31–35.

32. Brown, S. A. (1997). Basic anatomy, physiology, and husbandry. In E. V. Hillyer & K. E. Quesenberry (Eds.), *Ferrets, Rabbits, and Rodents: Clinical Medicine and Surgery* (pp. 3–13). Saunders.

33. Brown, S. A. (1997). Neoplasia. In E. V. Hillyer & K. E. Quesenberry (Eds.), *Ferrets, Rabbits, and Rodents: Clinical Medicine and Surgery* (pp. 99–114). Saunders.

34. Budd, J. (1981). Distemper. In J. W. Davis, L. H. Karstad, & D. O. Trainer (Eds.), *Infectious Diseases of Wild Mammals* (pp. 123–129). Iowa State University Press.

35. Burr, D. H., Rollins, D., Lee, L. H., Pattarini, D. L., Walz, S. S., & Tian, J. H. (2005). Prevention of disease in ferrets fed an inactivated whole cell *Campylobacter jejuni* vaccine. *Vaccine, 23*(34), 4315–4321.

36. Caplan, E. R., Peterson, M. E., Mullen, H. S., Quesenberry, K. E., Rosenthal, K. L., & Hoefer, H. L. (1996). Diagnosis and treatment of insulin-secreting pancreatic islet cell tumors in ferrets: 57 cases (1986–1994). *Journal of the American Veterinary Medical Association, 209*, 1741–1745.

37. Carpenter, J. W., Appel, M. J. G., & Erickson, R. C. (1976). Fatal vaccine-induced canine distemper virus infection in black-footed ferrets. *Journal of the American Veterinary Medical Association, 169*, 961–964.

38. Carroll, R. S., Erskine, M. S., & Baum, M. J. (1987). Sex difference in the effect of mating on the pulsatile secretion of luteinizing hormone in a reflex ovulator, the ferret. *Endocrinology, 121*, 1349–1359.

39. Carroll, R. S., Erskine, M. S., Doherty, P. C., Lundell, L. A., & Baum, M. J. (1985). Coital stimuli controlling luteinizing hormone secretion and ovulation in the female ferret. *Biology of Reproduction, 32*, 925–933.

40. Carter, G. R., Chengapa, M. M., & Roberts, A. W. (1995). *Essentials of Veterinary Microbiology* (5th ed.). Williams & Wilkins. (pp. 214–215). Campylobacter and Helicobacter.

41. Chivers, S. M., & Einon, D. F. (1982). Effects of early social experience on activity and object investigation in the ferret. *Developmental Psychobiology, 15*(1), 75–80.

42. Clingerman, K. J., Fox, J. G., & Walke, M. (1991). *Ferrets as Laboratory Animals: A Bibliography*. National Agricultural Library.

43. Coburn, D. R., & Morris, J. A. (1949). The treatment of *Salmonella typhimurium* infection in ferrets. *Cornell Veterinarian, 39*, 183–192.

44. Coleman, G. D., Chavez, M. A., & Williams, B. H. (1998). Cystic prostatic disease associated with adrenocortical lesions in the ferret (*Mustela putorius furo*). *Veterinary Pathology, 35*, 547–549.

45. Creed, J. E., & Kainer, R. A. (1981). Surgical extirpation and related anatomy of anal sacs of the ferret. *Journal of the American Veterinary Medical Association, 179*, 575–577.

46. Cross, M. L., Labes, R. E., & Mackintosh, C. G. (2000). Oral infection of ferrets with virulent *Mycobacterium bovis* or *Mycobacterium avium*: Susceptibility, pathogenesis and immune response. *Journal of Comparative Pathology, 123*(1), 15–21.

47. Dalrymple, E. F. (2004). Pregnancy toxemia in a ferret. *Canadian Veterinary Journal, 45*, 150–152.

48. Davidson, A., Birks, J., Griffiths, H., Kitchener, A., & Biggens, D. (1999). Hybridization and the phylogenetic relationship between polecats and the domestic ferret in Britain. *Biological Conservation, 87*, 155–161.

49. de Lisle, G. W., Kawakami, R. P., Yates, G. F., & Collins, D. M. (2008). Isolation of *Mycobacterium bovis* and other mycobacterial species from ferrets and stoats. *Veterinary Microbiology, 132*, 402–407.

50. Desmukh, D. R. (1987). Reye's syndrome. *Comparative Pathology Bulletin, 19*, 2.

51. Desmukh, D. R., Thomas, P. E., & McArthur, M. B. (1985). Serum glutamate-dehydrogenase and ornithine carbamyl transferase in Reye's syndrome. *Enzyme, 33*, 171–174.

52. Dietz, H. H., Chriél, M., Andersen, T. H., Jørgensen, J. C., Torpdahl, M., & Pedersen, H. (2006). Outbreak of *Salmonella Dublin*-associated abortion in Danish fur farms. *Canadian Veterinary Journal, 47*(12), 1201–1205.

53. Dillberger, J. E. (1985). Polycystic kidneys in a ferret. *Journal of the American Veterinary Medical Association, 186*, 74–75.
54. Dinsdale, J. R., & Rest, J. R. (1995). Yeast infection in ferrets [letter]. *Veterinary Record, 16*, 647.
55. Dunn, D. G., Harris, R. K., Meis, J. M., & Sweet, D. E. (1991). A histomorphologic and immunohistochemical study of chordoma in 20 ferrets (*Mustela putorius furo*). *Veterinary Pathology, 28*, 467–473.
56. Ehrhart, N., Withrow, S. J., Ehrhart, E. J., & Wimsatt, J. H. (1996). Pancreatic beta cell tumor in ferrets: 20 cases (1986–1994). *Journal of the American Veterinary Medical Association, 209*, 1737–1740.
57. Erdman, S. E., Brown, S. A., Kawasaki, T. A., Moore, F. M., Li, X., & Fox, J. G. (1996). Clinical and pathologic findings in ferrets with lymphoma: 60 cases (1982–1994). *Journal of the American Veterinary Medical Association, 208*, 1285–1289.
58. Erdman, S. E., Correa, P., Li, X., Coleman, L. A., & Fox, J. G. (1997). *Helicobacter mustelae*-associated gastric mucosa-associated lymphoid tissue (MALT) lymphoma in ferrets. *American Journal of Pathology, 151*, 273–280.
59. Erdman, S. E., Kanki, P. J., Moore, F. M., Brown, S. A., Kawasaki, T. A., & Mikule, K. W. (1996). Clusters of malignant lymphoma in ferrets. *Cancer Investigation, 14*, 225–230.
60. Erdman, S. E., Li, X., & Fox, J. G. (1998). Hematopoietic diseases. In J. G. Fox (Ed.), *Biology and Diseases of the Ferret* (2nd ed., pp. 231–246). Williams & Wilkins.
61. Erdman, S. E., Reimann, K. A., Moore, F. M., Kanki, P. J., Yu, Q. C., & Fox, J. G. (1995). Transmission of a chronic lymphoproliferative syndrome in ferrets. *Laboratory Investigation, 72*, 539–546.
62. Eshar, D., Wyre, N. R., & Brown, D. C. (2012). Urine specific gravity values in clinically healthy young pet ferrets (*Mustela furo*). *Journal of Small Animal Practice, 53*(2), 115–119.
63. Evans, H. E., & An, N. Q. (1998). Anatomy of the ferret. In J. G. Fox (Ed.), *Biology and Diseases of the Ferret* (pp. 19–69). Williams & Wilkins.
64. Fekete, S. Gy., Fodor, K., Proháczik, A., & Andrásofszky, E. (2005). Comparison of feed preference and digestion of three different commercial diets for cats and ferrets. *Journal of Animal Physiology and Animal Nutrition, 89*, 199–202.
65. Ferguson, D. C. (1985). Idiopathic hypersplenism in a ferret. *Journal of the American Veterinary Medical Association, 186*, 693–695.

66. Field and Laboratory Service Veterinary Staff. (1984). Diseases of the fitch. *Surveillance, 11*, 18.

67. Fisher, M., Beck, W., & Hutchinson, M. J. (2007). Efficacy and safety of selamectin (Stronghold®/Revolution™) used off-label in exotic pets. *International Journal of Applied Research in Veterinary Medicine, 5*(3), 87–96.

68. Forester, N. T., Parton, K., Lumsden, J. S., & O'Toole, P. W. (2000). Isolation of *Helicobacter mustelae* from ferrets in New Zealand. *New Zealand Veterinary Journal, 48*, 65–69.

69. Fox, J. G. (1998). Bacterial and mycoplasma diseases. In J. G. Fox (Ed.), *Biology and Diseases of the Ferret* (2nd ed., pp. 321–354). Williams & Wilkins.

70. Fox, J. G. (1998). Diseases of the gastrointestinal system. In J. G. Fox (Ed.), *Biology and Diseases of the Ferret* (2nd ed., pp. 273–290). Williams & Wilkins.

71. Fox, J. G. (1998). Housing and management. In J. G. Fox (Ed.), *Biology and Diseases of the Ferret* (2nd ed., pp. 173–182). Williams & Wilkins.

72. Fox, J. G. (1998). Mycotic diseases. In J. G. Fox (Ed.), *Biology and Diseases of the Ferret* (2nd ed., pp. 393–404). Williams & Wilkins.

73. Fox, J. G. (1998). Normal clinical and biologic parameters. In J. G. Fox (Ed.), *Biology and Diseases of the Ferret* (2nd ed., pp. 183–210). Williams & Wilkins.

74. Fox, J. G. (1998). Other systemic diseases. In J. G. Fox (Ed.), *Biology and Diseases of the Ferret* (2nd ed., pp. 307–320). Williams & Wilkins.

75. Fox, J. G. (1998). Parasitic diseases. In J. G. Fox (Ed.), *Biology and Diseases of the Ferret* (2nd ed., pp. 375–392). Williams & Wilkins.

76. Fox, J. G., & Bell, J. A. (1998). Growth, reproduction, and breeding. In J. G. Fox (Ed.), *Biology and Diseases of the Ferret* (2nd ed., pp. 211–227). Williams & Wilkins.

77. Fox, J. G., & Marini, R. P. (1998). Diseases of the endocrine system. In J. G. Fox (Ed.), *Biology and Diseases of the Ferret* (2nd ed., pp. 291–305). Williams & Wilkins.

78. Fox, J. G., & McLain, D. E. (1998). Nutrition. In J. G. Fox (Ed.), *Biology and Diseases of the Ferret* (2nd ed., pp. 149–172). Williams & Wilkins.

79. Fox, J. G., Ackerman, J. I., Taylor, N., Claps, M., & Murphy, J. C. (1987). *Campylobacter jejuni* infection in the ferret: An animal model of human campylobacteriosis. *American Journal of Veterinary Research, 48*, 85–90.

80. Fox, J. G., Ackerman, J., & Newcomer, C. E. (1983). Ferret as a potential

reservoir for human campylobacteriosis. *American Journal of Veterinary Research, 44*, 1049–1052.

81. Fox, J. G., Adkins, J. A., & Maxwell, K. O. (1988). Zoonoses in ferrets. *Laboratory Animal Science, 38*, 500.
82. Fox, J. G., Andrutis, K., & Yu, J. (1994). Animal models for *Helicobacter*-induced gastric and hepatic cancer. In R. H. Hunt & G. N. J. Tytgat (Eds.), *Helicobacter pylori: Basic Mechanisms to Clinical Cure* (pp. 504–522). Kluwer Academic Publishers.
83. Fox, J. G., Blanco, M., Yan, L., Shames, B., Polidoro, D., & Dewhirst, F. E. (1993). Role of gastric pH in isolation of *Helicobacter mustelae* from the feces of ferrets. *Gastroenterology, 104*, 86–92.
84. Fox, J. G., Cabot, E. B., Taylor, N. S., & Laraway, R. (1988). Gastric colonization of *Campylobacter pylori* subsp. *mustelae* in ferrets. *Infection and Immunity, 56*, 2994–2996.
85. Fox, J. G., Chilvers, T., Goodwin, C. S., Taylor, N. S., Edmonds, P., & Sly, L. I. (1989). *Campylobacter mustelae*, a new species resulting from the elevation of *Campylobacter pylori* subsp. *mustelae* to species status. *International Journal of Systematic Bacteriology, 39*, 301–303.
86. Fox, J. G., Correa, P., Taylor, N. S., Lee, A., Otto, G., & Murphy, J. C. (1990). *Helicobacter mustelae*-associated gastritis in ferrets: An animal model of *Helicobacter pylori* gastritis in humans. *Gastroenterology, 99*, 352–361.
87. Fox, J. G., Dangler, C. A., Sager, W., Borkowski, R., & Gliatto, J. M. (1997). *Helicobacter mustelae*-associated gastric adenocarcinoma in ferrets (*Mustela putorius furo*). *Veterinary Pathology, 34*, 225–229.
88. Fox, J. G., Edrise, B. M., Cabot, E., Beaucage, C., Murphy, J. C., & Prostak, K. S. (1986). *Campylobacter*-like organisms isolated from gastric mucosa of ferrets. *American Journal of Veterinary Research, 47*, 236–239.
89. Fox, J. G., Hotaling, L., Ackerman, J. B., & Hewes, K. (1986). Serum chemistry and hematology reference values in the ferret (*Mustela putorius furo*). *Laboratory Animal Science, 36*, 583.
90. Fox, J. G., Murphy, J. C., Ackerman, J. I., Prostak, K. S., Gallagher, C. A., & Rambow, V. J. (1982). Proliferative colitis in ferrets. *American Journal of Veterinary Research, 43*, 858–864.
91. Fox, J. G., Murphy, J. C., Otto, G., Pecquet-Goad, M. E., Lawson, G. H. K., & Scott, J. A. (1989). Proliferative colitis in ferrets: Epithelial dysplasia and translocation. *Veterinary Pathology, 26*, 515–517.
92. Fox, J. G., Otto, G., Murphy, J. C., Taylor, N. S., & Lee, A. (1991). Gastric colonization of the ferret with *Helicobacter* species: Natural and experimental infections. *Reviews of Infectious Diseases, 13*, S671–S680.

93. Fox, J. G., Otto, G., Taylor, N. S., Rosenblad, W., & Murphy, J. C. (1991). *Helicobacter mustelae*-induced gastritis and elevated gastric pH in the ferret (*Mustela putorius furo*). *Infection and Immunity, 59*, 1875–1880.

94. Fox, J. G., Palley, L. S., & Rose, R. (1992). Eosinophilic gastroenteritis with Splendore–Hoeppli material in the ferret (*Mustela putorius furo*). *Veterinary Pathology, 29*, 21–26.

95. Fox, J. G., Paster, B. J., Dewhirst, F. E., Taylor, N. S., Yan, L. L., & Macuch, P. J. (1992). *Helicobacter mustelae* isolation from feces of ferrets: Evidence to support fecal–oral transmission of a gastric *Helicobacter*. *Infection and Immunity, 60*, 606–611.

96. Fox, J. G., Pearson, R. C., & Bell, J. A. (1998). Diseases of the genitourinary system. In J. G. Fox (Ed.), *Biology and Diseases of the Ferret* (2nd ed., pp. 247–272). Williams & Wilkins.

97. Fox, J. G., Pearson, R. C., & Gorham, J. R. (1998). Viral diseases. In J. G. Fox (Ed.), *Biology and Diseases of the Ferret* (2nd ed., pp. 355–374). Williams & Wilkins.

98. Frederick, K. A., & Babish, J. G. (1985). Compendium of recent literature on the ferret. *Laboratory Animal Science, 35*, 298–318.

99. Garcia, A., Erdman, S. E., Xu, S., Feng, Y., & Rogers, A. B. (2002). Hepatobiliary inflammation, neoplasia, and argyrophilic bacteria in a ferret colony. *Veterinary Pathology, 39*, 173–179.

100. Garner, M. M., Ramsell, K., Morera, N., Juan-Salles, C., Jimenez, J., & Ardiaca, M. (2008). Clinicopathologic features of a systemic coronavirus-associated disease resembling feline infectious peritonitis in the domestic ferret (*Mustela putorius*). *Veterinary Pathology, 45*, 236–246.

101. Gigliotti, F., Harmsen, A. G., Haidaris, C. G., & Haidaris, P. J. (1993). *Pneumocystis carinii* is not universally transmissible between mammalian species. *Infection and Immunity, 61*, 2886–2890.

102. Gómez-Villamandos, J. C. (1995). Fatal cryptosporidiosis in ferrets (*Mustela putorius furo*): A morphopathologic study. *Journal of Zoo and Wildlife Medicine, 26*(4), 539.

103. Gorham, H. R., & Brandly, C. A. (1953). The transmission of distemper among ferrets and mink. Paper presented at the 90th Meeting of the American Veterinary Medical Association.

104. Gorham, J. R., Gordy, D. R., & Quortrup, E. R. (1949). Salmonella infections in mink and ferrets. *American Journal of Veterinary Research, 10*, 183–192.

105. Gorham, J. R., Leader, R. W., & Henson, K. B. (1964). The experimental transmission of a virus causing hypergammaglobulinemia in mink:

Sources and modes of infection. *Journal of Infectious Diseases, 114*, 341.

106. Greenlee, P. G., & Stephens, E. (1984). Meningeal cryptococcosis and congestive cardiomyopathy in a ferret. *Journal of the American Veterinary Medical Association, 184*, 840–841.
107. Haas, L., Lochelt, M., & Kaaden, O. R. (1988). Detection of Aleutian disease virus DNA in tissues of naturally infected mink. *Journal of General Virology, 69*, 705–710.
108. Hahn, E. W., & Wester, R. C. (1969). *The Biomedical Use of Ferrets in Research*. Marshall Farms Animals.
109. Hamir, A. N., Niezgoda, M., & Rupprecht, C. E. (2011). Recovery from and clearance of rabies virus in a domestic ferret. *Journal of the American Association for Laboratory Animal Science, 50*(2), 248–251.
110. Hammond, J., & Chesterman, F. C. (1972). The ferret. In *UFAW* (Ed.), *The Universities Federation for Animal Welfare Handbook on the Care and Management of Laboratory Animals* (5th ed.). Churchill Livingstone.
111. Hanley, C. S., MacWilliams, P., Giles, S., & Paré, J. (2006). Diagnosis and successful treatment of *Cryptococcus neoformans* variety *grubii* in a domestic ferret. *Canadian Veterinary Journal, 47*(10), 1015–1017.
112. Hart, J. E. (1985). Endocrine factors in hematological changes seen in dogs and ferrets given estrogens. *Medical Hypotheses, 16*, 159–163.
113. Heatley, J. J. (2011). Ferret cardiomyopathy. *Compendium on Continuing Education for the Practicing Veterinarian, 8*, 7–11.
114. Hebeler, D., & Wolf, P. (2001). Fütterung von Frettchen in der Heimtierhaltung. *Kleintierpraxis, 46*, 225.
115. Hein, J., Spreyer, F., Sauter-Louis, C., & Hartmann, K. (2012). Reference ranges for laboratory parameters in ferrets. *Veterinary Record*, *171(9), 218.*
116. Hill, S. L., & Lappin, M. R. (1995). Cryptosporidiosis in the dog and cat. In J. D. Bonagura & R. W. Kirk (Eds.), *Kirk's Current Veterinary Therapy* (pp. 728–731). Saunders.
117. Hillyer, E. V., & Quesenberry, K. E. (1997). Dermatologic diseases. In E. V. Hillyer & K. E. Quesenberry (Eds.), *Ferrets, Rabbits, and Rodents: Clinical Medicine and Surgery* (pp. 119–120). Saunders.
118. Hillyer, E. V., & Quesenberry, K. E. (1997). *Ferrets, Rabbits, and Rodents: Clinical Medicine and Surgery*. Saunders.
119. Hoffmann, K. P., Garipis, N., & Distler, C. (2004). Optokinetic deficits in albino ferrets (*Mustela putorius furo*): A behavioural and electrophysiological study. *Journal of Neuroscience, 24*, 4061–4069.

120. Hoppmann, E., & Barron, H. W. (2007). Ferret and rabbit dermatology. *Journal of Exotic Pet Medicine, 16*(4), 225–237.
121. Howard, J., Marinari, P. E., & Wildt, D. E. (2003). Black-footed ferret: Model for assisted reproductive technologies contributing to *in situ* conservation. In W. V. Holt, A. Pickard, J. C. Rodger, & D. E. Wildt (Eds.), *Reproductive Sciences and Integrated Conservation* (pp. 249–266). Cambridge University Press.
122. Hupfeld, D., Distler, C., & Hoffmann, K. P. (2006). Motion perception deficits in albino ferrets (*Mustela putorius furo*). *Vision Research, 46*(18), 2941–2948.
123. Hutchinson, M. J., Jacobs, D. E., & Mencke, N. (2001). Establishment of the cat flea (*Ctenocephalides felis felis*) on the ferret (*Mustela putorius furo*) and its control with imidacloprid. *Medical and Veterinary Entomology, 15*(2), 212–214.
124. Jackson, C. N., Rogers, A. B., Maurer, K. J., Lofgren, J. L. S., Fox, J. G., & Marini, R. P. (2008). Cystic renal disease in the domestic ferret. *Comparative Medicine, 58*, 161–167.
125. Jenkins, J. R. (2000). Ferret metabolic testing. In A. M. Fudge (Ed.), *Laboratory Medicine: Avian and Exotic Pets* (pp. 305–309). WB Saunders.
126. Jenkins, J. R., & Brown, S. A. (1993). *A Practitioner's Guide to Rabbits and Ferrets*. American Animal Hospital Association.
127. Jones, J. W., Pether, J. V. S., Rainey, H. A., & Swinburn, C. R. (1993). Recurrent *Mycobacterium bovis* infection following a ferret bite [letter]. *Journal of Infection, 26*, 225–226.
128. Kamphues, J., Schneider, D., & Leibetseder, J. (1999). *Supplemente zu Vorlesungen und Übungen in der Tierernährung. M & H Schaper*; Alfeld-Hannover: 262 pp.
129. Kauffman, C. A. (1981). Cell mediated immunity in ferrets: Delayed dermal hypersensitivity, lymphocyte transformation, and macrophage migration inhibitory factor production. *Developmental and Comparative Immunology, 5*, 125–134.
130. Kendrick, R. E. (2000). Ferret respiratory diseases. *Veterinary Clinics of North America: Exotic Animal Practice, 2*, 453–464.
131. Kenyon, A. J., Helmboldt, C. F., & Nielson, S. W. (1963). Experimental transmission of Aleutian disease with urine. *American Journal of Veterinary Research, 24*, 1066.
132. Kidder, J. D., Foote, R. H., & Richmond, M. E. (1998). Transcervical artificial insemination in the domestic ferret (*Mustela putorius furo*). *Zoo Biology, 17*, 393–404.

133. Kim, Y., Chongviriyaphan, N., Liu, C., Russell, R. M., & Wang, X. D. (2006). Combined antioxidant (beta-carotene, alpha-tocopherol, and ascorbic acid) supplementation increases the levels of lung retinoic acid and inhibits the activation of mitogen-activated protein kinase in the ferret lung cancer model. *Carcinogenesis, 27*(7), 1410–1419.

134. Kim, Y., Chongviriyaphan, N., Liu, C., Russell, R. M., & Wang, X. D. (2012). Combined alpha-tocopherol and ascorbic acid protects against smoke-induced lung squamous metaplasia in ferrets. *Lung Cancer, 75*(1), 15–23.

135. Kim, Y., Liu, X. S., Liu, C., Smith, D. E., Russell, R. M., & Wang, X. D. (2006). Induction of pulmonary neoplasia in the smoke-exposed ferret by 4-(methylnitrosamino)-1-(3-pyridyl)-1-butanone (NNK): A model for human lung cancer. *Cancer Letters, 209*, 209–219.

136. Kirkpatrick, C. E., & Dubey, J. P. (1987). Enteric coccidial infections. *Veterinary Clinics of North America: Small Animal Practice, 17*, 1414–1416.

137. Kiupel, M., Desjardins, D. R., Lim, A., Bolin, C., Johnson-Delaney, C. A., & Resau, J. H. (2012). Mycoplasmosis in ferrets. *Emerging Infectious Diseases, 18*(11), 1763–1770.

138. Knight, D. (1987). Heartworm infections. *Veterinary Clinics of North America: Small Animal Practice, 17*, 1463–1518.

139. Korhonen, H. (1965). The effects of environmental enrichment in ferrets. In C. P. Smith & V. Taylor (Eds.), *Environmental Enrichment Information Resources for Laboratory Animals: 1965–1995: Birds, Cats, Dogs, Farm Animals, Ferrets, Rabbits, and Rodents* (AWIV Resource Series No. 2). U.S. Department of Agriculture.

140. Krueger, K. L., Murphy, J. C., & Fox, J. G. (1989). Treatment of proliferative colitis in ferrets. *Journal of the American Veterinary Medical Association, 194*, 1435–1436.

141. Kwochka, K. W. (1986). Canine demodicosis. In R. Kirk (Ed.), *Current Veterinary Therapy 11* (pp. 531–537). Saunders.

142. Le Sueur, C., Bour, S., & Schaper, R. (2011). Efficacy and safety of the combination imidacloprid 10% / moxidectin 1.0% spot-on (Advocate® Spot-on for Small Cats and Ferrets) in the treatment of ear mite infection (*Otodectes cynotis*) in ferrets. *Parasitology Research, 109*(1 Suppl), 149–156.

143. Lee, E. J., Moore, W. E., Fryer, H. C., & Minocha, H. C. (1982). Haematological and serum chemistry profiles of ferrets (*Mustela putorius furo*). *Laboratory Animals, 16*, 133–137.

144. Li, X., & Fox, J. G. (1998). Neoplastic diseases. In J. G. Fox (Ed.), *Biology and Diseases of the Ferret* (2nd ed., pp. 405–448). Williams & Wilkins.

145. Li, X., Fox, J. G., & Padrid, P. A. (1998). Neoplastic diseases in ferrets: 574 cases (1968–1997). *Journal of the American Veterinary Medical Association, 212*, 1402–1406.

146. Li, X., Fox, J. G., Erdman, S. E., & Aspros, D. G. (1995). Cutaneous lymphoma in a ferret (*Mustela putorius furo*). *Veterinary Pathology, 32*, 55–56.

147. Li, X., Fox, J. G., Erdman, S. E., Lipman, N. S., & Murphy, J. C. (1996). Cystic urogenital anomalies in the ferret (*Mustela putorius furo*). *Veterinary Pathology, 33*, 150–158.

148. Li, X., Pang, J., & Fox, J. G. (1996). Coinfection with intracellular *Desulfovibrio* species and coccidia in ferrets with proliferative bowel disease. *Laboratory Animal Science, 46*, 569–571.

149. Liberson, A. J., Newcomer, C. E., Ackerman, J. I., Murphy, J. C., & Fox, J. G. (1983). Mastitis caused by hemolytic *Escherichia coli* in the ferret. *Journal of the American Veterinary Medical Association, 183*, 1179–1181.

150. Lipman, N. S., Marini, R. P., Murphy, J. C., Zhibo, Z., & Fox, J. G. (1993). Estradiol-17β-secreting adrenocortical tumor in a ferret. *Journal of the American Veterinary Medical Association, 203*, 1552–1555.

151. Lipman, N. S., Murphy, J. C., & Fox, J. G. (1987). Clinical, functional, and pathologic changes associated with a case of dilatative cardiomyopathy in a ferret. *Laboratory Animal Science, 37*, 210–212.

152. Liu, C., & Coffin, D. L. (1957). Studies on canine distemper infection by means of fluorescein-labeled antibody. 1. The pathogenesis, pathology, and diagnosis of the disease in experimentally infected ferrets. *Virology, 3*, 115.

153. Loeb, W., & Quimby, F. (1999). *The Clinical Chemistry of Laboratory Animals*. Taylor and Francis.

154. Lucas, J., Lucas, A., & Furber, H. (2000). *Mycobacterium genavense* infection in two aged ferrets with conjunctival lesions. *Australian Veterinary Journal, 78*, 685–689.

155. Ludlow, M., Nguyen, D. T., Silin, D., Lyubomska, O., de Vries, R. D., & von Messling, V. (2012). Recombinant canine distemper virus strain Snyder Hill expressing green or red fluorescent proteins causes meningoencephalitis in the ferret. *Journal of Virology, 86*(14), 7508–7519.

156. Ludwig, E., Reischl, U., Holzmann, T., Melzl, H., Janik, D., & Gilch, C. (2011). Risk for *Mycobacterium celatum* infection from ferrets. *Emerging Infectious Diseases, 17*(3), 553–555.

157. Lugton, I. W., Wobeser, G., Morris, R. S., & Caley, P. (1997). Epidemiology of *Mycobacterium bovis* infection in feral ferrets (*Mustela furo*) in New Zealand: II. Routes of infection and excretion. *New Zealand Veterinary Journal, 45*(4), 151–157.

158. Lunn, J. A., Martin, P., & Zaki, S. (2005). Pneumonia due to *Mycobacterium abscessus* in two domestic ferrets (*Mustela putorius furo*). *Australian Veterinary Journal, 83*, 542–546.

159. Malakoff, R. L., Laste, N. J., & Orcutt, C. J. (2012). Echocardiographic and electrocardiographic findings in client-owned ferrets: 95 cases (1994–2009). *Journal of the American Veterinary Medical Association, 241*, 1484–1489.

160. Manning, D. D., & Bell, J. A. (1990). Lack of detectable blood groups in domestic ferrets: Implications for transfusion. *Journal of the American Veterinary Medical Association, 197*, 84–86.

161. Manning, D., & Bell, J. A. (1990). Derivation of gnotobiotic ferrets: Perinatal diet and hand-rearing requirements. *Laboratory Animal Science, 40*, 51–55.

162. Marini, R. P., Callahan, R. J., Jackson, L. R., Jyawook, S., Esteves, M. I., & Fox, J. G. (1997). Distribution of technetium 99m-labeled red blood cells during isoflurane anesthesia in ferrets. *American Journal of Veterinary Research, 58*, 781–785.

163. Marini, R. P., Fox, J. G., Taylor, N. S., Yan, L., McColm, A., & Williamson, R. (1999). Ranitidine-bismuth citrate and clarithromycin, alone or in combination, for eradication of *Helicobacter mustelae* infection in ferrets. *American Journal of Veterinary Research, 60*, 1280–1286.

164. Marini, R. P., Jackson, L. R., Esteves, M. I., Andrutis, K. A., Goslant, G. M., & Fox, J. G. (1994). The effect of isoflurane on hematologic variables in ferrets. *American Journal of Veterinary Research, 55*, 1479–1483.

165. Marini, R. P., Taylor, N. S., Liang, A. Y., Knox, K. A., & Pena, J. A. (2004). Characterization of hemolytic *Escherichia coli* strains in ferrets: Recognition of candidate virulence factor CNF1. *Journal of Clinical Microbiology, 42*, 5904–5908.

166. Marshall, K. R., & Marshall, G. W. (1973). *The Biomedical Use of Ferrets in Research*. Marshall Farms Animals.

167. Martínez, J., Martorell, J., Abarca, M. L., Olvera, A., Ramis, A., & Woods, L. (2012). Pyogranulomatous pleuropneumonia and mediastinitis in ferrets (*Mustela putorius furo*) associated with *Pseudomonas luteola* infection. *Journal of Comparative Pathology, 146*, 4–10.

168. Martinez, J., Ramis, A. J., Reinacher, M., & Perpiñán, D. (2006). Detection

of feline infectious peritonitis virus-like antigen in ferrets. *Veterinary Record, 158*, 523.

169. Mayer, J., & Burgess, K. (2012). An update on ferret lymphoma: A proposal for a standardized classification of ferret lymphoma. *Journal of Exotic Pet Medicine, 21*, 123–125.
170. Mayer, J., Wagner, R., Mitchell, M. A., & Fecteau, K. (2013). Use of recombinant human thyroid-stimulating hormone for thyrotropin stimulation test in euthyroid ferrets. *Journal of the American Veterinary Medical Association, 243*(10).
171. McCullers, J. A., McAuley, J. L., Browall, S., Iverson, A. R., & Boyd, K. L. (2010). Influenza enhances susceptibility to natural acquisition of and disease due to *Streptococcus pneumoniae* in ferrets. *Journal of Infectious Diseases, 202*, 1287–1295.
172. McLain, D. E., Harper, S. M., Roe, D. A., Babish, J. G., & Wilkinson, C. (1985). Congenital malformations and variations in reproductive performance in the ferret: Effects of maternal age, color, and parity. *Laboratory Animal Science, 35*, 251–255.
173. Miller, D. S., Eagle, R. P., Zabel, S., Rosychuk, R., & Campbell, T. W. (2006). Efficacy and safety of selamectin in the treatment of *Otodectes cynotis* infestation in domestic ferrets. *Veterinary Record, 159*, 748. https://doi.org/10.1136/vr.159.22.748.
174. Miller, L. A., Fagerstone, K. A., Wagner, R. A., & Finkler, M. (2013). Use of a GnRH vaccine, GonaConTM, for prevention and treatment of adrenocortical disease (ACD) in domestic ferrets. *Vaccine, 31*(41), 4619–4623.
175. Miller, W. R., & Merton, D. A. (1982). Dirofilariasis in a ferret. *Journal of the American Veterinary Medical Association, 180*, 1103–1104.
176. Mitchell, M. J. (2008). *Exotic Pet Medicine, 17*(3), 228–229.
177. Moore, D. R. (1982). Late onset of hearing in the ferret. *Brain Research, 235*(1, 2), 309–311.
178. Moore, D. R., & Kowalchuk, N. E. (1988). Auditory brainstem of the ferret: Effects of unilateral cochlear lesions on cochlear nucleus volume and projections to the inferior colliculus. *Journal of Comparative Neurology, 272*, 503–515.
179. Moore, G. E., Glickman, N. W., Ward, M. P., Engler, K. S., & Lewis, H. B. (2005). Incidence of and risk factors for adverse events associated with distemper and rabies vaccine administration in ferrets. *Journal of the American Veterinary Medical Association, 226*, 909–912.
180. Moreland, A. F., Battles, A. H., & Nease, J. H. (1986). Dirofilariasis in a

ferret. *Journal of the American Veterinary Medical Association, 188*, 864.

181. Morgan, J. P., & Travers, K. E. (1998). Use of the ferret in cardiovascular research. In J. G. Fox (Ed.), *Biology and Diseases of the Ferret* (2nd ed., pp. 499–510). Williams & Wilkins.
182. Morrisey, J. K., & Kraus, M. S. (2011). Cardiovascular and other diseases. In K. Quesenberry & J. Carpenter (Eds.), *Ferrets, Rabbits, and Rodents* (3rd ed., pp. 71–73). Saunders.
183. Morrow, D. A. (1980). *Current Therapy in Theriogenology*. Saunders.
184. Mullen, H. (1997). Soft tissue surgery. In E. V. Hillyer & K. E. Quesenberry (Eds.), *Ferrets, Rabbits, and Rodents: Clinical Medicine and Surgery* (pp. 131–144). Saunders.
185. Murray, J., Kiupel, M., & Maes, R. K. (2010). Ferret coronavirus-associated diseases. *Veterinary Clinics of North America: Exotic Animal Practice, 13*, 543–560.
186. Naismith, D. J., & Cursiter, M. C. (1972). Is there a specific requirement for carbohydrate in the diet? *Proceedings of the Nutrition Society, 31*, 94A.
187. Nemelka, K. W., Brown, A. W., Wallace, S. M., Jones, E., Asher, L. V., & Pattarini, D. (2009). Immune response to and histopathology of *Campylobacter jejuni* infection in ferrets (*Mustela putorius furo*). *Comparative Medicine, 59*(4), 363–371.
188. Niezgoda, M., Briggs, D. J., & Shadduck, J. (1998). Viral excretion in domestic ferrets (*Mustela putorius furo*) inoculated with a raccoon rabies isolate. *American Journal of Veterinary Research, 59*, 1629–1632.
189. Noli, C., van der Horst, H., & Wjillemse, T. (1996). Demodicosis in ferrets (*Mustela putorius furo*). *Veterinary Quarterly, 18*, 28–31.
190. Norrby, E., & Oxman, M. N. (1990). Measles virus. In B. N. Fields, D. M. Knipe, & R. M. Chanock (Eds.), *Fields Virology* (pp. 1013–1044). Raven Press.
191. Nowak, R. M. (1999). *Walker's Mammals of the World* (6th ed.). Johns Hopkins University Press.
192. O'Donnell, C. D., & Subbarao, K. (2011). The contribution of animal models to the understanding of the host range and virulence of influenza A viruses. *Microbes and Infection, 13*(5), 502–515.
193. Ohshima, K., Shen, D. T., Henson, J. B., & Gorham, J. R. (1978). Comparison of the lesions of Aleutian disease in mink and hypergammaglobulinemia in ferrets. *American Journal of Veterinary Research, 39*, 653–657.
194. Orcutt, C. (1997). Dermatologic diseases. In E. V. Hillyer & K. E. Quesenberry (Eds.), *Ferrets, Rabbits, and Rodents: Clinical Medicine and Surgery* (pp. 115–125). Saunders.

195. O'Toole, P. W., Snelling, W. J., Canchaya, C., Forde, B. M., & Hardie, K. R. (2010). Comparative genomics and proteomics of *Helicobacter mustelae*, an ulcerogenic and carcinogenic gastric pathogen. *BMC Genomics, 11*, 164.

196. Otto, G., Fox, J. G., Wu, P. Y., & Taylor, N. S. (1990). Eradication of *Helicobacter mustelae* from the ferret stomach: An animal model of *Helicobacter* (*Campylobacter* pylori) chemotherapy. *Antimicrobial Agents and Chemotherapy, 34*, 1232–1236.

197. Oxenham, M. (1990). Aleutian disease in the ferret. *Veterinary Record, 126*, 585.

198. Palley, L. S., Corning, B. F., Fox, J. G., Murphy, J. C., & Gould, D. H. (1992). Parvovirus-associated syndrome (Aleutian disease) in two ferrets. *Journal of the American Veterinary Medical Association, 201*, 100–106.

199. Parker, G. A., & Picut, C. A. (1993). Histopathologic features and post-surgical sequelae of 57 cutaneous neoplasms in ferrets (*Mustela putorius furo* L.). *Veterinary Pathology, 30*, 499–504.

200. Parrott, T. Y., Greiner, E. C., & Parrott, J. D. (1984). *Dirofilaria immitis* in three ferrets. *Journal of the American Veterinary Medical Association, 184*, 582–583.

201. Patterson, M. M., Kirchain, S. M., & Whary, M. T. (1999). Clinical trial to control ear mites in a ferret colony. *Laboratory Animal Science, 49*, 437.

202. Patterson, M. M., Rogers, A. B., Schrenzel, M. D., Marini, R. P., & Fox, J. G. (2003). Alopecia attributed to neoplastic ovarian tissue in two ferrets. *Comparative Medicine, 53*, 213–217.

203. Peltola, V. T., Boyd, K. L., McAuley, J. L., Rehg, J. E., & McCullers, J. A. (2006). Bacterial sinusitis and otitis media following influenza virus infection in ferrets. *Infection and Immunity, 74*, 2562–2567.

204. Pennick, K. E., Stevenson, M. A. M., Latimer, K. S., Ritchie, B. W., & Gregory, C. R. (2005). Persistent viral shedding during asymptomatic Aleutian mink disease parvoviral infection in a ferret. *Journal of Veterinary Diagnostic Investigation, 17*, 594–597.

205. Perpiñán, D., & López, C. (2008). Clinical aspects of systemic granulomatous inflammatory syndrome in ferrets (*Mustela putorius furo*). *Veterinary Record, 162*, 180–184.

206. Perpiñán, D., Ramis, A., Tomás, A., Carpintero, E., & Bargalló, F. (2008). Outbreak of canine distemper in domestic ferrets (*Mustela putorius furo*). *Veterinary Record, 163*, 248–252.

207. Peter, A. T., Bell, J. A., Manning, D. D., & Bosu, W. T. (1990). Real-time ultrasonography determination of pregnancy and gestational age in ferrets. *Laboratory Animal Science, 40*, 91–92.

208. Phillips, P. H., O'Callaghan, M. G., Moore, E., & Baird, R. M. (1987). Pedal *Sarcoptes scabiei* infestation in ferrets (*Mustela putorius furo*). *Australian Veterinary Journal, 64*, 289–290.
209. Poessel, S. A., Biggins, D. E., Santymire, R. M., Livieri, T. M., Crooks, K. R., & Angeloni, L. (2011). Environmental enrichment affects adrenocortical stress responses in the endangered black-footed ferret. *General and Comparative Endocrinology, 172*(3), 526–533.
210. Porter, D. D., Larsen, A. E., & Cox, N. A. (1975). Isolation of infectious bovine rhinotracheitis virus from Mustelidae. *Journal of Clinical Microbiology, 1*, 112–113.
211. Porter, H. G., Porter, D. D., & Larsen, A. E. (1982). Aleutian disease in ferrets. *Infection and Immunity, 36*, 379–386.
212. Powell, D. A., Gonzales, C., & Gunnels, R. D. (1991). Use of the ferret as a model for pediatric endotracheal intubation training. *Laboratory Animal Science, 41*, 86–89.
213. Price, D. J., & Morgan, J. E. (1987). Spatial properties of neurons in the lateral geniculate nucleus of the pigmented ferret. *Experimental Brain Research, 68*(1), 28–36.
214. Prohaczik, A., Kulcsar, M., & Huszenicza, G. (2009). Metabolic and endocrine characteristics of pregnancy toxemia in the ferret. *Veterinary Medicine, 54*(2), 75–80.
215. Pyle, N. J. (1940). Use of ferrets in laboratory work and research investigations. *American Journal of Public Health, 30*, 787.
216. Quesenberry, E., & Carpenter, J. (2012). *Ferrets, Rabbits, and Rodents: Clinical Medicine and Surgery* (3rd ed.). Saunders.
217. Quesenberry, K. E., & Rosenthal, K. L. (1997). Endocrine diseases. In E. V. Hillyer & K. E. Quesenberry (Eds.), *Ferrets, Rabbits, and Rodents: Clinical Medicine and Surgery* (pp. 85–98). Saunders.
218. Qureshi, T., Labes, R. E., Lambeth, M., Montgomery, H., Griffin, J. F. T., & Mackintosh, C. G. (2000). Transmission of *Mycobacterium bovis* from experimentally infected ferrets to non-infected ferrets (*Mustela furo*). *New Zealand Veterinary Journal, 48*(4), 99–104.
219. Ragg, J. R., Moller, H., & Waldrup, K. A. (1995). The prevalence of bovine tuberculosis (*Mycobacterium bovis*) infections in feral populations of cats (*Felis catus*), ferrets (*Mustela furo*), and stoats (*Mustela erminea*) in Otago and Southland, New Zealand. *New Zealand Veterinary Journal, 43*(7), 333–337.
220. Ragg, J. R., Walderup, K. A., & Moller, H. (1995). The distribution of gross lesions of tuberculosis caused by *Mycobacterium bovis* in feral

ferrets (*Mustela furo*) from Otago, New Zealand. *New Zealand Veterinary Journal, 43*, 338–341.

221. Raila, J., Gomez, C., & Schweigert, F. J. (2002). The ferret as a model for vitamin A metabolism in carnivores. *Journal of Nutrition, 132*(6 Suppl 2), 1787S–1789S.
222. Raj, V. S., Smits, S. L., Pas, S. D., Provacia, L. B., Moorman-Roest, H., & Osterhaus, A. D. (2012). Novel hepatitis E virus in ferrets, the Netherlands. *Emerging Infectious Diseases, 18*, 1369–1370.
223. Ramsell, K. A., & Garner, M. M. (2010). Disseminated idiopathic myofasciitis in ferrets. *Veterinary Clinics of North America, 13*, 561–575.
224. Rassnick, K. M. (1995). Use of a vascular access system for administration of chemotherapeutic agents to a ferret with lymphoma. *Journal of the American Veterinary Medical Association, 206*, 500–504.
225. Regh, J. E., Gigliotti, F., & Stokes, D. (1988). Cryptosporidiosis in ferrets. *Laboratory Animal Science, 38*, 155–158.
226. Rosenbaum, M. R., Affolter, V. K., Usborne, A. L., & Beeber, N. L. (1996). Cutaneous epitheliotropic lymphoma in a ferret. *Journal of the American Veterinary Medical Association, 209*, 1441–1444.
227. Rosenthal, K. (1994). Ferrets. *Veterinary Clinics of North America, 24*(1), 1–23.
228. Rosenthal, K. L. (1997). Respiratory diseases. In E. V. Hillyer & K. E. Quesenberry (Eds.), *Ferrets, Rabbits, and Rodents: Clinical Medicine and Surgery* (pp. 77–84). Saunders.
229. Rosenthal, K. L., & Peterson, M. E. (1996). Evaluation of plasma androgen and estrogen concentrations in ferrets with hyperadrenocorticism. *Journal of the American Veterinary Medical Association, 209*, 1097–1102.
230. Rosenthal, K. L., Peterson, M. E., & Quesenberry, K. E. (1993). Hyperadrenocorticism associated with adrenocortical tumor or nodular hyperplasia of the adrenal gland in ferrets: 50 cases (1987–1991). *Journal of the American Veterinary Medical Association, 203*, 271–275.
231. Rupprecht, C. E., Gilbert, J., Pitts, R., Marshall, K. R., & Koprowski, H. (1990). Evaluation of an inactivated rabies virus vaccine in domestic ferrets. *Journal of the American Veterinary Medical Association, 196*, 1614–1616.
232. Ryan, K. D. (1984). Hormonal correlates of photoperiod-induced puberty in a reflex ovulator, the female ferret. *Biology of Reproduction, 31*, 925.
233. Ryland, L. M., & Gorham, J. R. (1978). The ferret and its diseases. *Journal of the American Veterinary Medical Association, 173*, 1154–1158.

234. Ryland, L. M., Bernard, S. L., & Gorham, J. R. (1983). A clinical guide to the pet ferret. *Compendium on Continuing Education for the Practicing Veterinarian, 5*, 25.

235. Saifuddin, M., & Fox, J. G. (1996). Identification of a DNA segment in ferret Aleutian disease virus similar to the hypervariable capsid region of mink Aleutian disease parvovirus. *Archives of Virology, 141*, 1329–1336.

236. Sasai, H., Kato, K., Sasaki, T., Koyama, S., Kotani, T., & Fukata, T. J. (2000). Echocardiographic diagnosis of dirofilariasis in a ferret. *Small Animal Practice, 41*(4), 172–174.

237. Saunders, G. K., & Thomsen, B. V. (2006). Lymphoma and *Mycobacterium avium* infection in a ferret (*Mustela putorius furo*). *Journal of Veterinary Diagnostic Investigation, 18*, 513–515.

238. Schulman, F. Y., Montall, R. J., & Hauer, P. J. (1993). Gastroenteritis associated with *Clostridium perfringens* type A in black-footed ferrets (*Mustela nigripes*). *Veterinary Pathology, 30*, 308–310.

239. Schultheiss, P. C., & Dolginow, S. Z. (1994). Granulomatous enteritis caused by *Mycobacterium avium* in a ferret. *Journal of the American Veterinary Medical Association, 204*, 1217–1218.

240. Scott, D. W., Miller, W. H., & Griffin, C. E. (1995). *Small Animal Dermatology*. Saunders.

241. Shen, D. T., & Gorham, J. R. (1980). Survival of pathogenic distemper virus at 5°C and 25°C. *Veterinary Medicine and Small Animal Clinic, 75*, 69–72.

242. Shen, D. T., Gorham, J. R., Ryland, L. M., & Strating, A. (1981). Using jet injection to vaccinate mink and ferrets against canine distemper, mink viral enteritis, and botulism, type C. *Veterinary Medicine and Small Animal Clinic, 76*, 856–859.

243. Sherrill, A., & Gorham, J. R. (1985). Bone marrow hypoplasia associated with estrus in ferrets. *Laboratory Animal Science, 35*, 280–286.

244. Shimbo, F. M. (1992). *A Tao Full of Detours, the Behavior of the Domestic Ferret*. Ministry of Publications.

245. Shump, A. U., & Shump, K. A., Jr. (1978). Growth and development of the European ferret (*Mustela putorius*). *Laboratory Animal Science, 28*, 89–91.

246. Shump, A., et al. (1974). *A Bibliography of Mustelids. Part 1: Ferrets and Polecats*. Michigan Agricultural Experiment Station 6977.

247. Sinclair, D. G., Evans, E. V., & Sibbald, I. R. (1962). The influence of apparent digestible energy and apparent digestible nitrogen in the diet on weight gain, feed consumption, and nitrogen retention of growing mink.

Canadian Journal of Biochemistry and Physiology, 40, 1375–1389.

248. Sledge, D. G., Bolin, S. R., Lim, A., Kaloustian, L. L., Heller, R. L., & Carmona, F. M. (2011). Outbreaks of severe enteric disease associated with *Eimeria furonis* infection in ferrets (*Mustela putorius furo*) of 3 densely populated groups. *Journal of the American Veterinary Medical Association, 239*(12), 1584–1588. https://doi.org/10.2460/javma.239.12.1584.

249. Smith, L. (1975). *The Pathogenic Anaerobic Bacteria*. Charles C. Thomas.

250. Smith, P. C. (1978). Experimental infectious bovine rhinotracheitis virus infection of English ferrets. *American Journal of Veterinary Research, 39*, 1369–1372.

251. Smith, W., & Stuart-Harris, C. H. (1936). Influenza infection of man from the ferret. *Lancet, 2*, 21.

252. Stamoulis, M. E., Miller, M. S., & Hillyer, E. V. (1997). Cardiovascular diseases. In E. V. Hillyer & K. E. Quesenberry (Eds.), *Ferrets, Rabbits, and Rodents: Clinical Medicine and Surgery* (pp. 63–70). Saunders.

253. Stauber, E., Robinette, J., & Basaraba, R. (1990). Mast cell tumors in three ferrets. *Journal of the American Veterinary Medical Association, 196*, 766–767.

254. Stephensen, C. B., Welter, J., Thaker, S. R., Taylor, J., Tartaglia, J., & Paoletti, E. (1997). Canine distemper virus (CDV) infection of ferrets as a model for testing Morbillivirus vaccine strategies: NYVAC- and ALVAC-based CDV recombinants protect against symptomatic infection. *Journal of Virology, 71*, 1506–1513.

255. Stokes, D. C., Gigliotti, F., Rehg, J. E., Snellgrove, R. L., & Hughes, W. T. (1987). Experimental *Pneumocystis carinii* pneumonia in the ferret. *British Journal of Experimental Pathology, 68*, 267–276.

256. Straube, E. F., & Walden, N. B. (1981). Zinc poisoning in ferrets (*Mustela putorius furo*). *Laboratory Animals, 15*, 45–47.

257. Supakorndej, P., McCall, J. W., & Lewis, R. E. (1992). Biology, diagnosis, and prevention of heartworm infection in ferrets. In *American Heartworm Society* (pp. 59–69). Batavia, IL.

258. Sweet, C., Bird, R. A., Jakeman, K., Coates, D. M., & Smith, H. (1987). Production of passive immunity of neonatal ferrets following maternal vaccination with killed influenza A virus vaccines. *Immunology, 60*, 83–89.

259. Symmers, W., & Thomson, A. (1953). Observations on tuberculosis in the ferret (*Mustela furo*). *Journal of Comparative Pathology, 63*, 20–29.

260. Thomas, P. E., & Desmukh, D. R. (1986). Effect of arginine-free diet on ammonia metabolism in young and adult ferrets. *Journal of Nutrition,*

116, 545–551.

261. Thomson, A. (1951). A history of the ferret. *Journal of the History of Medicine and Allied Sciences, 6*, 471.
262. Thornton, P. C., Wright, P. A., Sacra, P. J., & Goodier, T. E. (1979). The ferret, *Mustela putorius furo*, as a new species in toxicology. *Laboratory Animals, 13*, 119–124.
263. Timm, K. I. (1988). Pruritis in rabbits, rodents, and ferrets. *Veterinary Clinics of North America, 18*, 1088–1089.
264. Tompkins, D. S., Watt, J. I., Rathbone, B. J., & West, A. P. (1988). The characterization and pathological significance of gastric *Campylobacter*-like organisms in the ferret: A model of chronic gastritis? *Epidemiology and Infection, 101*, 269–278.
265. Torres-Medina, A. (1987). Isolation of atypical rotavirus causing diarrhea in neonatal ferrets. *Laboratory Animal Science, 37*, 167.
266. Trautwein, G. W., & Helmboldt, C. F. (1966). Mastitis in mink due to *Staphylococcus aureus* and *Escherichia coli*. *Journal of the American Veterinary Medical Association, 149*, 924–928.
267. USDA APHIS. (2013). *Pseudorabies (Aujeszky's disease) and its eradication*. Retrieved from: http://www.aphis.usda.gov/publications/animal_health/content/printable_version/pseudo_rabies_report.pdf
268. USDHHS. (1996). *Guide for the Care and use of Laboratory Animals* (p. 32). National Academy Press.
269. Valheim, M., Djønne, B., Heiene, R., & Caugant, D. A. (2001). Disseminated *Mycobacterium celatum* (type 3) infection in a domestic ferret (*Mustela putorius furo*). *Veterinary Pathology, 38*(4), 460–463.
270. Viallier, J., Viallier, G., & Prave, M. (1983). Place de *Mycobacterium avium* dans l'épidémiologie mycobactérienne actuelle chez les animaux domestiques et sauvages. *Science et Médicine Vétérinaire Comparée, 85*, 103–109.
271. Vinegar, A., Sinnett, E. E., Kosch, P. C., & Miller, M. L. (1985). Pulmonary physiology as a model for inhalation toxicology. *Laboratory Animal Science, 35*, 246–250.
272. von Messling, V., Springfeld, C., Devaux, P., & Cattaneo, R. (2003). A ferret model of canine distemper virus virulence and immunosuppression. *Journal of Virology, 77*, 12579–12591.
273. Wagner, R. (2012). Hypothyroidism in ferrets. *Association of Exotic Mammal Veterinarians 11th Annual Conference*, Oakland, CA, October 25, pp. 29–31.

274. Wagner, R. A. (2009). Ferret cardiology. *Veterinary Clinics of North America: Exotic Animal Practice, 12*, 115–134.
275. Wagner, R. A., & Dorn, D. P. (1994). Evaluation of serum estradiol concentrations in alopecic ferrets with adrenal gland tumors. *Journal of the American Veterinary Medical Association, 205*, 703–707.
276. Wagner, R. A., et al. (2014). The treatment of adrenal cortical disease in ferrets with 4.7-mg deslorelin acetate implants. *Journal of Exotic Pet Medicine, 18*(2), 146–152.
277. Wagner, R. A., Piché, C. A., & Jöchle, W. (2005). Clinical and endocrine responses to treatment with deslorelin acetate implants in ferrets with adrenocortical disease. *American Journal of Veterinary Research, 66*, 910–914.
278. Wang, X. D., Krinsky, N. L., Marini, R. P., Tang, G., Yu, J., & Hurley, R. (1992). Intestinal uptake and lymphatic absorption of β-carotene in ferrets: A model for human β-carotene metabolism. *American Physiological Society, 263*, G480–G486.
279. Wang, X. D., Marini, R. P., Hebuterne, X., Fox, J. G., Krinsky, N. I., & Russell, R. M. (1995). Vitamin E enhances the lymphatic transport of β-carotene and its conversion to vitamin A in the ferret. *Gastroenterology, 108*, 719–726.
280. Wang, X. D., Russell, R. M., Marini, R. P., Tang, G., Dolnikowski, G., & Fox, J. G. (1993). Intestinal perfusion of β-carotene in the ferret raised retinoic acid level in portal blood. *Biochimica et Biophysica Acta, 1167*, 159–164.
281. Weiss, C. A., Williams, B. H., & Scott, J. B. (1999). Surgical treatment and long-term outcome of ferrets with bilateral adrenal tumors or adrenal hyperplasia: 56 cases (1994–1997). *Journal of the American Veterinary Medical Association, 215*, 820–823.
282. Weiss, C. A., Williams, B. H., & Scott, M. V. (1998). Insulinoma in the ferret: Clinical findings and treatment comparison of 66 cases. *Journal of the American Animal Hospital Association, 34*, 471–475.
283. Weiss-Buerger, M. (1981). An investigation of the influence of exploration and playing on learning by polecats, *Mustela putorius* × *Mustela furo*. *Zeitschrift für Tierpsychologie, 551*, 33–62.
284. Welchman, D. D. B., Oxenham, M., & Done, S. H. (1993). Aleutian disease in domestic ferrets: Diagnostic findings and survey results. *Veterinary Record, 132*, 479–484.
285. Whisson, D., & Moore, T. (1997). An annotated bibliography on the ferret (*Mustela putorius furo*). *California Department of Fish and Game,*

Wildlife Management Division, Bird and Mammal Conservation Program Report 97-3, Sacramento, CA, 37 pp.

286. Wildt, D. E., Bush, M., & Morton, C. (1989). Semen characteristics and testosterone profiles in ferrets kept in a long-day photoperiod, and the influence of hCG timing and sperm dilution medium on pregnancy rate after laparoscopic insemination. *Journal of Reproduction and Fertility, 86*, 349–358.
287. Williams, B. H., Chimes, M. J., & Gardiner, C. H. (1996). Biliary coccidiosis in a ferret (*Mustela putorius furo*). *Veterinary Pathology, 33*, 437–439.
288. Williams, B. H., Eighmy, J. J., Berbert, M. H., & Dunn, D. G. (1993). Cervical chordoma in two ferrets (*Mustela putorius furo*). *Veterinary Pathology, 30*, 204–206.
289. Williams, B. H., Kiupel, M., West, K. H., Raymond, J. T., Grant, C. K., & Glickman, L. T. (2000). Coronavirus-associated epizootic catarrhal enteritis in ferrets. *Journal of the American Veterinary Medical Association, 217*, 526–530.
290. Williams, E. S., & Barker, I. K. (2001). *Infectious diseases of wild mammals* (3rd ed.). Iowa State University Press. ISBN: 9780813825564.
291. Williams, E. S., Thorne, E. T., & Kwiatkowski, D. R. (1992). Comparative vaginal cytology of the estrous cycle of black-footed ferrets (*Mustela nigripes*), Siberian polecats (*M. eversmanni*), and domestic ferrets (*M. putorius furo*). *Journal of Veterinary Diagnostic Investigation, 4*, 38–44.
292. Willis, L. S., & Barrow, M. V. (1971). The ferret (*Mustela putorius furo*) as a laboratory animal. *Laboratory Animal Science, 21*, 712–716.
293. Wise, A. G., Kiupel, M., & Maes, R. K. (2006). Molecular characterization of a novel coronavirus associated with epizootic catarrhal enteritis (ECE) in ferrets. *Virology, 349*, 164–174.
294. Wise, A. G., Kiupel, M., Garner, M. M., Clark, A. K., & Maes, R. K. (2010). Comparative sequence analysis of the distal one-third of the genomes of a systemic and an enteric ferret coronavirus. *Virus Research, 149*, 42–50.
295. Wise, A. G., Smedley, R. C., Kiupel, M., & Maes, R. K. (2009). Detection of group C rotavirus in juvenile ferrets (*Mustela putorius furo*) with diarrhea by reverse transcription polymerase chain reaction: Sequencing and analysis of the complete coding region of the VP6 gene. *Veterinary Pathology, 46*(5), 985–991.
296. Wyre, N. R., Michels, D., & Chen, S. (2013). Selected emerging diseases in ferrets. *Veterinary Clinics of North America: Exotic Animal Practice, 16*(2), 469–493.

297. Yu, J., Russell, R. M., Salomon, R. N., Murphy, J. C., Palley, L. S., & Fox, J. G. (1995). Effect of *Helicobacter mustelae* infection on epithelial cell proliferation in ferret gastric tissues. *Carcinogenesis, 16*, 1927–1931.

CHAPTER 9

INFECTIOUS DISEASES OF GUINEA PIGS

Contents

Introduction.. 398

9.1. Overview of Guinea Pigs.. 398

9.2. Diseases in Guinea Pigs.. 407

References.. 427

INTRODUCTION

Infectious diseases in guinea pigs are a significant concern for pet owners and breeders due to their susceptibility to various pathogens, including bacteria, viruses, fungi, and parasites. Guinea pigs are social animals that thrive in groups, but this close contact can facilitate the spread of infections, making early detection and prevention critical for maintaining their health.

Bacterial infections are among the most common issues in guinea pigs. *Bordetella bronchiseptica* and *Streptococcus pneumoniae* are two primary respiratory pathogens that can cause severe illnesses. Symptoms such as sneezing, coughing, nasal discharge, and lethargy are common, and if untreated, these infections can lead to pneumonia and death. Antibiotic treatment is often necessary, but prevention through proper hygiene and avoiding exposure to infected animals is key.

Viral infections, though less common, can still pose risks. Guinea pigs are susceptible to viral infections like *Lymphocytic choriomeningitis virus* (LCMV), which can cause symptoms ranging from mild respiratory distress to severe neurological issues. Viral infections are challenging to treat, so prevention through strict biosecurity measures is essential. Fungal infections, such as ringworm, are also a concern, particularly in young or immunocompromised guinea pigs. Ringworm, caused by dermatophyte fungi, leads to hair loss and skin lesions and is highly contagious to both other animals and humans. Treatment typically involves antifungal medications and maintaining a clean environment. Parasitic infestations, including mites and lice, can lead to intense itching, hair loss, and secondary bacterial infections due to scratching. Regular monitoring and prompt treatment with appropriate medications can help control these infestations. Preventing infectious diseases in guinea pigs involves maintaining a clean environment, providing a balanced diet, and ensuring regular veterinary check-ups. Isolation of new or sick guinea pigs, good husbandry practices, and early intervention are critical to keeping these animals healthy and preventing the spread of infections.

9.1. OVERVIEW OF GUINEA PIGS

Like chinchillas, guinea pigs are rodents classified as hystricognaths. They are members of the family Cavidae, which includes the 14 species of animals known as Patagonian hares (or maras) and cavies. Cavidae are distinguished by having four digits on the forepaw and three on the hindfoot. A guinea pig's features include a large head, short legs, unfurred ears, and a stocky build. It weighs 500–1500 g, has a head and body length of 200–400 mm, and no external tail. Wild cavies live in rocky areas, savannas, forest edges, and swamps throughout South America, extending from Venezuela and Columbia to Brazil and northern Argentina. They inhabit burrows dug by themselves or by other animals, and they can number up to 10 people in a group. They forage for a variety of plant materials at night, when they are most active. The guinea pig was first domesticated at least 900 BCE, and it may have started as early as 5000 BCE.

Members of the Caviidae family, which also includes several South American rodent species, are guinea pigs, also known as cavies. Since cavies are social creatures, they all like living in groups. There is no wild species of *Cavia porcellus*, the domestic guinea pig. Related species consume grass and leaves while living in the wild on grassy plains. Domestic guinea pigs are valued research animals as well as well-liked pets. Domestication of cavies began at least 900 BCE, and possibly as early as 5000 BCE. It is most likely from the highlands of Bolivia, Ecuador, and Peru that domesticated cavies first appeared. Native Americans in these regions used them for sustenance and spiritual purposes. These tamed animals were discovered by Europeans not long after they arrived in South America. In the 1500s, cavies were first brought to Europe. They are kept as pets all over the world nowadays.

9.1.1. Description and Physical Characteristics of Guinea Pigs

Like other cavies, guinea pigs are stocky and have short legs.

Table 9.1. *Cardiorespiratory and Metabolic Measurements in Guinea Pigs*

Parameters	**Guinea pig normoxia**	**Guinea pig acute hypoxia**	**Rat normoxia**	**Rat acute hypoxia**
MABP (mm Hg)	50 ± 3	$34 \pm 2^{***}$	115 ± 14	$53 \pm 4^{**}$
pO_2 (mm Hg)	60 ± 3	$20 \pm 1^{***}$	85 ± 3	$39 \pm 2^{***}$
pCO_2 (mm Hg)	32 ± 1	36 ± 3	32 ± 2	30 ± 3
Glucose (mg/dL)	140 ± 11	$195 \pm 17^{*}$	147 ± 16	$229 \pm 20^{**}$
Lactate (mmol/L)	0.92 ± 0.06	$1.87 \pm 0.38^{*}$	1.66 ± 0.17	$4.8 \pm 0.4^{***}$

Mean arterial blood pressure (MABP) and arterial blood levels of pO_2, pCO_2, glucose, and lactate from guinea pigs and rats in normoxia (air) and acute hypoxia (10% O_2; 10 min) conditions. Data are mean ± SEM; n = 8.

$^{*}p < 0.05$;

$^{**}p < 0.01$;

$^{***}p < 0.001$ *vs. normoxia.*

Source: Gonzalez-Obeso E, Docio I, Olea E, Cogolludo A, Obeso A, Rocher A and Gomez-Niño A (2017) Guinea Pig Oxygen-Sensing and Carotid Body Functional Properties. Front. Physiol. 8:285. doi: 10.3389/fphys.2017.00285.

Adults that are fully grown typically measure between 8 and 19 inches (20 and 50 cm) in length. The average adult guinea pig's height is five inches (13 cm). An adult weighs 30 to 35 ounces (850–1000 g) on average. Their typical body temperature ranges from 39□ to 40□, or 102°F to 104°F. Although guinea pig lifespans vary, they typically last six to eight years (Table 9.1).

As of right now, guinea pigs are recognized in at least 13 breeds. The American breed, which has short, smooth hair, the Abyssinian breed, which has short hair that grows in whorls, and the Peruvian breed, which has longer, silkier hair, are some of the more popular breeds. There have also been developed breeds without hair. A variety of colors and color combinations are available for guinea pigs, such as black, tan, cream, brown, and white.

Guinea pigs can see both forward and backward thanks to the eyes on either side of their heads, though they might have problems looking straight ahead. Their ears are small, and their heads are blunt. Each of the guinea pigs' four front toes and three hind toes are called digits. Every toe has an extremely sharp claw. They have longer hind legs than front legs. They don't have an outward tail.

20 teeth total upper and lower incisors for cutting and tearing, premolars, and molars are present in guinea pigs. The teeth of guinea pigs are "open rooted," meaning they continue to grow throughout their lives. Guinea pigs grind, chew, and consume food, which wears down their teeth. For this reason, it's critical that your guinea pig eats enough hay, grass, and abrasive foods to keep their teeth healthy and at the right length. The normal cheek fat pads can make it challenging to examine guinea pig teeth.

9.1.2. Special Considerations for Guinea Pigs

Only the small temperature range of 65°F to 75°F (18□ to 23□) is comfortable for guinea pigs. Additionally, they favor low relative humidity (less than 50%). If the temperature rises above 85°F (29□), they could get heat stroke. Therefore, extra caution must be taken when transferring these animals on extremely hot or cold days. They should not be left in direct sunlight all the time and do not do well in drafty areas.

Guinea pigs are tiny animals that frequently sustain injuries when left outside of their cages. If they fall off a table or are stepped on, they could sustain severe injuries. Guinea pigs are drawn to anything that appears intriguing to chew on, just like many other rodents. These animals frequently encounter electrical cords as hazards. Guinea pig injuries are easily caused by other pets, including dogs, cats, and ferrets. Furthermore, a lot of common houseplants are poisonous to guinea pigs. To ensure its safety, your guinea pig needs to be closely watched whenever it is outside of its cage.

Guinea pigs are primarily diurnal, meaning they typically wake up during the day and go to sleep at night, in contrast to hamsters and rats. This makes them ideal for keeping as pets because during the day they will be active and you can watch over and interact with them without disturbing them while they sleep.

Having multiple guinea pigs offers your pet company and companionship. It is impossible for guinea pigs to mate if you only keep them together. Male guinea pigs should either be neutered or introduced to one another before finishing their weaning process from their mothers if you intend to keep them together. This will assist in averting fights. Keep guinea pigs away from dogs, cats, and other pets as they may become frightened or stressed out if they come into contact with them. Additionally, this will aid in stopping the spread of contagious illnesses.

Guinea pigs are extremely sensitive to a wide range of widely used antibiotics, whether they are administered orally, intravenously, or topically as creams. Your guinea pig may be poisoned by several of the most commonly used antibiotics. A veterinarian should always be consulted before giving your guinea pig antibiotics or products containing antibiotics.

9.1.3. Selecting a Guinea Pig

Look for a guinea pig that seems alert, plump, and healthy when selecting one for a pet. Although the animal might be wary or fearful at first, it should soon warm up to gentle caresses. Examine the animal thoroughly. The eyes should not be lustrous or crusty, but rather clear and bright. There should be no discharge or discoloration and clear visibility in the nose, eyes, ears, and anus. The long incisors in the upper jaw should overlap and barely touch the bottom incisors, and the teeth should be spotless and clean. Make sure the feet are well-formed and have easy mobility. There should be no wounds, red areas, or scars on the feet.

Additionally, take a peek at the guinea pigs' housing. The enclosure or cage needs to be odor free and spotless. Overcrowding can cause stress in the animals and lower their resistance to illness, so it's best to keep them apart. Male and female groups are frequently used to house guinea pigs. It is common for a fairly young female guinea pig to be pregnant at the time of purchase because female guinea pigs can become pregnant as early as two months of age.

Concerning a prepurchase veterinary check, ask the seller. A conscientious dealer will consent to either a third-party veterinarian's inspection or the return of an animal that is ill or pregnant after it has been checked out right away. Before bringing your new guinea pig home, or as soon as possible after, try to get it checked out by a veterinarian (Collins, 1988; Harkness and Wagner, 1989).

9.1.4. Home for a Guinea Pig

Make sure the cage is ready, that the right food is on hand, and that you have already bought and set up the cage before bringing your guinea pig home. A suitable brush and some bedding for the cage should also be available.

9.1.4.1. Housing

Guinea pig commercial housing is frequently undersized. The majority of guinea pig cages on the market are made for just one animal, not two or more. A single guinea pig should have a space of seven square feet (0.7 m^2) (a 42 by 24 inch [106 by 61 cm] cage would be an example of this). For every additional guinea pig, you should allot two to four square feet (0.2 to 0.4 m^2) of floor space. It is ideal to have a cage height of 18 inches (46 cm). These sized cages give your guinea pig plenty of room for play and other activities that enhance its quality of life.

You can build cages out of wire, glass, plastic, or metal. Guinea pigs need good ventilation to stay healthy, so if a cage with solid sides is chosen, the top should have a wire mesh that lets in lots of air. Since guinea pigs cannot jump, a lid is not necessary to restrict their activity. To shield the guinea pig from other household pets, a lid might be required.

Guinea pigs' small feet are prone to injury from stepping on wire mesh and other narrow objects, so the bottom of their cage should be smooth and free of any uneven surfaces like mesh shelves, floor grids, or ramps. A little wooden or cardboard box placed inside the cage will give the animal a feeling of security and a comfortable place to sleep.

Your guinea pig's house should be placed somewhere between 65°F and 75°F (18□ and 24□) that is free from drafts and direct sunlight. This needs to be done away from places with a lot of moisture, like damp basements or laundry rooms. It is not advisable to keep guinea pigs and rabbits in the same cage or near each other as certain infectious diseases can spread between the two species.

It is not appropriate to let your pet's bedding become wet or to retain moisture in it. The best options for bedding are kiln-dried pine or commercially available chapter-based bedding. Avoid using cedar and raw pine because they contain oils that irritate the lungs of guinea pigs. Chapter bedding is frequently chosen over wood shavings because it is more easily digested than wood, which can clog the digestive system. Both sawdust and cat litter are bad options because they can be eaten by children and inhaled, damaging the lungs. Over the bedding, high-quality, soft grass may be provided for your guinea pig to chew on, but high-quality grass hay—timothy and orchard grass are popular is usually a better option because it will give your pet more tooth abrasion and wearing down that are necessary for good dental health.

9.1.4.2. Diet

Guinea pigs are plant-eating animals or herbivores. The main ingredients of a high-quality guinea pig diet are commercial pellets, hay, fruits, and vegetables. A good "dinner plate" is a small, heavy, smooth-surfaced ceramic dish because they are easy to clean and are not easily toppled.

While most guinea pigs are happy eating nearly any kind of vegetable, they have a preference for leafy greens like lettuce and carrot tops. Instead of iceberg lettuce, which is less nutrient-dense, use romaine or green leaf lettuce. Guinea pigs also frequently take pleasure in nibbling on tiny carrot pieces and occasionally in slicing up high-fiber fruits like apples or pears. After a few hours, any uneaten fresh food should be disposed of. All the time, there should be access to high-quality grass hay.

The majority of your pet's diet should consist of guinea pig pellets, which are easily found in pet stores. Scanning the label is a good idea before purchasing pellets. Pellets of high quality should have little corn content and be free of animal byproducts. A lot of guinea pig pellets contain a lot of alfalfa. These pellets are good for guinea pigs that are growing, young, and pregnant; however, a pellet made of timothy, which has less calcium, is a better option for animals that are fully grown. See your veterinarian for recommendations regarding pellet content. If given sufficient access to hay and fresh produce, adult guinea pigs can consume approximately 1/8 cup of pellets daily. But you should modify the amount of food according to how much your pet eats.

Like humans, guinea pigs are incapable of synthesizing vitamin C. To prevent scurvy, they need to consume a lot of this vitamin in their diets. Pellets designed for other animals (like rabbits) shouldn't be used in place of guinea pig pellets as they won't have enough vitamin C. Vitamin C levels in commercially prepared guinea pig pellets should be adequate. If the pellets are kept in storage for longer than a few months, a significant portion of the vitamin C may be lost. It's a good idea to get enough vitamin C from other sources, like fresh fruits and vegetables that are high in the nutrient. Additionally, your guinea pig might require a daily vitamin C supplement. Ask your veterinarian what's best for your animal companion.

Always have access to fresh water to avoid dehydration. Minimum contamination can be achieved by using a drip bottle that is affixed to the cage's side at a height that is convenient for your pet to reach. Every day, replace the water and clean the drip bottle. Avoid using distilled water and water with a high mineral content, particularly water with a high calcium content, and do not add any supplements to the water. Guinea pigs have a finicky appetite. They frequently form food preferences at a very young age and are resistant to dietary changes. Finding out what foods the guinea pig is used to eating from the retailer or previous owner is a smart idea when buying or adopting a new one. To expand the variety of foods your young guinea pig will eat, try small amounts of various pellets, fruits, and vegetables if you are buying or raising one.

9.1.4.3. Exercise

Regular exercise is essential to your guinea pig's health. A well-thought-out cage arrangement will promote activity in your pet. Activity is encouraged by placing the food in one area of the cage, the water in another, and the sleeping area (a tiny wooden or cardboard box) in yet another. A tiny barrier or an additional level with a solid floor can be added to promote exercise even more. A few pieces of PVC pipe, no more than five inches (13 cm) in diameter, can be added to your pet's cage to give them the extra exercise they need. Exercise wheels are not advised because they can harm guinea pigs.

9.1.4.4. Temperament

Guinea pigs are kind, sociable, amiable, and frequently very talkative animals. They prefer to live in small groups—with no more than one male in a mixed group; if a male is kept with females, it is best to neuter him to avoid unintentional breeding). A variety of sounds, including a "squeak," "chirp," and a "tooth-chatter" sound that resembles a purr, can be made by guinea pigs. Even when handled incorrectly, guinea pigs seldom bite. A guinea pig should be carefully picked up by placing one hand under its chest and the other supporting its hindquarters. When handled as young animals, guinea pigs are frequently very affectionate and love being held with care.

9.1.5. Routine Health Care of Guinea Pigs

Although there are no vaccinations needed for guinea pigs, it is advised that you take your pet in for a routine checkup with a veterinarian who has experience with guinea pigs at least once a year.

9.1.5.1. Signs of Illness

Daily care should be given to guinea pigs. This lets you spend quality time with your pet on a regular basis. Additionally, it gives you the chance to examine your pet for potential wounds, skin issues, unexpected weight gain or loss, dental issues, and other health issues. A guinea pig's lack of appetite, weight loss, hunched posture, abnormal gait or limp, abnormally large or skinny belly, alteration in the consistency of the hair coat, and difficulty breathing are some indicators that the animal is ill. A sick guinea pig may exhibit reduced energy or show no reaction to touch or noises.

These animals' digestive or respiratory systems are typically the targets of health issues. A sick guinea pig may exhibit nasal or eye discharge in addition to diarrhea. Examine your pet's mouth for signs of dental issues, such as swelling, overgrown teeth, or drooling. Additionally, you should look for sores or broken nails on your pet's feet and check its ears for leaks or irritation.

It is best to take the guinea pig to the vet right away if you see any of these symptoms. These tiny animals can get sick easily, so it's important to diagnose and treat any issues as soon as possible.

9.1.5.2. Grooming

Regular nail trims are necessary for guinea pigs (whenever your pet needs). Beginning at an early age, clip the nails of guinea pigs with a small animal or human nail clipper. The quick, which is where the blood vessels are located, should not be cut as this can result in bleeding and pain. It might be simpler to see the quick in a dark nail if a strong light is shone through it. Trim the nail neatly to a distance of approximately 0.3 cm, or an eighth of an inch, beyond the quick's end. Use styptic powder or a styptic pencil to stop the bleeding if you unintentionally cut into the quick. Your pet will grow acclimated to this necessary care if you begin clipping their nails at an early age and proceed with caution.

Bathing guinea pigs is rarely necessary. Use a shampoo designed specifically for small animals if necessary. Shampoos designed for humans will cause guinea pig skin to become dry. Avoid getting water or shampoo in the guinea pigs' eyes or ears when giving them a bath in a shallow bowl of warm water. To avoid giving your pet the chills, make sure you rinse them well and dry them right away. Long-haired guinea pigs require periodic brushing and grooming to keep their hair tidy and avoid matting (Collins, 1988; Harkness and Wagner, 1989).

9.1.6. Breeding and Reproduction of Guinea Pigs

Veterinarians generally advise against guinea pig breeding for individual pet owners. For young guinea pigs, it can be challenging to find homes with responsible and loving pet owners. Female guinea pigs typically live shorter lives as a result of breeding. Due to the normal stiffening of the pubic symphysis, which is a joint of tough fibrous cartilage between the two pubic bones of the pelvis, which occurs when the female reaches adulthood, breeding a female guinea pig after she reaches 8 months of age can be extremely dangerous. If a sow (a female pig reaches adulthood without having given birth before, she might not be able to give birth to her young in a normal way. Even when carried out by a veterinarian with guinea pig experience, cesarean sections are rarely successful in guinea pigs.

A veterinarian with experience in treating small animals can spay or neuter pet guinea pigs, but the procedure is not without risk and can be costly. In general, neutering a male is less complicated than spaying a female. Nevertheless, guinea pigs frequently have negative reactions to anesthesia, surgery, or being housed in strange environments. Separating male and female guinea pigs in their housing is the most secure, practical, and affordable method of preventing reproduction.

If you do choose to have your guinea pig neutered or spayed, look for a veterinarian who has performed many of these procedures successfully. As your pet heals from surgery, keep them quiet and away from other guinea pigs. It is effective to use a tiny, tidy cage with towels for bedding. At least two towel changes should be made each day. Verify that the guinea pig is getting enough food and liquids; additional vitamin C may also be beneficial.

Boars, or male guinea pigs, can mate as early as two to three months of age after reaching sexual maturity. Usually, female guinea pigs, or sows, reach sexual maturity at two months of age (55 to 70 days). But males and females can reach maturity and fertility even earlier. Estrous or fertile periods can occur in females at any time of the year, though they are most frequent in the spring. There are 16 days in the estrous cycle. A female's fertile period lasts between six and 11 hours, usually at night. A new estrous cycle starts soon after a female guinea pig gives birth.

Guinea pigs take between 59 and 72 days to gestate. Although a litter of two to four pups is more typical, litter sizes vary from one to eight. Every year, a female can give birth to up to five litters. Any male guinea pigs nearby will congregate and attempt to establish themselves as the dominant male to mate and protect the pregnant female. All of the males will attempt to mate with the new mother during the female's brief postpartum estrus, which lasts for only about 30 minutes.

9.1.7. Care of Newborns

A newborn guinea pig weighs approximately 3 oz or 100 g. They can see and run, and they have hair, even at such a small size. Although they will still nurse from their mother, after two days they can start to nibble on solid food (like moistened pellets). The infants will nurse from the mother's lone pair of mammary glands for approximately three weeks. The offspring will stop nursing and start eating the same foods as their parents when they weigh about six ounces (170 g).

It is crucial to start giving young guinea pigs careful handling when they are just two or three weeks old. This is the time when young guinea pigs learn to tolerate being held and form bonds with humans (Collins, 1988; Harkness and Wagner, 1989).

9.2. DISEASES IN GUINEA PIGS

New World, montane, hystricomorph rodents commonly kept as pets are called guinea pigs. Guinea pigs are superficially similar to mice and rats, but they differ in several anatomical and physiological ways that may make them more susceptible to certain diseases. The diagnosis and management of diseases in guinea pigs are covered in this chapter.

9.2.1. Gastrointestinal Diseases

9.2.1.1. Antibiotic-Associated Enterotoxemia

The gastrointestinal flora of guinea pigs is primarily gram-positive, and they are highly susceptible to antibiotics, which destroy this flora. The most vulnerable gram-positive bacteria are destroyed by medications like penicillin, ampicillin, chlortetracycline, clindamycin, erythromycin, and lincomycin, which allows *Clostridium difficile* to overgrow and develop its toxin. *Hemorrhagic typhlitis* and secretory diarrhea are brought on by the hyperactivity of secretomotor neurons brought on by toxin production. In clinical terms, the guinea pig exhibits anorexia, dehydration, and hypothermia along with diarrhea. Lesions, clinical symptoms, and history are typically used to make the diagnosis because *C. difficile* is challenging to separate. Enterotoxemia linked to antibiotics is managed symptomatically. To treat hypothermia, supportive therapy may involve putting the animal in an incubator or placing its cage on a recirculating water blanket. Restoring hydration can be accomplished by intravenous, intraperitoneal, or subcutaneous injection of lactated Ringer's solution or other crystalline fluids. An animal that is dehydrated should receive a daily maintenance dose of 10 ml/100 g until it is drinking normally. By giving a commercial Lactobacillus product, intestinal microflora can be restored. As an alternative, 5 ml of live-culture yogurt can be taken orally each day. Transfaunation using healthy guinea pig flora has shown to be more successful than refaunation using rodent fecal slurry, even though the former has been done. Chloramphenicol 50 mg/kg PO given every eight hours may help prevent additional clostridial overgrowth.

Antibiotic-associated enterotoxemia can be avoided by giving the right antibiotics to guinea pigs. Guinea pigs respond well to trimethoprim-sulfa (30 mg/kg SC, IM, or PO q12h × seven days), chloramphenicol (50 mg/kg PO q12h × seven days), and enrofloxacin (10 mg/kg PO q12h). Although well tolerated at 6 mg/kg SC given every eight hours for five days, sodium ampicillin at 8 mg/kg is toxic. Cefazolin at 50 mg/kg IM, SC is well tolerated; however, it would need to be given every 30 minutes to be effective against Bordetella, a common pathogen in guinea pigs. Elevated cefazolin dosages result in injection site irritation and fatality. When a guinea pig is treated with antibiotics, the animal should also take a Lactobacillus supplement both during the course of the treatment and for five days after the antibiotics are stopped.

9.2.1.2. Bacterial Enteritis

For guinea pigs, *Salmonella typhimurium* and *Salmonella enteritidis* are the most frequent causes of bacterial enteritis. Feed contaminated by feces is the mode of transmission. Stressed animals are more vulnerable, especially weanlings, pregnant sows, elderly animals, and animals with nutritional deficiencies. Abortion, conjunctivitis, weakness, weight loss, and a scruffy coat are among the symptoms. There may or may not be diarrhea. The liver and spleen are enlarged at necropsy, and there may be yellow necrotic foci in the viscera. A stool sample is used for sensitivity testing and culture to make the diagnosis. Given the possibility that the afflicted animal could develop into an asymptomatic carrier, treatment consists of administering the proper amount of antibiotics and fluids. By maintaining a clean environment, storing food in airtight containers, and giving all fresh fruits and vegetables given to guinea pigs a thorough wash, the disease can be avoided. The disease salmonellosis is a zoonotic one.

Yersinia pseudotuberculosis, *Clostridium perfringens*, *Escherichia coli*, *Pseudomonas aeruginosa*, and *Listeria monocytogenes* are additional causes of bacterial diarrhea in guinea pigs. Similar to Salmonella, these are acquired by eating contaminated food. *Y. Pseudotuberculosis* can result in intestinal and local lymph node abscesses. *E. coli* can lead to depression, wasting, and even death in weanlings. Yellow fluid may be found in the intestines. Culture and sensitivity analysis are the cornerstones of antibiotic treatment. Furthermore recommended is supportive therapy. In guinea pigs, Tyzzer's disease has been documented as a result of *Clostridium piliforme* (previously *Bacillus piliformis*). The fecal-oral route is the means by which the organism is spread, and young, stressed animals are more susceptible. An unthrifty appearance, diarrhea, and sudden death are symptoms of Tyzzer's disease. Dependent subcutaneous edema and an excess of serous fluid were noted in one case. Intestinal inflammation and focal hepatic necrosis are among the lesions found during necropsy. Since *C.*

piliforme is an intracellular bacterium, standard culture media will not support its growth. The identification of the organisms on hematoxylin and eosin- or silver-stained sections of the liver or intestine provides the final diagnosis during necropsy. The care has been unsatisfactory. Good husbandry practices and stress reduction, especially during weaning, can prevent this disease.

9.2.1.3. Parasitic Diarrhea

Cryptosporidium wrairi is the cause of protozoal diarrhea in guinea pigs, a condition that affects the small intestine. The fecal-oral route is the mode of transmission. Although immunocompetent adults can be impacted, weanlings, and immunosuppressed animals are more vulnerable. Nine indicators include diarrhea, losing weight, not gaining weight, and passing away. Immunocompetent guinea pigs are resistant to reinfection and recover in four weeks. The organisms can be detected histopathologically on the brush border of mucosal epithelial cells or observed during fecal examination. There is no proven treatment. The oocysts of *C. wrairi* can be killed in the environment by heating it to above 149°F (65□), freezing it to below 32°F (0□), or using a 5% ammonia solution. Cryptosporidiosis is a disease that can spread to animals.

Paraspidodera uncinata (a roundworm) and *Balantidium caviae* (a protozoan) are residents of guinea pigs. Rarely, Diarrhea is caused by *B. caviae*, and *P. uncinata* does not normally cause any clinical illness, although it can induce bronchoalveolar eosinophilia.

9.2.1.4. Viral Diarrhea

One case of wasting syndrome in weanling guinea pigs has been documented, involving diarrhea, anorexia, rapid weight loss, and acute death. Transmission electron microscopy was used to identify coronavirus-like particles in the feces; no other organisms were isolated.

9.2.1.5. Malocclusion

Because the hypsodontic (open-rooted) incisors, premolars, and molars of guinea pigs grow continuously, it is possible for any of these teeth to overgrow. While malocclusion has a strong genetic predisposition, other factors such as diet, trauma, or infection may also be involved. Mandibular teeth overgrow medially, occluding the oropharynx and abrading the tongue, while maxillary premolars and molars overgrow laterally, abrading the cheeks. A physical examination can be used to identify overgrown incisors. Evaluation of the cheek teeth should always be a part of a physical examination because premolar malocclusion usually goes unnoticed until the animal appears anorectic, prehends but cannot swallow food, and loses weight. Other indications of malocclusion include excessive salivation (slobbers), wetting of the chin and forepaws, and secondary moist dermatitis (Collins, 1988; Harkness and Wagner, 1989) (Figure 9.1).

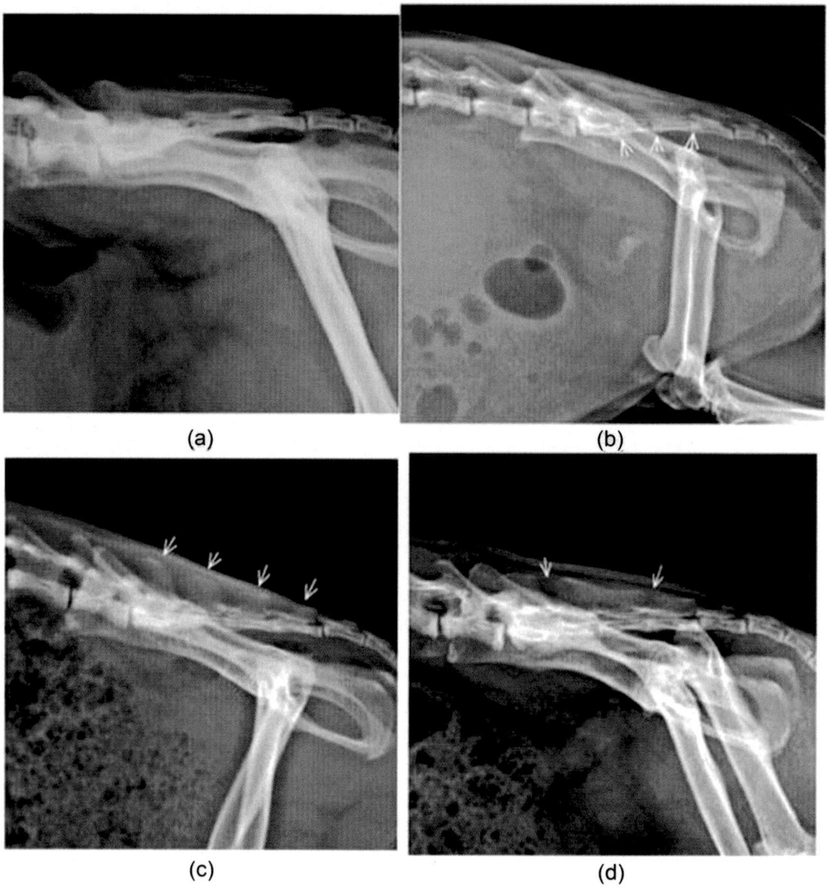

Figure 9.1. *Lateral radiographs of the lumbosacral vertebral column demonstrating morphological variants of the sacral (S) bone-in guinea pigs: (**a**) Complete fusion of the vertebral bodies and fusion of spinous processes S1–S4 creates the median sacral crest; (**b**) Fusion of the vertebral bodies with vestigial intervertebral space (arrows) and fusion of spinous processes S1–S4 represents the most common variant; (**c**) Fusion of vertebral bodies and separation of all spinous processes (arrows); (**d**) Fusion of vertebral bodies and separation of first and fourth spinous processes.*

Source: Proks P, Johansen TM, Nývltová I, Komenda D, Černochová H, Vignoli M. Vertebral Formulae and Congenital Vertebral Anomalies in Guinea Pigs: A Retrospective Radiographic Study. Animals. *2021; 11(3):589. https://doi.org/10.3390/ani11030589.*

A thorough oral examination can be used to diagnose malocclusion. Since the guinea pig has a narrow mouth cavity and many buccal folds, it is easier to perform this examination after the animal has been given isoflurane anesthesia. A vaginal speculum or otoscope can be used to carefully examine the premolars and molars, and gauze strips looped over the incisors can be used to open the mouth. Pediatric rongeurs, dental drills, or files made specifically for this purpose can all be used to trim teeth. If you use rongeurs, take caution not to split your teeth. For the duration of the animal's life, regular trimming is required (typically every four to six weeks). To stop decreased collagen formation and subsequent tooth movement in the socket, guinea pigs can be fed 15–25 mg of vitamin C per day. By giving a healthy diet and refraining from breeding afflicted animals, the disease can be avoided.

9.2.2. Respiratory Diseases

9.2.2.1. Bacterial Pneumonia

Among guinea pigs, pneumonia is most likely the most common illness, especially in humid or wet conditions. Guinea pigs are particularly vulnerable to respiratory infections brought on by *Streptococcus pneumoniae* and *Bordetella bronchiseptica*. Borrelia is a gram-negative rod that is frequently carried by nonhuman primates, dogs, and rabbits. Guinea pigs and rabbits can infect each other with Bordetella. Many species of asymptomatic carriers also spread the gram-positive bacterium streptococcus. Stress makes people more prone to illness, with young guinea pigs being the most commonly impacted. The organisms are spread through fomites, aerosolization, and direct contact. Inappetence, ocular, and nasal discharge, and dyspnea are among the warning signs. Bordetella causes purulent bronchopneumonia characterized by tympanic bullae exudate and lung lobe consolidation. Abortions and metritis have also been reported. In addition to pericarditis and fibrinopurulent pleuritis, streptococcus also causes bronchopneumonia. Streptococcus can also cause abortions and tortticollis. Septic arthritis is the result of a vitamin C deficiency and *S. pneumoniae* infection.

Bacterial pneumonia is diagnosed by culture and exudate sensitivity analysis. Indirect immunofluorescence and the enzyme-linked immunosorbent assay (ELISA) are two methods for detecting Bordetella antibodies; however, sensitivity, and culture studies should be utilized to verify active cases. Consolidated lung lobes or an opacity in the tympanic bullae may be seen on radiographs. The most often prescribed antibiotics for treating bacterial pneumonia in guinea pigs are enrofloxacin (5–10 mg/kg q12h × 7–21 days), trimethoprim-sulfa (30–50 mg/kg PO q12h × seven days), and chloramphenicol palmitate (30–50 mg/kg PO q12h × 7–21 days). Effective treatment options include a combination of enrofloxacin and doxycycline (2.5 mg/kg PO q12h) administered for 7–21 days. Any of these antimicrobial agents may require long-term (4–6 weeks) therapy. The novel quinolone trovafloxacin works well against *S. pneumoniae* and was effectively used to treat guinea pigs that had been experimentally induced to contract Legionnaires' disease. Lactobacillus should be administered both during and for about five days following antibiotic treatment. Supplemental treatment options include oxygen therapy, vitamin C, and lactated Ringer's solution. Pneumonia can be controlled by providing good care, avoiding stress, and keeping guinea pigs apart from dogs and rabbits.

Streptobacillus moniliformis and *Haemophilus* species are two more bacteria that are important for clinical settings. Haemophilus is a bacteria that can cause subcutaneous abscesses and is easily transmitted to rats. It has been found in the respiratory tracts of guinea pigs and rabbits. *Streptobacillus moniliformis* has been known to cause granulomatous pneumonia in guinea pigs and is typically linked to cervical lymphadenitis.

9.2.2.2. Viral Pneumonia

In guinea pigs, a form of necrotizing bronchopneumonia has been reported. The illness has an incubation period of 5 to 10 days, and animals under stress are more susceptible to contracting the disease. While the morbidity is minimal, the mortality rate is significantly high, and animals affected by this condition often succumb to it rapidly. Characteristic lesions include necrotizing, exfoliative bronchiolitis. The sloughed epithelium contains substantial, round to oval, basophilic intranuclear inclusions. There's a belief that an adenovirus is the cause of this disease.

9.2.2.3. Pulmonary Neoplasia

Guinea pigs are frequently affected by a tumor called bronchiogenic pulmonary adenoma. It happened to about 30% of the animals older than three years in one study. Sometimes this illness is misdiagnosed as pneumonia (Collins, 1988; Harkness and Wagner, 1989).

9.2.3. Urogenital Diseases

9.2.3.1. Urinary Calculi

Guinea pig urinary tract calculi are frequently seen. Congealed ejaculum in elderly boars can cause proteinaceous urethral blockages, which can lead to urethritis. It seems that females who are older than three years old are more likely to develop cystitis and cystic calculi. In one report, several guinea pigs with *Streptococcus pyogenes* cystitis were found to have calcium oxalate uroliths. Stones are formed when calcium deposits around the nidi that the bacteria create. Anorexia, hematuria, dysuria, and a hunched, huddled posture are symptoms of urolithiasis. Hemorrhagic exudate inside the bladder, calculi attached to the bladder wall, and bleeding or congestion of the bladder mucosa are examples of lesions. *Streptococcus pyogenes* results in yellowish-gray deposits inside the lumen and hyperemia of the bladder mucosa. Although urolithiths can occasionally be palpated, radiography is typically used to make a conclusive diagnosis. Due to the difficulty of removing an acid load with guinea pigs, urinary acidifiers are typically not recommended. Although potassium citrate is occasionally used to aid in the prevention of calculi formation, this medication's effectiveness has not been established. Retrograde flushing may help clear the obstruction in guinea pigs with urethral calculi. Surgery should be considered if cystic or ureteral calculi are present, or if this procedure fails. For guinea pigs, isoflurane or sevoflurane is the preferred anesthetic. Antibiotics and fluid therapy are two aspects of supportive care for urinary calculi.

9.2.3.2. Chronic Interstitial Nephritis

When guinea pigs older than three years old are necropsied, chronic interstitial nephritis is frequently discovered. Additionally, reports of it in guinea pigs with hyperglycemia and diabetes mellitus have been made. As a result of *Staphylococcal pododermatitis*, animals may experience nephritis and chronic renal amyloidosis. Good hygiene, a balanced diet, and confinement can all help to prevent secondary interstitial nephritis by lowering the risk of pododermatitis.

9.2.3.3. Renal Parasitism

A renal coccidian called *Klossiella cobayae* may be present in guinea pigs. The epithelial cells that line the renal tubules are home to the organism. Urine contains sporozoites that are shed. An infection does not cause any clinical disease, and treatment is typically not advised. Renal parasitism may respond well to trimethoprim-sulfa or sulfadimethoxine treatment (Tables 9.2 and 9.3).

Table 9.2. *Association Between Patient and Diagnostic Features with the Guinea Pigs' Sex (Female Versus Male) in 117 Clinical Cases Diagnosed with Lower Urinary Tract Disease*

	Sex				
	Female		**Male**		***p*-Value**
	n	**%**	***n***	**%**	
Reproductive condition (n = 117)					$p < 0.001$ (S)
Intact	67	65.00%	36	35.00%	
Neutered	1	7.10%	13	92.90%	
Type of medical consultation (n = 117)					$p = 0.091$ (NS)
First opinion	43	65.20%	23	34.80%	
Referral	25	49.00%	26	51.00%	
Disease recurrence (n = 117)					$p = 0.038$ (S)
Yes	42	67.70%	20	32.30%	
No	26	47.30%	29	52.70%	
Family history of UD (n = 117)					$p < 0.001$ (S)
Yes	45	75.00%	15	25.00%	
No	23	40.40%	34	59.60%	
Reason for consultation (n = 117)					$p = 0.337$ (NS)
Urinary signs	49	59.80%	33	40.20%	
Urinary + Nonspecific signs	12	66.70%	6	33.30%	
Nonspecific signs	7	43.80%	9	56.20%	
Shock	0	0.00%	1	100.00%	
Dysuria (n = 117)					$p = 0.038$ (S)
Yes	54	64.30%	30	35.70%	
No	14	42.40%	19	57.60%	
Hematuria (n = 117)					$p = 0.075$ (NS)
Yes	50	64.10%	28	35.90%	
No	18	46.20%	21	53.80%	

Stranguria (n = 117)					$p = 0.038$ (S)
Yes	54	64.30%	30	35.70%	
No	14	42.40%	19	57.60%	
Abdominal pain (n = 117)					$p = 0.126$ (NS)
Yes	46	63.90%	26	36.10%	
No	22	48.90%	23	51.10%	
Diagnosis (n = 117)					$p < 0.001$ (S)
Urolithiasis	20	38.50%	32	61.50%	
Cystitis/UTI	48	73.80%	17	26.20%	

n—number of guinea pigs; *p*—statistical significance; S—statistically significant association; NS—non-statistically significant association; UD—urinary disease; US—Urinary signs; UTI—urinary tract infection; %—relative percentage of guinea pigs.

Source: Azevedo S, O'Malley B, Greene C, Moran H, Magalhães TR, Queiroga FL. Lower Urinary Tract Diseases in Guinea Pigs: A 14-Year Retrospective Study (2004–2018). Animals. *2023; 13(1):112. https://doi.org/10.3390/ani13010112.*

Table 9.3. *Association Between Diagnosis and Therapeutic Features with the Guinea Pigs' Response to Treatment in 114 Clinical Cases Diagnosed with Lower Urinary Tract Disease*

	Response to Treatment				
	Yes		**No**		***p*-Value**
	n	**%**	***n***	**%**	
Diagnosis (n = 114)					$p = 0.994$ (NS)
Urolithiasis	32	64.00%	18	36.00%	
Cystitis/UTI	41	64.10%	23	35.90%	
Type of treatment (n = 114)					$p = 0.520$ (NS)
Medical	50	61.70%	31	38.30%	
Surgical	23	69.70%	10	30.30%	
Use of glucosamine (n = 114)					$p = 0.408$ (NS)
Yes	52	66.70%	26	33.30%	
No	21	58.30%	15	41.70%	

Use of potassium citrate (n = 114)					$p = 0.004$ (S)
Yes	40	78.40%	11	21.60%	
No	33	52.40%	30	47.60%	
Type of surgery (n = 28)					$p = 0.010$ (S)
Cystotomy	10	58.80%	7	41.20%	
Urethrotomy	9	100.00%	0	0.00%	
Cystotomy + Urethrotomy	0	0.00%	2	100.00%	

n—number of guinea pigs; p—statistical significance; S—statistically significant association; NS—non-statistically significant association; UD—urinary disease; US—Urinary signs; UTI—urinary tract infection;%—relative percentage of guinea pigs.

Source: Azevedo S, O'Malley B, Greene C, Moran H, Magalhães TR, Queiroga FL. Lower Urinary Tract Diseases in Guinea Pigs: A 14-Year Retrospective Study (2004–2018). Animals. *2023; 13(1):112. https://doi.org/10.3390/ani13010112.*

9.2.3.4. Ovarian Cysts

In 76% of female guinea pigs between the ages of 18 months and five years, cystic rete ovarii have been found; these are most frequently found in animals between the ages of two and four years. Cysts grow larger as an animal ages, developing spontaneously and having a diameter of 0.5 to 7 cm.

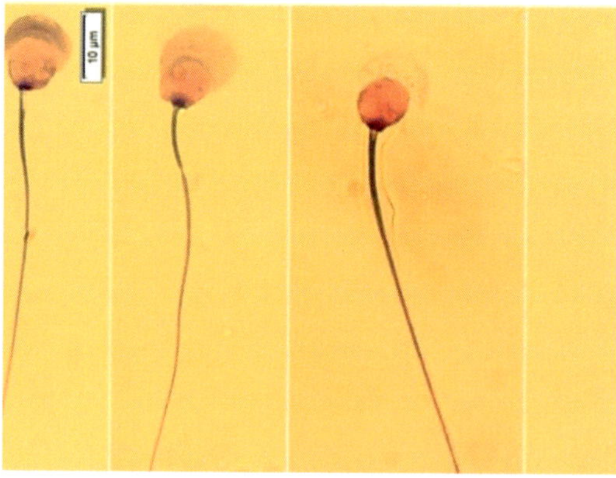

Figure 9.2. *Morphology of guinea pig spermatozoa with Hemacolor® staining (Merck, Darmstadt, Germany). Type 1: intact acrosome (a), Type 2: expansion by acrosomal reaction (b), Type 3: the onset of decomposition (c), and Type 4: loss of acrosomal contents and matrix (d). Bar represents 10 µm.*

Source: https://azeah.com/sites/default/files/gpigovcysts.png.

They are often filled with a clear fluid and can be single or multilocular. Usually, one ovary is affected, and if two ovaries are affected, it's usually the right ovary. Testosterone can promote the growth of ovarian epithelial cells, which can lead to the formation of cysts. In addition to occasional cases of anorexia, exhaustion, and depression, affected animals exhibit abdominal distention. In the flank area, there will be bilateral symmetric hair loss if the cysts are functioning. After 15 months of age, a decline in fertility is the most reliable indicator of cysts in breeding sows (Figures 9.2 and 9.3).

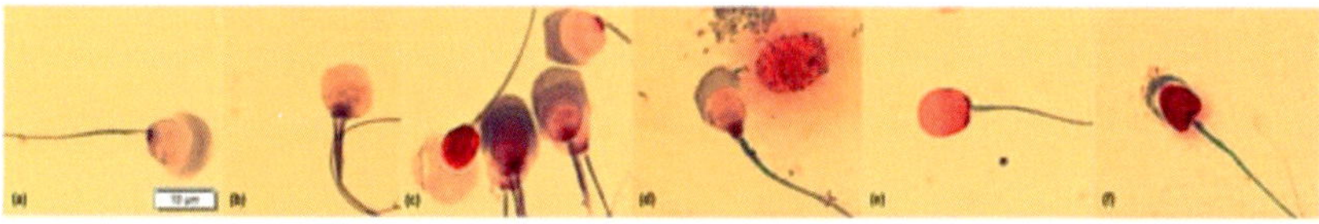

Figure 9.3. *Atypical forms in guinea pig spermatozoa with Hemacolor® staining (Merck, Darmstadt, Germany). Normal (a), spermatozoa with macronucleus (b), micronucleus (c), pyriform (d), elongated (e), and lanceolate (f). Bar represents 10 μm.*

Source: Quispe-Ccasa HA, Briceño-Mendoza YM, Cayo-Colca IS. Guinea Pig Sperm Morphology and Fertility under Different Photoperiod. Animals. *2023; 13(14):2249. https://doi.org/10.3390/ani13142249.*

Utilizing ultrasonography, an ovarian cyst can be diagnosed. Real-time diagnostic images can be obtained with a mechanical sector transducer operating at 6 or 10 MHz, oriented in the sagittal plane. The course of treatment is ovariohysterectomy. Additional issues linked to cystic rete ovarii comprise leiomyomas, endometritis, and cystic endometrial hyperplasia. Exposure to testosterone, diethylstilbestrol, or estradiol can cause uterine and ovarian neoplasms.

9.2.3.5. Vaginitis and Scrotal Plugs

It is possible for inguinal sebaceous secretions and wet, dirty bedding to stick to the penis, scrotum, or vulva, causing a secondary infection or obstructing the flow of urine and feces. The affected area can be gently soaked in a diluted chlorhexidine solution to treat the condition, and any debris can be carefully removed. Systemic antibiotics may be administered if necessary. Good husbandry techniques and proper sanitation can easily prevent this condition.

9.2.3.6. Pregnancy Toxemia

The last two weeks of pregnancy and the first week following delivery are when primiparous, obese sows are most likely to experience pregnancy toxemia, also known as pregnancy ketosis. Boars are also prone to toxemia, even though pregnancy is one of the contributing factors. The two most important risk factors are obesity and fasting. This disease can also be triggered by changes in the environment, dietary changes, and other stressors. The signs appear suddenly. The guinea pig stops drinking, develops anorexia, and within a day is dyspneic and prostrate. Death and convulsions may happen in two to five days. Clinically, the animal has aciduria (pH 5–6), ketonemic ketonesis, proteinuria, and hypoglycemia (less than 60 mg/dL). The stomach is empty and the liver is enlarged and fatty. Pregnant sows may experience disseminated intravascular coagulopathy, ischemia, and thromboplastin release as a result of their heavy, gravid uteri compressing their own vascular supply. Pregnancy toxemia is typically not well treated. Maintain the guinea pig's body temperature, give it intravenous or intraperitoneal fluids as needed, and give it glucose either intravenously or orally. Corticosteroids and calcium gluconate are further treatments that can be administered to an animal that is in shock. Anorexia, an empty stomach, and treatment stress can all result in a deadly case of enteritis. Pregnancy toxemia patients have a guarded to grave prognosis.

The best way to treat pregnancy toxemia is to prevent it. A high-quality diet and precautions against obesity should be implemented for every guinea pig. Giving animals fresh veggies will help them become accustomed to a varied diet and reduce the likelihood that they will reject new foods. Ad libitum access to fresh water and avoidance of environmental, physical, and social stressors are important, especially in late pregnancy.

9.2.3.7. Dystocia

Guinea pigs frequently have dystocia. The fibrocartilage of the pubic symphysis disintegrates in female guinea pigs during the final week of pregnancy due to the hormone relaxin, which is released from the pituitary and endometrium. This causes a 3-cm separation. A sow's pelvic symphysis will fuse in close apposition permanently if she is not bred before she is seven to eight months old, which will prevent the fetuses from passing through during parturition. Dystocia can also result from uterine inertia, obesity, and large fetuses (Figure 9.4).

Puppies born to guinea pigs typically arrive within 30 minutes, with a five minute interval in between for resting. Dystocia is characterized by ineffective contractions and straining. Animals may exhibit bloody or greenish-brown vulvar discharge in addition to depression. Signs and history are used to make the diagnosis. To induce contractions, intramuscular injections of 0–3 IU/kg of oxytocin can be used if the pelvic symphysis is sufficiently wide and the cause is uterine inertia. When manually extracting pups, it's important to quickly remove the fetal membranes. An emergency cesarean section is recommended if oxytocin is unable to induce contractions or if the pubic symphysis is not more than 20 to 25 mm dilated.

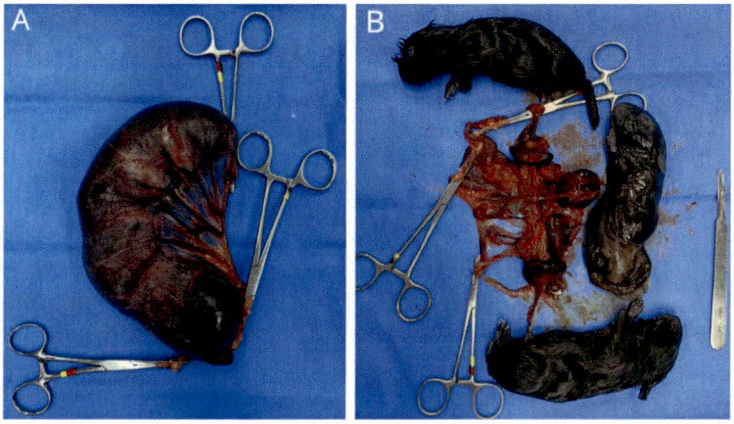

Figure 9.4. *Dystocia.*

Source: de Andrade, J. P., Miranda, S. de M., Soares, C. F., Castanheira, T. L. L., Ferrante, B., and de Carvalho, M. P. N. (2024). Fetal dystocia in guinea pigs: A case report. Brazilian Journal of Veterinary Medicine, 46, e002024. https://doi.org/10.29374/2527-2179.bjvm002024.

.Buprenorphine (0.05 mg/kg SC q8–12h) can be administered to the female, and then mask induction and maintenance of isoflurane or sevoflurane can be performed. The ventral midline approach is employed as usual. The uterus is longitudinally incised and exteriorized. As soon as the pups are taken out, they need to be cleaned up and given breathing stimulation. An inverted suture pattern is used to close the uterus. Both the skin and the linea are closed using nonabsorbable suture and absorbable suture, respectively, in a straightforward interrupted pattern. The prognosis is guarded to poor because guinea pigs do not handle the stress of anesthesia and surgery well. The best ways to avoid dystocia are to avoid obesity and to avoid breeding guinea pigs before they are six months old.

9.2.3.8. Mastitis

Guinea pig mastitis is brought on by *E. coli*, *Streptococcus, Pseudomonas, Pasteurella, Klebsiella,* and *Staphylococcus* species. Sows are more susceptible to infection due to unclean, damp cages and stress from nursing pups. Infections occur through bite wounds on the teats or through the teat canal. Clinically, the glands can appear cyanotic, cool, and swollen at first, then red and warm. Oftentimes, the milk is bloody. Both the mother and the pups may die from an infection that spreads throughout the body. The treatment of mastitis in guinea pigs can involve the use of systemic antibiotics, specifically enrofloxacin, trimethoprim-sulfa, and chloramphenicol palmitate. The glands that are impacted can be treated with hot packs. It is advisable to clean the surroundings and possibly wean the puppies early. The mammary glands can be removed if required.

9.2.3.9. Mammary Gland Tumors

The majority of guinea pig mammary gland tumors are benign fibroadenomas. Adenocarcinomas, which are locally invasive but rarely metastasize, make up about 30% of cases. If a tumor is removed, a significant portion of normal tissue should be removed as well as the tumor. Removal of local lymph nodes is also recommended.

9.2.4. Dermatologic Diseases

9.2.4.1. Dermatophytosis

In guinea pigs, ringworm can occur. Because of their immature immune systems and lower levels of fungistatic fatty acids in their sebum, young animals are more vulnerable. *Trichopyton mentagrophytes* is the most commonly isolated dermatophyte, though *M. canis* has also been found. Disease typically develops as a result of overcrowding, subpar husbandry, and other stressors, though animals may be asymptomatic carriers. The organism is easily spread through fomites and direct contact. The guinea pig exhibits pruritic lesions that are focal, circular patches of crusting alopecia. Usually, lesions begin on the forehead, ears, and face before moving to the back and down the limbs.

Hairs and crusts from the edge of the lesion are scraped off or picked, and these are then placed on dermatophyte test medium (DTM) or another suitable culture medium to help make the diagnosis. Miconazole is an external treatment for ringworm (q24h × 2–4 weeks). Effective topical use of butenafine as a 1% cream has been reported (q24h × 10–20 days). Fluconazole can be given as 16 mg/kg PO every 24 hours for a period of 14 days. Although griseofulvin has been shown to be effective when given in pediatric doses of 25 mg/100 g PO q10 days × three treatments, the recommended dosage is typically 25 mg/kg PO q24h × 14–28 days. This medication should not be given to pregnant animals due to its teratogenic properties. One potentially zoonotic illness is ringworm.

9.2.4.2. Ectoparasites

Mites (*Trixacarus caviae, Chorodiaides caviae*), lice (*Gliricola porcelli, Gyropus ovalis*), and fleas (*Ctenocephalides felis*) can infest guinea pigs.

The most important of these is the sarcoptid mite, Trixacarus. Trixacarus is a burrowing mite that can infect people momentarily. Severe pruritus with excoriations and secondary infection are indicators of Trixacarus infestation. Some animals experience pruritus so intense that it can result in seizures. Lesions can spread to the shoulders and neck and are seen on the thighs and back. There may be yellowish crusts on the afflicted skin. Under a microscope, necrotic patches, eosinophilic microabscesses, and orthokeratotic hyperkeratosis are visible. The stratum corneum may contain mite cross-sections. Hematologic abnormalities such as leukocytosis, monocytosis, eosinophilia, and basophilia are linked to Trixacarus infestation. Compared to Trixacarus, the nonburrowing fur mite Chirodiscoides causes far less trouble. The perineal and hip regions may become lesions due to Chirodiscoides, but the infestation is typically asymptomatic. Both adults and eggs of the debris-feeding lice Gyropus and Grimicola cling to hair shafts. Alopecia, rough hair coats, and crusts are signs of lice infestations. These parasites are all spread through direct touch.

An ectoparasitic infection can be diagnosed visually (fleas), by scraping the skin (mites), or by plucking or combing hairs to check for eggs and adults (lice). To see Trixacarus, skin scrapings may need to be digested with potassium hydroxide.

It is effective to treat trichomoniasis with ivermectin (0.5–0.8 mg/kg SC, repeat in seven days). Ivermectin is also used to treat lice and schirodicoides (0–3 mg/kg SC, PO, repeat in 10 days). Cat flea powder with a pyrethrin base can be used to treat fleas. A topical parasiticide called selamectin (6 mg/kg; Revolution, Pfizer, Exton, PA) has also shown promise in guinea pig experiments. To stop reinfestation, the environment should be thoroughly cleaned repeatedly after treating any ectoparasite.

9.2.4.3. Cervical Lymphadenitis

Guinea pigs are frequently afflicted with *Cervical lymphadenitis*, or lumps. The main culprit is the gram-positive coccus *Streptococcus zooepidemicus*, also known as Lancefield's group C. On rare occasions, *S. moniliformis* is likewise implicated. In guinea pigs, zooepidemicus typically resides in the nasal cavity and conjunctiva. Bacteria infiltrate the cervical lymph nodes and result in abscessation if the animal's oral mucosa is worn down by biting, malocclusion, or dietary roughage (hay stems). The disease can also develop in animals whose nasal and conjunctival mucosa are intact. Stress makes one more prone to infection. The guinea pig typically exhibits pus-filled ventral cervical masses at the time of diagnosis. Bacteria can occasionally spread throughout the body and cause necrotizing bronchopneumonia or septicemia. The diagnosis is made using clinical signs, impression smear results, pus Gram stains, culture, and sensitivity testing. The abscess can either be surgically removed (which is preferred) or drained, and the wound must be thoroughly flushed. It is best to administer a systemic antibiotic. By feeding the guinea pig a healthy diet and providing a clean, stress-free environment, this condition can be avoided. *S. moniliformis* is a potentially zoonotic illness.

9.2.4.4. Bite Wounds

Despite their social nature, guinea pigs living in groups may sustain bite wounds that occasionally become infected. To treat the wound, saline solution must be drained and irrigated, and systemic antibiotics must be given in accordance with the findings of sensitivity testing.

9.2.4.5. Pododermatitis

Bumblefoot, or pododermatitis, is a common condition in guinea pigs. Obese animals kept in cages with wire floors or rough bedding are usually the ones with the disease. On the palmar and plantar surfaces of the feet, hyperkeratosis areas form. These become ulcerated, allowing *Staphylococcus aureus* to invade again. Osteomyelitis can arise from an infection that penetrates deeply into the tissues of the feet, ascending tendons, and entering bone. Podocetodermatitis causes severe pain, frequent vocalizations, and reluctance to walk in guinea pigs. Clinical signs and lesion identification are used to make the diagnosis. Osteomyelitis may be identified with radiography. The course of treatment includes bandaging the wound, using the proper analgesic medication, surgically debridement of the lesions, foot soaks with chlorhexidine, and appropriate systemic antibiotics. The prognosis is uncertain, just as it is with pododermatitis in most species. The best defenses against the illness are to keep the surroundings dry and clean, to house the animals on plush, nonabrasive bedding, and to steer clear of obesity.

9.2.4.6. Tumors

Trichofolliculoma, a benign basal cell epithelioma, is the most common skin tumor in guinea pigs. Usually over the lumbosacral region, trichomefolliculomas manifest as solid or cystic masses. Surgical removal of them is simple.

9.2.4.7. Alopecia

In guinea pigs, alopecia due to noninfectious causes is fairly common. Barbering, endocrine alopecia, and vitamin deficiencies fall into this category. A detailed inspection of the hair-loss area is necessary to identify barbering. Broken hair shafts can be seen in barbering, but the underlying skin is not red or itchy. Alopecia across the flanks is a sign of self-barbering, which can happen as a result of boredom. Providing chew toys, hay, or other roughage may help to solve this issue. When guinea pigs live in groups, the dominant animal frequently grooms the subordinates. Late-gestation guinea pig sows may develop a temporary endocrine alopecia. Weanlings with partial alopecia have been reported. It has been shown that adding hay to the diet can lower trichophagia.

9.2.5. Musculoskeletal Diseases

9.2.5.1. Scurvy

Guinea pigs are unable to produce l-gulono-γ-lactone oxidase due to a mutated gene. As a result, they are unable to produce vitamin C internally or convert glucose to ascorbic acid. Because of this, pregnant animals need to consume 30 mg of vitamin C per day, while guinea pigs only need 15 to 25 mg per day. Collagen synthesis requires ascorbic acid; deficiency in vitamin C from the diet leads to defective type IV collagen, laminin, and elastin, which weakens blood vessel integrity and causes bleeding in joints and gingiva. Additionally, collagen keeps teeth firmly in their sockets; without it, teeth become dislodged and malocclusion happens. Furthermore, vitamin C is required for the proper retention of vitamin E.

Rough hair coat, anorexia, diarrhea, teeth grinding, vocalizing in pain, delayed wound healing, lameness, and heightened susceptibility to bacterial infections are all indicators of vitamin C deficiency. Long bone epiphyses and the ribs' costochondral junctions are enlarged on radiographs. Fractures that are pathological may also be visible. Scurvy is more common in young, growing animals, and clinical illness can appear as soon as two weeks after ascorbic acid deficiency.

Radiographic lesions, clinical signs, and history are used to diagnose vitamin C deficiency. The diagnosis can be verified by measuring the levels of serum ascorbic acid. Parenteral ascorbic acid should be administered as soon as possible via IP, IM, or SC at a dose of 50 mg/day. The same dosage of vitamin C may be taken orally once the response has been observed. Following recuperation, the diet should include daily supplements of vitamin C. Use fresh, high-quality guinea pig (not rabbit) chow within 90 days of the milling date to ensure adequate vitamin C intake. Oranges, kale, and fresh cabbage are also good sources of vitamin C. For example, 50 g of cabbage has 30 mg of vitamin C, and 100 g of kale has 125 mg. 200 to 400 mg/L35 of vitamin C tablets can be added to the drinking water; the medicated water needs to be changed every day.

9.2.5.2. Osteoarthritis and Osteoarthrosis

In guinea pigs, spontaneous osteoarthritis has been reported. Guinea pig osteoarthritis mimics human osteoarthritis in both morphological and epidemiological aspects. One of the risk factors is obesity.

There have been reports of young guinea pigs developing osteoarthrosis and spontaneous cartilage degeneration in the femorotibial joint. The illness doesn't seem to be common, and no cause has been found.

9.2.5.3. Iatrogenic Muscle Necrosis

In guinea pigs, the injection site can develop muscle necrosis when fentanyl and droperidol are combined. Distal to the injection site, guinea pigs have also been linked to nerve damage and self-mutilation caused by ketamine and diazepam (Collins, 1988; Harkness and Wagner, 1989).

9.2.6. Neurologic Diseases

9.2.6.1. Lymphocytic Choriomeningitis Virus

Although it's more frequently associated with mice and hamsters, the arenavirus known as LCMV can cause meningitis and paralysis of the hind limbs in guinea pigs. The choroid plexus, ependyma, and meninges are among the lesions that have lymphocytic infiltrates. The virus is spread by direct contact, ingestion, or inhalation of contaminated urine, saliva, or feces. In addition to transplacental transmission, biting insects can spread LCMV. Humans can contract LCMV from one another. Human LCMV infections typically cause fever, vomiting, and headaches; fatalities are uncommon.

9.2.7. Ophthalmologic Diseases

9.2.7.1. Inclusion Body Conjunctivitis of Guinea Pigs

In guinea pigs, *Chlamydophila psittaci*-related conjunctivitis has been reported. The majority of affected animals are between one and three weeks old. We don't know the transmission mode. Serous ocular discharge and conjunctival reddening are symptoms. The detection of intracytoplasmic inclusions in conjunctival scrapings aids in the diagnosis. As the illness clears up on its own in two to three weeks, there's no need for treatment. Even so, *Chlamydophila psittaci* is a zoonotic pathogen; there are no known instances of guinea pig-to-human transmission.

9.2.8. Miscellaneous Diseases

9.2.8.1. Ototoxicity

Many medications have been known to cause ototoxicity in guinea pigs. When applied topically, gentamicin may be ototoxic. For guinea pigs, cortisporin otic suspension (Glaxo Wellcome, Research Triangle Park, NC) containing polymyxin B and neomycin is also ototoxic. In guinea pigs, cisplatin at 7.5 mg/kg IM twice in a five day period caused ototoxicity.

9.2.8.2. Diabetes Mellitus

Male adult Abyssinian-Hartley guinea pigs have been reported to have spontaneous diabetes mellitus. The diabetes is comparable to adult-onset diabetes in humans and is not insulin-dependent. The affected animals live for approximately five years and develop bladder hypertrophy and voiding dysfunction. The diagnosis of diabetes mellitus was made for an adult female guinea pig who had small, frequent urination and cystitis. The best insulin for the guinea pig was Caninsulin (Intervet Canada, Whitby, Ontario); it responded to NPH insulin as well. A proper low-fat, high-fiber diet is crucial for both prevention and treatment of diabetes mellitus, even if it is a temporary condition (Vannevel, 1998).

9.2.8.3. Adrenal Tumors

In guinea pigs, adrenal tumors have been reported. An adult male guinea pig with obesity, bilateral alopecia, hepatomegaly, and depression was the subject of one case report. Ultrasonography was used to diagnose an adrenal tumor and ureterolith. The guinea pig passed away following surgery to remove both, though.

9.2.8.4. Heat Stress

Heat stress can affect guinea pigs. Heat stress can occur in guinea pigs kept outside, even in low temperatures of 70° to 75°F (21° to 24□). When guinea pigs experience heat stress, they will salivate a lot to regulate their body temperature. They have pale mucous membranes, shallow, rapid breathing, and a raised rectal temperature. These symptoms could be followed by death or a coma. Parenteral fluids, corticosteroids, and cool water baths are all part of the supportive treatment. The prognosis is quite cautious.

9.2.8.5. Metastatic Mineralization

In most cases, metastatic mineralization in guinea pigs is discovered by accident during necropsy. The cause is unknown, although it may be associated with subclinical mineral imbalances, dehydration, or excessive dietary vitamin D_3 or mineral supplementation. Though it usually shows no symptoms, the illness can cause muscle rigidity and kidney problems. When an animal reaches the age of one year, it can sustain injuries such as kidney, heart, vessel, stomach, and colon mineralization. There isn't a cure.

9.2.8.6. Cavian Leukemia

The most frequent type of tumor in guinea pigs is lymphosarcoma. The type C retrovirus is the cause of the illness. The animals exhibit lymphadenopathy and a disheveled coat. Rarely, mediastinal masses, splenomegaly, and hepatomegaly are observed. The findings of a complete blood count and cytologic analysis of aspirates of enlarged nodes or fluids from the abdomen or pleura are used to make the diagnosis. The total white blood cell count in leukemic animals ranges from 25,000 to 500,000 cells/mm^3. Visceral organs and lymph nodes may be enlarged at necropsy, with proliferating lymphoblasts infiltrating the area. The illness takes two to five weeks to complete. Although some animals have responded to chemotherapy at first, the prognosis is not good (Collins, 1988; Harkness and Wagner, 1989).

REFERENCES

1. Arika, T., Tokoo, M., & Hase, T. (1990). Effects of butenafine hydrochloride, a new benzylamine derivative, on experimental dermatophytosis in guinea pigs. *Antimicrobial Agents and Chemotherapy, 34*, 2250–2253.
2. Badia, J. M., Torres, J. M., & Tur, C. (1996). Saline wound irrigation reduces the postoperative infection rate in guinea pigs. *Journal of Surgical Research, 63*, 457–459.
3. Belis, J. A., Curley, R. M., & Lang, C. M. (1996). Bladder dysfunction in the spontaneously diabetic male Abyssinian-Hartley guinea pig. *Pharmacology, 53*, 66–70.
4. Bendele, A. M., White, S. L., & Hulman, J. F. (1989). Osteoarthritis in guinea pigs: Histopathologic and scanning electron microscopic features. *Laboratory Animal Science, 39*, 115–121.
5. Bendele, A., McComb, J., & Gould, T. (1999). Animal models of arthritis: Relevance to human disease. *Toxicologic Pathology, 27*, 134–142.
6. Beregi, A., Zorn, S., & Felkai, F. (1999). Ultrasonic diagnosis of ovarian cysts in ten guinea pigs. *Veterinary Radiology & Ultrasound, 40*, 74–76.
7. Boot, R., Thuis, H. C. W., & Veenema, J. L. (1999). Serological relationship of some V-factor dependent *Pasteurellaceae* (Haemophilus sp.) from guinea pigs and rabbits. *Laboratory Animals, 33*, 91–94.
8. Boot, R., Thuis, H., & Wieten, G. (1995). Multifactorial analysis of antibiotic sensitivity of *Bordetella bronchiseptica* isolates from guinea pigs, rabbits, and rats. *Laboratory Animals, 29*, 45–49.
9. Chrisp, C. E., Suckow, M. A., & Fayer, R. (1992). Comparison of the host ranges and antigenicity of *Cryptosporidium parvum* and *Cryptosporidium wrairi* from guinea pigs. *Journal of Protozoology, 39*, 406–409.
10. Collins, B. (1988). Common diseases and medical management of rodents and lagomorphs. In E. R. Jacobson & G. V. Kollias (Eds.), *Exotic Animals* (pp. 261–316). Churchill Livingstone.
11. Conder, G. A., Richards, I. M., & Jen, L. W. (1989). Bronchoalveolar eosinophilia in guinea pigs harboring inapparent infections of *Paraspidodera uncinata. Journal of Parasitology, 75*, 144–146.
12. Conlon, B. J., McSwain, S. D., & Smith, D. W. (1998). Topical gentamicin and ethacrynic acid: Effects on cochlear function. *Laryngoscope, 108*, 1087–1089.
13. Deeb, B. J., DiGiacomo, R. F., & Wang, S. P. (1989). Guinea pig inclusion conjunctivitis (GPIC) in a commercial colony. *Laboratory Animals, 23*, 103.

14. Earnest-Koons, K. A., Griffith, J. W., & Lang, C. M. (1988). Incidence and classification of thyroid lesions in 210 guinea pigs. *Laboratory Animal Science, 38*, 514.
15. Feldman, S. H., Richardson, J. A., & Clubb, F. J., Jr. (1990). Necrotizing viral bronchopneumonia in guinea pigs. *Laboratory Animal Science, 40*, 82–83.
16. Field, K. J., Griffith, J. W., & Lang, C. M. (1989). Spontaneous reproductive tract leiomyomas in aged guinea-pigs. *Journal of Comparative Pathology, 101*, 287–294.
17. Finnie, J. W., Noonan, D. E., & Swift, J. G. (1999). Adenovirus pneumonia of guinea pigs. *Australian Veterinary Journal, 77*, 191, 192.
18. Fish, R. E., & Besch-Williford, C. (1992). Reproductive disorders in the rabbit and guinea pig. In R. W. Kirk & J. D. Bonagura (Eds.), *Kirk's Current Veterinary Therapy* (Vol. 11, pp. 1175–1179). WB Saunders.
19. Fritz, P. E., Hurst, W. J., & White, W. J. (1987). Pharmacokinetics of cefazolin in guinea pigs. *Laboratory Animal Science, 37*, 646–651.
20. Fuentealba, C., & Hanna, P. (1996). Mange induced by *Trixacarus caviae* in a guinea pig. *Canadian Veterinary Journal, 37*, 749–750.
21. Gaschien, L., Ketz, C., & Lang, J. (1998). Ultrasonographic detection of adrenal gland tumor and ureterolithiasis in a guinea pig. *Veterinary Radiology & Ultrasound, 39*, 43–46.
22. Gerold, S., Huisinga, E., & Iglauer, F. (1997). Influence of feeding hay on the alopecia of breeding guinea pigs. *Journal of Veterinary Medicine, 44*, 341–348.
23. Gibson, S. V., & Wagner, J. E. (1986). Cryptosporidiosis in guinea pigs: A retrospective study. *Journal of the American Veterinary Medical Association, 189*, 1033–1034.
24. Goldsmith, L. T., Weiss, G., & Steinetz, B. G. (1995). Relaxin and its role in pregnancy. *Endocrine Disorders in Pregnancy, 24*, 171–186.
25. Griffith, J. W., Brasky, K. M., & Lang, C. M. (1997). Experimental pneumonia virus of mice infection of guinea pigs spontaneously infected with *Bordetella bronchiseptica. Laboratory Animals, 31*, 52–57.
26. Harkness, J. E., & Wagner, J. E. (1989). *The Biology and Medicine of Rabbits and Rodents* (3rd ed.). Lea and Febiger.
27. Hurley, R. J., Murphy, J. C., & Lipman, N. S. (1995). Diagnostic exercise: Depression and anorexia in recently shipped guinea pigs. *Laboratory Animal Science, 45*, 305–308.
28. Jaax, G. P., Jaax, N. K., & Petrali, J. P. (1990). Coronavirus-like virions

associated with a wasting syndrome in guinea pigs. *Laboratory Animal Science, 40*, 375–378.

29. Johnson-Delaney, C. A. (1998). Small mammal endocrinology. In *Proceedings of the Annual Conference of the Association of Avian Veterinarians, Small Mammal and Reptile Program* (pp. 99–111).
30. Keller, L. S. F., Griffith, J. W., & Lang, C. M. (1987). Reproductive failure associated with cystic rete ovarii in guinea pigs. *Journal of Veterinary Pathology, 24*, 335–339.
31. Kirchner, B. K., Lake, S. G., & Wightman, S. R. (1992). Isolation of *Streptobacillus moniliformis* from a guinea pig with granulomatous pneumonia. *Laboratory Animal Science, 42*, 519–521.
32. Kunstyr, I., Maess, J., & Naumann, S. (1984). Adenovirus pneumonia in guinea pigs: An experimental reproduction of the disease. *Laboratory Animals, 18*, 55–60.
33. Liu, J. F., & Lee, Y. W. (1998). Vitamin C supplementation restores the impaired vitamin E status of guinea pigs fed oxidized frying oil. *Journal of Nutrition, 128*, 116–122.
34. Mahmoodian, F., & Peterkofsky, B. (1999). Vitamin C deficiency in guinea pigs differentially affects the expression of type IV collagen, laminin, and elastin in blood vessels. *Journal of Nutrition, 129*, 83–91.
35. Manning, P. J., Wagner, J. E., & Harkness, J. E. (1984). Biology and diseases of guinea pigs. In J. G. Fox, B. J. Cohen, & F. M. Loew (Eds.), *Laboratory Animal Medicine* (pp. 149–181). Academic Press.
36. Murphy, J. C., Ackerman, J. I., & Marini, R. P. (1991). Cervical lymphadenitis in guinea pigs: Infection via intact ocular and nasal mucosa by *Streptococcus zooepidemicus*. *Laboratory Animal Science, 41*, 251–254.
37. Nagino, K., Shimohira, H., & Ogawa, M. (2000). Comparison of the therapeutic efficacy of oral doses of fluconazole and griseofulvin in a guinea pig model of dermatophytosis. *Journal of Antibiotics (Tokyo), 53*, 207–210.
38. Nishikimi, M., Kawai, T., & Yagi, K. (1992). Guinea pigs possess a highly mutated gene for l-gulono-γ-lactone oxidase, the key enzyme for l-ascorbic acid biosynthesis missing in this species. *Journal of Biological Chemistry, 267*, 21967–21972.
39. Okewole, P. A., Odeyemi, P. S., & Oladummade, M. A. (1991). An outbreak of *Streptococcus pyogenes* infection associated with calcium oxalate urolithiasis in guinea pigs (*Cavia porcellus*). *Laboratory Animal Science, 25*, 184–186.

40. Pechere, J. C., & Gootz, T. D. (1998). Bacteriological activity of trovafloxacin, a new quinolone, against respiratory tract pathogens. *European Journal of Clinical Microbiology & Infectious Diseases, 17*, 405–412.

41. Peng, X., Griffith, J. W., & Lang, C. M. (1987). Cystitis and cystic calculi in aged guinea pigs. *Laboratory Animal Science, 34*, 527.

42. Perry, B. P., & Smith, D. W. (1996). Effect of Cortisporin otic suspension on cochlear function and efferent activity in the guinea pig. *Laryngoscope, 106*, 1557–1561.

43. Porter, K. B., Tsibris, J. C. M., & Porter, G. W. (1997). Use of endoscopic and ultrasound techniques in the guinea pig leiomyoma model. *Laboratory Animal Science, 47*, 537–539.

44. Rothwell, T. L. W., Pope, S. E., & Rajczyk, Z. K. (1991). Haematological and pathological responses to experimental *Trixacarus caviae* infection in guinea pigs. *Journal of Comparative Pathology, 104*, 179–185.

45. Saito, T., Zhang, Z. J., & Yamada, T. (1997). Similar pharmacokinetics and differential ototoxicity after administration with cisplatin and transplatin in guinea pigs. *Acta Otolaryngologica (Stockholm), 117*, 61–65.

46. Silva, E. G., Tornos, C., & Deavers, M. (1998). Induction of epithelial neoplasms in the ovaries of guinea pigs by estrogenic stimulation. *Gynecologic Oncology, 71*, 240–246.

47. Silva, E. G., Tornos, C., & Fritsche, H. A. (1997). The induction of benign epithelial neoplasms of the ovaries of guinea pigs by testosterone stimulation: A potential animal model. *Modern Pathology, 10*, 879–883.

48. Spink, R. R. (1978). Urolithiasis in a guinea pig. *Veterinary Medicine and Small Animal Clinic, 73*, 501–502.

49. Suckow, M. A., Chrisp, C. E., & Rush, H. G. (1989). Cryptosporidiosis in immunocompetent guinea pigs. *Laboratory Animal Science, 39*, 470.

50. Timm, K. I. (1988). Pruritus in rabbits, rodents, and ferrets. *Veterinary Clinics of North America: Small Animal Practice, 18*, 1077–1091.

51. Vangeel, I., Pasmans, F., & Vanrobaeys, M. (2000). Prevalence of dermatophytes in asymptomatic guinea pigs and rabbits. *Veterinary Record, 146*, 440–441.

52. Vannevel, J. (1998). Diabetes in the guinea pig: Not uncommon. *Canadian Veterinary Journal, 40*, 613.

53. Vannevel, J. (1998). Diabetes mellitus in a 3-year-old, intact, female guinea pig. *Canadian Veterinary Journal, 39*, 503–504.

54. Waggie, K. S., Wagner, J. E., & Kelley, S. T. (1986). Naturally occurring

Bacillus piliformis infection (Tyzzer's disease) in guinea pigs. *Laboratory Animal Science, 36*, 504.

55. Wallach, J. D., & Boever, W. J. (1983). *Diseases of Exotic Animals*. WB Saunders.
56. Williams, C. S. F. (1976). *Practical Guide to Laboratory Animals*. CV Mosby.
57. Witt, W. M., Hubbard, G. B., & Fanton, J. W. (1988). *Streptococcus pneumoniae* arthritis and osteomyelitis with vitamin C deficiency in guinea pigs. *Laboratory Animal Science, 38*, 192–194.
58. Wullenweber, M., & Boot, R. (1994). Interlaboratory comparison of enzyme-linked immunosorbent assay (ELISA) and indirect immunofluorescence (IIF) for detection of *Bordetella bronchiseptica* antibodies in guinea pigs. *Laboratory Animals, 28*, 335–339.
59. Xia, Y., Hu, H. Z., & Liu, S. (2000). *Clostridium difficile* toxin A excites enteric neurons and suppresses sympathetic neurotransmission in the guinea pig. *Gut, 46*, 481–486.
60. Young, J. D., Hurst, W. J., & White, W. J. (1987). An evaluation of ampicillin pharmacokinetics and toxicity in guinea pigs. *Laboratory Animal Science, 37*, 652–656.

INDEX

A

AAFP (American Association of Feline Practitioners) 31
AAHA (American Animal Hospital Association) 31
Abdominal masses 127
Abnormal behavior 40
Absorbency 275
Adrenal Gland Disease 370
Airborne bacteria 112
Airborne transmission 2, 50, 61
Alanine aminotransferase (ALT) 90
Alert management 12
American Animal Hospital Association (AAHA) 139
American Board of Veterinary Practitioners (ABVP) 6
American Society for Prevention of Cruelty to Animals (ASPCA) 7
Animal agriculture 5
Animal behavior 6
Animal control protocols 47
Animal experiments 110
Animal Health 18, 20, 304, 326
Animal husbandry 3, 46, 136
Animal management 6
Animal mortality 19
Animal populations 11, 21, 23
Animal turnover 2
Animal welfare 7, 8, 10, 21, 22, 23, 40, 44, 47, 64, 85
Anthropophilic species 57
Association of Shelter Veterinarians (ASV) 6
Atherosclerosis 247

B

Bacterial Diseases in Ferrets 358
Bacterial infections 270, 346, 398
Bacterial urinary tract infection 116
Behavioral expressions 41
Behavioral health 21, 37, 39, 47
Behavioral indicators 41
Behavioral professionals 23
Biochemical analysis 123
Biosecurity xv, 2, 25, 48, 218, 260, 261, 263, 264, 270, 398
Biosecurity procedures 2, 263
Blood flow 125, 226, 302, 363
Blood pressure 125, 324, 399
Blood sample analysis 123
Blood tests 116
Body posture 41
Bordetella 31, 33, 136, 137, 139, 147, 148, 154, 167, 169, 170,

172, 176, 177, 179, 180, 182, 183, 185, 186, 187, 188, 193, 196, 207, 208, 209, 211, 212, 213, 214, 270, 310, 313, 337, 398, 408, 411, 427, 428, 431
Brachycephalic dogs 38
Breeding and Reproduction of Ferrets 357
Breed Specific Care 38

C

Canine adenovirus (CAV-2) 31
Canine distemper (CDV) 134
Canine distemper virus (CDV) 31
Canine parainfluenza (CPI) 137, 146
Canine parvovirus (CPV-2) 134
Canine respiratory disease complex (CRDC) 136
Cardboard boxes 45, 107, 274
Cardiorespiratory xiii, 399
Care of Newborns 284, 406
Cellular immunity 138
Chlamydophila 139
Chlorhexidine 93
Chronic progressive nephrosis 116
Chronic respiratory 104, 112
Circadian cycle 108
Clean environment 113, 273, 398, 408
Climbazole 85
Clinic environment 122
Clostridial Enterotoxaemia 314
Colony forming units (CFU) 77
Community programs 8
Complete blood count (CBC) 123
Corona virus 327
Culture monitoring 76

D

Database collection 25
Data Collection 29
Dehydration 39, 117, 121, 184, 232, 351, 359, 360, 403, 407, 426
Dermatologic Diseases 420
Dermatophyte infection 67, 76, 81, 95
Dermatophytes 60, 78
Dermatophyte Test Medium (DTM) 75
Dermatophyte transmission 61
Dermatophytosis xi, 56, 57, 58, 60, 61, 62, 63, 64, 65, 66, 67, 68, 69, 70, 73, 74, 76, 77, 81, 82, 83, 84, 85, 86, 88, 89, 90, 91, 92, 95, 96, 97, 98, 99, 270, 328, 420, 427, 429
Diagnostic testing 11, 28, 81, 84, 122, 123, 124, 127, 153, 154, 166, 188, 202, 260, 291, 310, 367
Dietary imbalance 108
Disease Transmission 14, 24, 84
Domestic Rabbits 269, 285, 342

E

Ectoparasites 285, 420
Effective disease management xv
Electric clippers 84
Emergencies of Ferrets 355
Emotional and physical development 36
Endoparasites 289, 346
Enilconazole 93
Environmental contamination 60, 61, 62, 83, 86, 87, 88, 92, 97
Environmental cultures 80, 94
Environmental factors 23, 45, 59, 312
Environmental surfaces 61
Environmental wellness 21
Epidermophyton 57

Erroneous diagnostics 96
Erythema 62, 91, 250, 287, 359
Euthanasia 17, 18, 23, 46, 262, 264

F

Facial expression 41
Feline herpesvirus type 1 (FHV-1) 138
Feline immunodeficiency virus (FIV) 90
Feline leukemia virus (FeLV) 90
Feline panleukopenia (parvovirus) (FPV) 134
Feline respiratory disease complex (FRDC) 136
Female hamsters 111
Fenugreek 278
Financially sustainable 3
Flat tumors 117
Food intake 107
Food preparation 16, 83
Food security xv
Food supplement 128
Funding agencies 12
Fungal culture 60, 61, 65, 69, 70, 74, 75, 76, 77, 78, 79, 80, 81, 82, 85, 92, 94, 95, 98, 100
Fungal Diseases of Ferrets 363
Fungi 73, 115

G

Gastrointestinal 2, 112, 128, 160, 163, 228, 229, 230, 231, 233, 266, 270, 276, 322, 346, 355, 359, 362, 378, 407
Gastrointestinal Diseases 407
Gastrointestinal infections 2
Genetic heritage 116
Geographic region 48
Gerbils 106, 108
Get infections 112
Glass microscope slides 69
Glomerulonephrosis 116
Grooming behavior 59, 113
Grooming program 37
Growth performance 278, 279

H

Hairballs 272, 275
Hair follicles 114
Hand hygiene 2
Harderian glands 113
Health assessments 2
Health management 5, 134, 204, 214
Health monitoring systems 20
Hepatitis 31, 34, 136, 146, 169, 247, 323, 326, 340, 390
Herpes virus (FHV-1) 134
High-efficiency particulate air (HEPA) 16
High-quality nutrition 279
Hospitalizations 127
Housing environment 60
Humane education 8
Humane societies 9, 10, 273
Hygienic practices 14
Hyperpigmentation 62
Hypocalcemia 248
Hypothermia 87, 110, 407
Hypovitaminosis A 248

I

Illness management 8
Immune responses 32, 97, 177, 300, 305
Immune system 2, 59, 136, 137, 139, 153, 156, 157, 172, 218, 326
Indirect transmission 112
Induce air exchange 49
Infected animals 24, 60, 66, 92, 94, 113, 293, 302, 398

Infectious diseases xv, xvi, 2, 3, 6, 17, 21, 22, 25, 28, 29, 30, 34, 47, 48, 105, 159, 191, 213, 218, 226, 253, 260, 263, 264, 346, 398, 402
Inflammatory skin 66, 115
Insulinomas 357, 369
International Committee on Taxonomy of Viruses (ICTV) xiii, 237

K

Keratoacanthomas 117
Ketoconazole 85, 89, 90, 91

L

Lameness 128, 182, 232, 247, 423
Length of stay (LOS) 3, 76
Leptospirosis 31, 116
Lime sulfur 86
Listeriosis 315
Liver enzyme 89, 147
Livestock veterinarian 3
Lyme Disease 317

M

Malnourishment 110, 235
Mammal species 103
Managing wildlife 8
Mass treatment 24
Mass vaccination 24
Maternally derived antibodies (MDAs) 137
Medical devices 32
Medical Record Keeping 29
Medical records 29, 202
Mental depression 39
Mice's growth 107
Microbial protein 276
Microchip 27, 29, 32, 36, 37, 224
Micromedium plates 75
Microscopic analysis 67
Microscopic examination xi, 65, 69, 286
Microsporum 56, 57, 73, 86, 97, 98, 99, 363
Miscellaneous Diseases 425
Modified live virus (MLV) 33, 135, 195
Municipal Shelters 8
Muscular dystrophy 277
Musculoskeletal Diseases 423
Mycologic 91, 92
Mycotic Infections 328
Myomorpha 104
Myxomatosis xii, 270, 317, 318, 319, 320, 321, 322, 323, 339

N

Natural immunization 135
Natural infection 135, 150, 154, 299
Neoplasia 102, 227, 301, 364, 380, 383
Neurological symptoms 182, 270, 297, 310, 325
Neurologic Diseases 424
Normothermic patients 123

O

Ophthalmologic Diseases 424
Oral fluid administration 128
Oral terbinafine 89
Ovarian cancer 36

P

Panleukopenia 18, 31, 34, 48, 52, 134, 136, 137, 138, 148, 149, 150, 161, 169, 170, 195, 196, 212
Papillomatosis 327

Paradoxically 3
Parasite Control 35
Parasite infections 111
Parasites of Rabbits 285, 306
Parasitic diseases 270, 378
Parasitic Diseases of Ferrets 363
Pasteurellosis 270, 306, 307, 308, 309, 310, 311, 312, 339, 342, 343, 344
PCV/TS (packed cell volume and total plasma solids) 25
Periodontal disease 30, 38
Periodontal Disease Prevention 38
Personal protective equipment (PPE) 2, 218
Pet rabbits 280, 285, 287, 289, 317, 322
Physical examination 20, 25, 26, 27, 64, 65, 74, 108, 224, 225, 227, 248, 263, 301, 368, 370, 409
Physiological stress 57
Policy and Protocol Development 30
Polymerase chain reaction (PCR) 81, 166, 177, 263, 310
Polyvinyl chloride (PVC) 107
Population control 6
Population density 3, 46, 56
Population healthy 20
Preventive management 95
Private Shelters 8
Protozoa 112, 270, 291, 293
Public education 20
Public health protection 7

R

Rabbit Hemorrhagic Disease (RHD) 270
Rabies vaccination 31, 151, 152, 153
Record-keeping 6, 29
Renal Disease 247
Reproductive performance 278
Reproductive stress 36
Respiratory diseases xvi, 17, 32, 33, 107, 185, 194, 199, 211, 212, 382
Respiratory disorders 112, 119, 188
Respiratory infections 2, 9, 17, 33, 48, 83, 112, 121, 150, 181, 183, 185, 189, 194, 199, 200, 207, 208, 225, 411
Respiratory mycoplasmosis 112
Respiratory system 49, 190, 194, 346, 350
Rhinotracheitis 31, 34, 40, 52, 150, 176, 208, 209, 210, 214, 389, 392
Ringtail 115
Rodent patients 123, 127, 128
Rodents xvi, 102, 103, 104, 105, 107, 108, 109, 110, 111, 112, 114, 116, 117, 120, 121, 122, 123, 124, 125, 126, 127, 128, 178, 232, 262, 293, 295, 305, 313, 314, 316, 339, 342, 393, 398, 400, 407, 427, 430
Routine Health Care for Ferrets 352

S

Salmonellosis 117, 314
Semiaquatic 103
Shelter Administration 12
Shelter Design 15
Shelter environment xvi, 21, 24, 32, 46, 49, 60, 134, 149, 150, 204, 209, 260
Shelter management 5, 6, 20, 202
Shelter medicine 4, 14
Shelter Medicine 4, 6, 53, 208, 209, 210, 211, 212, 213
Shelter Resources 11
Sick animals 2, 11, 12, 16, 17, 19, 38, 48, 49, 50, 56, 57, 89, 111,

201
Skin disorders 113
Social and emotional stress 36
Social companionship 41
Social environment 44
Sodium hypochlorite 93, 189
Staff members 2, 4, 13, 26, 32, 42, 43, 44, 46, 49, 51, 61, 83, 146, 152, 177, 189, 191, 192, 195, 200, 201, 205, 261, 263, 264, 265
Staphylococcus bacteria 115
Stereotypical behaviors 41
Streptococcus pyogenes 110, 413, 429
Stress management 6
Subcutaneous tissues 128

T

Terrestrial habitat 103
Treponematosis 315
Trichophyton xi, 56, 57, 64, 75, 76, 78, 80, 85, 91, 93, 98
Trichophyton species 57, 85
Tularaemia 316
Tyzzer's Disease 313

U

Urogenital Diseases 412
U.S. Department of Agriculture (USDA) 104

V

Vaccination 30, 134, 138, 139, 140, 141, 142, 145, 150, 153, 155, 165, 172, 195, 209, 262, 270, 326, 346, 352, 360
Vaccination clinics 135
Vaccination protocols 2, 194
Vaccine-preventable diseases 135
Ventilation systems 2, 16, 49, 194
Veterinary forensics 6
Veterinary medicine xv
Veterinary professionals 15, 184, 280
Viral Diseases of Ferrets 359
Viral Hemorrhagic Disease (VHD) 322
Viral infections 270, 346, 398
Visual examination 66, 68, 81
Visual (physical) examination 65
vitamins A 277

W

Water bowls 2, 221, 305
Watering system 274
Welfare services 11
White blood cell (WBC) 90

Z

Zoonotic xv, 6, 17, 22, 56, 82, 83, 92, 178, 208, 227, 231, 232, 233, 253, 260, 261, 263, 287, 299, 306, 316, 339, 408, 420, 422, 424
Zoonotic disease xv, 6, 83, 227, 233, 260, 263